Natural History

Development in Japan and Its Future

Kunio Iwatsuki

Translated by Tadao Hikihara

Kyoto Tsushinsha Press

Published in Japan by Kyoto Tsushinsha Press,
309, Oikeno-cho, Muromachi Oike, -agaru,
Nakagyo-ku, Kyoto 604-0022.
https://www.kyoto-info.com/

Part of the production expenses is defrayed from the International Cosmos Prize Award to the author.

Designed by Nakasone Design.

Printed in Japan by Kyodo Printing Industry.

Bound by Otakeguchi Shiko.

Originally published in Japanese by the University of Tokyo Press.

Translated from Japanese by Tadao Hikihara.

ISBN978-4-903473-11-6

Foreword to the English Version

This is an English version of Iwatsuki, K., 'Natural History' published in Japanese by the University of Tokyo Press in 2018.

The translation was by Mr. Tadao Hikihara, to whom I am grateful for his comprehensive and patient contribution as it is quite difficult to render Japanese sentences in this kind of book into English. As Mr. Hikihara is not a biologist, I have reviewed some particularly technical terms, although their expression is mostly based on his interpretations. Some of the problems faced in this translation will be explained at the end of this book in the post-script, which was newly prepared for the English version of this book.

Introduced by Mr. Shoichi Mitani, the Expo '90 Foundation, the translation and printing of this book were contributed by Kyoto Tsushinsha Press. I owe very much to this publisher for issuing the English version of the book. Mr. F.Kobayashi took care of English editing, to whom I am deeply grateful.

The English in the postscript, two forewords, and the table of contents were carefully checked by Dr. David E. Boufford, Harvard University Herbaria, who has always helped with linguistic checks in addition to our discussions on taxonomic topics. I wish to express my cordial thanks to him for his patient and long-term instruction in my English expressions.

Knowing that the direct English version of a book originally written in Japanese is difficult to be digested by non-Japanese readers, I cordially expect them to follow the sentences patiently so that they can gain an understanding of the development of natural history studies in Japan. The descriptions and discussions in this book are mostly based on the natural history in Japan, and I fear that many English-speaking people may not be very interested in the facts about Japan that are introduced in this book. Still, I expect that a wider recognition of natural history as understood in Japan may contribute something to the promotion of a general understanding of Japanese knowledges in this domain to the international society. I hope the English version of this book, originally issued in Japanese, will contribute to and promote natural history studies and research.

In compiling the ideas of natural history, especially in Japan, I learned from a variety of people: teachers, seniors, and all kinds of colleagues, and I thank all of them on this occasion, although all the shortcomings and potential misunderstandings in this book are my own responsibility.

Preface

The University of Tokyo Press began publishing the Natural History series in 1993. This book is the last of this fifty-volume series. For the final volume, I was asked by the Press to provide an overview of natural history research conducted at Japanese universities, a topic not previously covered in writing.

I have already contributed two volumes to this series: 'Natural History of the Ferns' (1996) and 'Botanical Gardens in Japan' (2004). As I turned eighty around the time when I began to write this volume, I desired to summarize my experiences on this occasion.

My career as a researcher began as a post-graduate student at Kyoto University. Since then I have pursued research interests at various types of universities, including the University of Tokyo (a national university), Rikkyo (Sophia) University (a private university), and the Open University (at that time under the name of the University of the Air). At the University of Tokyo, I worked as a professor and as the director of the Botanical Gardens, which contributes to the area of lifelong learning. The Open University is a university that advocates lifelong learning. While serving as a faculty member and conducting research in my area of specialization, I was also deeply involved in natural history-related organizations.

After working at these universities, I participated in the activities of a museum, serving for ten years as the director of the Museum of Nature and Human Activities, Hyogo. As I have strived to promote natural history throughout my career, I developed a desire to summarize my views on natural history in Japan as a whole. This book discusses natural history from the perspective of a researcher who worked primarily at several Japanese universities.

While natural history covers an extensive range of subjects, the authors of this series have largely focused on the natural history of living organisms. I am no exception. My specialized area of research covers vascular plants, specifically ferns. As I hammer my thoughts into shape, topics naturally center around my research area. The aim of natural history studies is to delineate nature as a whole, as I repeatedly state in this book, although it is practical to first look at individual parts. Accordingly, the question posed is how to position and understand the parts of an integrated whole. I have strived to narrate this using the subjects of my research as examples. I await the assessment of the readers as to whether my efforts have been successful.

The way Japanese society views lifelong learning has changed substantially in recent years. Japanese natural history museums are facing increasing difficulties in terms of the scale of their activities and the support they receive, despite never being sufficient in the past. Nonetheless, they have achieved solid results owing to the positive attitudes of the staff toward their work.

At one time in Japan, museum curator jobs were regarded as a reluctant choice for those who could not hold a research post at a university, despite a desire to develop a research career. In recent years, some of the younger generation have wanted to find jobs at natural science museums, not giving universities a second glance. This is good news for individuals like me who experienced the older days. I have no intention to blame universities as research institutions of natural history. I know that the challenges needing to be addressed in natural history at universities have become increasingly significant. At the same time, my expectations are high for an increasing number of talented young scientists to be engaged in suitable activities at museums and similar facilities throughout Japan.

Natural history leaders in Japan have built a sound theoretical foundation with the publication of the Natural History series. Since the commencement of these publications a quarter century has passed, during which the field of natural history has received increasingly enthusiastic attention and the activities of natural history museums have become vigorous. Without a doubt, this results from the peaceful cultural practices on the Japanese archipelago. As I was involved in the management of the Museum of Nature and Human Activities, I sense afresh the significance of, and the backdrop to, the development of natural history during this period.

The Natural History series is completed with the current volume. This does not mean that the study of natural history is complete. Rather, hopes for the future rest on radical advances in natural history based on the foundation provided by this series. As the author of this book, the final volume of the series, it would be my greatest pleasure if it plays a part in this role.

CONTENTS

[1] Japanese characters may be pronounced the same, but have different meanings.

Chapter
1

Back through Natural History: Chronological Observations of Nature

In contemplating natural history, Chapter 1 traces the past and present usage, or evolution, of the term "natural history," exploring the history of this area of study. The definition of the term does coincide with the activities indicated by the term. Nevertheless, considering both definition and usage is useful in delineating the concept of the term, thereby exploring contemporary issues facing natural history.

1.1 The Category of Natural History

(1) The designation "natural history"

The term "natural history" originates from the Latin *historia naturalis*, derived from 'Naturalis historia,' a work completed in 77 (A.D.) by Gaius Plinius Secundus, or Pliny the Elder (22/23–79). (This does not imply that the term was coined by Pliny. It is simply the designation of a work having survived today.) The work is huge, consisting of 37 volumes that present a general view of human knowledge at the time. It is an encyclopedia covering everything from the makeup of the universe and movements of the planets to the science of nature and technology, as well as currency, paintings, and human-made structures. The work was dedicated to the emperor Titus F. V. (39–81). Pliny published 10 books himself. The rest are said to have been published by his nephew and adopted son, Gaius Plinius Caecilius Secundus, or Pliny the Younger (61–112).

Plinius goes down in posterity as a military man and politician, rather than a natural historian. He is said to have written over 100 books, of which only the aforementioned tome has survived until today. Given that the book title 'Naturalis historia' is the most ancient record of the term natural history, it is appropriate to historically trace the concept of natural history to this book's content.

(2) 'Bowuzhi' (Records of Diverse Matters)

In slightly more recent years in China, 'Bowuzhi' was compiled by Zhang Hua (232–300), a Jin dynasty man.

Zhang was a Western Jin scholar who flourished in the third century. Celebrities of those times who are still known today were either excellent politicians or successful warriors. Zhang was also a high official who served in the Wei and Jin dynasties during the Three States and Jin periods, which demonstrates his abilities in political, diplomatic, and military affairs. Despite being from an impoverished family, Zhang Hua could join the central government because of his literary talent and broad knowledge. His work 'Jiao-Liao Fu' also met with acclaim from a high official (Tanaka, 1986). Biographies of Zhang Hua exclusively describe his work in the political arena.

The assigned duty of Zhang, as a government official, was to compile an authentic history. However, he became deeply involved in politics. Consequently, Chen Shou (233–297), who was recommended by Zhang to that duty, wrote a representative historical text from that time instead, 'Records of the Three Kingdoms.' Moreover, 'Bowuzhi,' known to be a work of Zhang, has not remained in its original form. The readable form surviving today is a result of editing by Chang Jing (?–550) and others later in the Bei-Wei period. The original work was a compilation of miscellaneous and fragmentary topics amounting to about 100 volumes. It has been ascertained that years later people edited this down to 10 volumes by revisions such as deleting absurd topics.

'Bowuzhi' also presents a general view of all kinds of knowledge from that time, collecting topics of interest from within the country and abroad that include zoology, botany, and myths. Topics found in the work surviving today comprise general geography, landscapes, products, rare animals, birds, insects, plants, medicinal plants, regimen, books, rituals, music, garments, and historical facts, as well as foreign people, manners, customs, and products. In sum, 'Bowuzhi' aims to thoroughly collect every kind of knowledge equivalent to, from a contemporary perspective, an encyclopedia.

The records say that 'Bowuzhi' by Zhang Hua was brought to Japan as early as the country's Nara period (710–794). The book must have had some influence on the Japanese concept of nature. Of course, I am aware that "nature" in this context does not directly lead to natural history, as this book intends to discuss.

In China, 'Bowuzhi' by Zhang Hua served as a model for other works published thereafter. 'Xu bowuzhi' (Continuation to the 'Bowuzhi'), written in 10 volumes by Li Shi and published in the Sung period (960–1279), follows the style of Zhang's work. In Japan, a reprinted edition of 'Xu bowuzhi' was published in the mid-Edo period.

Evolutionary extensions of Zhang's work from the Ming period (1368–1644) include 'Bowuzhi bu' (Supplement to the 'Bowuzhi') by You Qian (recruited to the government in 1501, according to records) and 'Guang bowuzhi' (Enlargement of the 'Bowuzhi') by Dong Sizhang (1587–1628). These works are indicative of the significant influence exerted by Zhang Hua on the history of natural history in China.

Pliny's 'Naturalis historia,' as with Zhang's 'Bowuzhi,' covers somewhat different areas of natural history than this present book. Their works are records that provide a comprehensive range of the knowledge that circulated in their time, which, in contemporary terms, performed the role of an encyclopedia. Demand for such comprehensive surveys of knowledge still exists today. The influence of these predecessors lingers on in current encyclopedias, whose recent evolutionary form may take the electronic-based Wikipedia. In the Western world, the term "natural history" has also gradually approached its contemporary usage. In France, led by Denis Diderot (1713–1784) and Jean Le Rond d'Alembert (1717–1783), the Encyclopediasts' attempt to compile extensive knowledge before the French Revolution flourished. This effort was joined by other influential members, including Voltaire (Francois-Marie Arouet, 1694–1778) and Jean-Jacques Rousseau (1712–1778).

Two attitudes still exist towards natural history: 1) to understand it in the framework of the works of Pliny and Zhang Hua and 2) to grasp it under evolutionary names, such as the *study of natural history* and *natural history and science*. The latter framework, as described later, exists in contemporary Japan, where it is assumed that natural history was transformed during the developmental stage of science.

The biological aspect of natural history that emerged in the time of ancient Greece developed into practical compilations in Europe as *materia medica*, inherited by the concepts of fauna and flora. In that process, minerals were similarly considered as products. This resembles the development of herbalism in China, which took place in parallel with the evolution of the tradition of Zhang's 'Bowuzhi.'

It is worthwhile to note that both in Europe and China, although natural things were recorded in the form of natural history, almost no wild creatures appear in literary arts. In contrast, diverse wild animals and plants appear in poems in the 'Man'yoshu,' the oldest collection of Japanese poetry. Nonetheless, very few classical writings in the East and the West, such as the Bible, 'Iliad,' 'Odyssey,' the 'Book of Songs,' and 'Ramayana,' refer to wild animals or plants, although names of some bred animals and cultivated plants that are familiar in everyday life occasionally appear.

Lamarck (see Tea Time 2), who made his debut as an inconspicuous French Encyclopediast, eventually served as a driving force for the concept of evolution by coining the term "biology." In later chapters, this book reflects on the fact that young Lamarck, who had been following a down-to-earth approach as a botanist, contributed to compiling a separate botanical volume of the 'Encyclopédie.' As exemplified in Section 2.3, he presented the first recorded description of *Asplenium unilaterale* in the 'Encyclopédie' in 1786, the scientific name long-used for the Japanese species *Hymenasplenium hondoense* Murakami & Hatanaka.

Pliny and Zhang Hua alike were politicians by profession, rather than botanists in the contemporary sense. Politicians in those days, whether in the East or the West, were successful military men. At the same time, politicians were people of letters. In the very early stages of recording human knowledge, terms such as *naturalis historia* and *bowuzhi* seem to have been understood as a collection of knowledge. The titles of these books were, at that time, used for a collection of knowledge then available. The literary abilities of politicians were indispensable for collecting and compiling human knowledge. Strong driving forces behind human history are politics and military affairs. In these areas, information gathering must have been a basic task since the time that humans were living in the wilderness.

Nevertheless, a collection of knowledge in this context was something that resembled extensive reading and good memory. It was not akin to modern science, through which scholars pursue universal principles lying behind phenomena, as guided by their intellectual curiosity. Another speculation is that, in terms of human intelligence, before the emergence of modern science curiosity might have been biased towards extensive reading and good memory.

Many superb studies present historical evidence of how the terms *natural history* and *bowuzhi* were used in ancient times. This book refrains from again presenting these achievements. Ascertaining the etymology of the word *history* and its contemporary meaning is certainly significant. However, it somewhat deviates from the purpose of Chapter 1. In line with this book, this chapter will simply trace that achievements that have accumulated and evolved using the term *natural history* or *bowuzhi* since around the beginning of the Christian era (or, in Japan, before the emergence of the first state). This process uncovers a history, whether in the East or the West, that began with the collection of intellectual accomplishments in documents reported to a person of power. Similarly, in Japan, descriptions of nature have been recorded in various forms since the country's first production of documents. Nonetheless, the situation in Japan was somewhat different from that in

Greece or China.

One cannot go without mentioning that the title 'Naturalis historia' by Pliny was succeeded centuries later in 'L'Histoire Naturelle' by Georges-Louis Leclerc, Comte de Buffon (1707–1788). Its subtitle explains "general and particular, with the description of the King's Cabinet (present National Museum of Natural History)." The work was written mostly by Buffon and partly by contributors, with publication beginning in 1749. After Buffon died at the time of the publication of volume 36, his successor, La Cepede (B.-G.-E. de La V.-s.-I., comte de La Cepede, 1756–1825), added eight volumes to complete the work in 1804. In the form of a list of items at the natural history museum, the work describes how animals, plants, and minerals exist in nature, while also presenting them as specimens. It was truly an accomplishment that deserves the name "natural history." The first edition printed in full color is said to have been a best-seller. This may imply that Paris, specifically the aristocratic circle of the city in the early years of the 19th century, was willing to absorb scientific knowledge about nature's creations.

Outside the domain of science, a well-known example of natural history is 'Histoires naturelles' (1896) by Jules Renard (1864–1910). The book contains this exquisite writer's record of nature observations, which made a strong impression on me when I was a child. Joseph-Maurice Ravel (1875–1937) composed vocal works based on 'Histoires naturelles' in 1908, which has also contributed to an increasing number of natural history lovers.

Most recently, the term *natural history* appears in 'The Natural History of Iblard' by Japanese painter Naohisa Inoue, which depicts his unique fantastic scenes. The term has thus come to have flexibly and intuitively developed implications.

1.2 Natural History and Science: Creation and Transmission of Knowledge

(1) Aristotle's natural philosophy

The term *naturalis historia* emerged in the first century and *bowuzhi* in the third century. Without being particular about the term natural history, this section looks back at how humans have viewed nature in their culture and system of knowledge. Next, this section focuses on what contributions the field known as natural history has made to culture and academia. For these purposes, the road to the history of natural history begins with looking at the burgeoning of Western science, the mainstream of contemporary natural science.

Before the emergence of human culture, in other words, before the evolution of *Homo sapiens* from a species of wild animal to culture-bearing humans, discerning natural objects was vital for survival. Humans produced basic living needs, such as clothing, food, and housing, from natural objects. Their life has been sustained by natural objects.

Humans gave names to animals and plants. Individual human beings began to exchange information among them via words. The amount of information stored in society increased. Subsequently, the distinction of animal and plant species, the most important information for human survival, also became a subject of human intellectual curiosity beyond the simply practical. Wild animals learn from their parents distinctions between edible and inedible things and between friends and foes. Since the development of language in human society, notions about natural objects have been put into words and transmitted between individuals. Society began to share discovered knowledge, such as, for example, that millet is edible, and roots of *Aconitum* must not be eaten but can be used as a fish poison.

Before long, the accumulated knowledge of natural things allowed for human intellectual intrigue into the relationships between things. Intellectual curiosity towards relationships between things significant for human life and towards those things irrespective of utility began to develop in a delicately mixed manner.

Along with human understanding of natural objects and the frequent exchange of information practical for living, people began to live affluent lives. It was not long before human society began to accumulate knowledge that can be described as culture. Their understanding of nature began to fit as a part of intellectual activities, not simply as a means of meeting urgent survival needs. Studies conducted to understand nature prior to modern natural science are collectively called "natural philosophy" by some scholars.

Aristotle (384–322 BC) is one of the first scholars to understand nature in a systemic framework and record this in written works. Accordingly, he is known as the founder of natural philosophy and the originator of the field of natural history. It is a historical fact that Aristotle's system of study was not properly transferred to the early Christian world and needed, prior to the Renaissance, to be rediscovered in Arabic literature inherited by the Islamic world. However, this book refrains from delving into this lineage further.

Aristotle is described more as a philosopher than as a scientist. His natural philosophy regards nature more speculatively. Of Aristotle's surviving works, those in the domains of logic, metaphysics, politics, and ethics are most popular. However, as an originator of natural science, he left many books on zoology, including zoography, as well as on natural philosophy, astronomy, and meteorology. (The collected works of Aristotle have been compiled repeatedly. To write this book, I primarily relied on the Japanese version of his collected works published during 1968–1973 by Iwanami Shoten Publisher.) Aristotle wrote no book on botany. Theophrastus (see Tea Time 1), his disciple and coeval, is known to be the founder of botany.

Although Aristotle's descriptions of and attitude towards nature are noticeably speculative due partly to his work in metaphysics, Aristotle examined material objects through his own observation and strived to write accurate descriptions. Among others, his descriptions of marine animals provide valuable examples. Sea urchins have a huge appetite due to a mouth with five strong tooth-like plates. The name of the structure is "Aristotle's lantern," dedicated to the founder of natural philosophy. His attitude of understanding individual facts observed in nature in the framework of laws that underlie living things shares commonalities with contemporary natural science.

Objects and phenomena in the realm of nature were individually observed and recorded. Likely, these observations produced abundant knowledge in human society. Aristotle made the first human attempt to collect and organize this vast amount of information into a system of knowledge. It is highly likely that by his time human society had advanced and accumulated enough knowledge for him to do so. Even so, Aristotle was presumably a man who read extensively and had a good memory because he compiled the human knowledge of his time in a comprehensive manner. Accordingly, he must have had collaborators in acquiring information.

If collected information was simply recited, he would not be able to frame any objective conclusion sufficient for satisfying his curiosity. That was why he attempted to sort out and systematize knowledge in an Aristotelian manner. Present-day people have remarkably advanced scientific information. However, fundamental questions remain, such as: What is the meaning of life, and how is it possible to make the environment comfortable for humans to live in? Even with the seemingly enormous amount of human knowledge, we are still far from reaching a scientific conclusion that precisely addresses these questions. With the fragmental scientific knowledge of his times, Aristotle was unable to derive a scientific and logical conclusion that would fulfill his intellectual curiosity. What he was able to do was to present inferences as conclusions derived through speculation. As a result, he is not the founder of natural science. He is defined within the category of philosophers, rather than scientists, by prevailing notions of the present-day. Nonetheless, he may be praised as the originator of very speculative natural philosophy.

Aristotle tutored Alexander III (356–323 BC), who expanded the territory of Macedonia to a global level between the age of 13 and his accession to the throne. As this fact suggests, Aristotle was a Macedonian. After the death of Alexander the Great, persecution of Macedonian people took place in Athens. Aristotle retired, likely due to this persecution, dying in the year following his retirement (some scholars say that he poisoned himself).

In Athens in Aristotle's time, Macedonian people were often persecuted. A situation that was certainly misfortunate for him. Although he founded the "Lyceum" intending to facilitate academic pursuits, he was not a very successful educator. (The Lyceum was a gymnasium located at the Temple of Apollo in Athens. Aristotle is known to have taught his disciples at the gymnasium. Hence the "Lyceum" is also used as a name to refer to his school.) Aristotle failed to train great successors who would inherit the tradition of Socrates and Plato and surpass Aristotle. Likely due to this, his works were treated as sacred books in the academic world. Through the Middle Ages, even his incorrect understanding was regarded as true and left uncriticized, which was an unfortunate outcome.

In Aristotle's system of studies (scholasticism), logic is fundamental, with natural philosophy and metaphysics categorized into theory, politics and ethics categorized into practice, and poetics categorized into production. Natural philosophy was placed in the domain of theory and understood as aligning with metaphysics. Of the intellectual activities, those involved in the growth of knowledge were included in theory, while the understanding of nature was placed in the domain covered by natural philosophy. In natural philosophy, all objects and phenomena in nature were subjects of intellectual curiosity, resulting in the study of topics ranging from celestial bodies to weather and living things, as well as humanity. In the end, Aristotle's scholastic view of the world collapsed with the establishment of modern physics in the 19th century, after having gone through the scientific revolution by Galileo and Descartes. The English term "physicist" appeared in 1840.

Unlike present-day natural science, natural philosophy placed emphasis on speculative elements rather than on simple depictions of observed objects, as seen with Aristotle's research and observation methodology. By present-day notions, natural science for the purpose of understanding nature is based on experiments and observations to organize concrete natural objects and phenomena into a mathematical framework. These efforts explore the objective universality inherent to natural objects and phenomena in an attempt to determine laws of causality.

However, in the 19th century, natural science began to distinguish itself from social science when the specialization of science progressed in the Western world. Each scientific field became independent of each other as it specialized. This led to a remarkable deepening of knowledge in each field. This trend has been and is increasingly apparent. Until very recently, natural science had been definitely divided into four branches: physics, chemistry, biology, and geology. Nowadays, whether these divisions are significant or not for research purposes is occasionally questioned. Moreover, extreme specialization of fields is countered by insistences, not few, that bridging humanities and natural science is one urgent issue facing contemporary sciences. (In fact, the clear distinction between humanities and science prevalent in Japan is not common.)

The Latin word for natural philosophy is *physica*, which is derived from *Ta Physika*, a title that encompasses Aristotle's entire body of work. According to his thoughts, natural philosophy is a study of the principles behind the stationary or moving states of objects, and phenomena emerge based on these principles. Accordingly, natural philosophy, in a wide sense, is no different from present-day natural science in that the subjects of study are all objects and phenomena in the natural world.

From the perspective of the modernization of science, the heliocentric theory of Copernicus and Galileo began by negating Aristotle's geocentric theory. Similarly, Descartes' throwing off the yoke of scholasticism was meaningful. Modern science started by refusing Aristotle; however, Aristotle is still regarded as the founder of science. Even today scientific knowledge is limited, and it is actually unavoidable for science to use not only empirical evidence, but also speculation.

Aristotle thought that, as seen in his work, objects and phenomena do not exist separately in nature. Rather, they, as a whole, constitute a single nature. No object or phenomenon exists or occurs separately. All objects and phenomena are mutually related in some manner. To what extent he was aware of this is not known. However, he was interested in, and sought knowledge about, all kinds of things and was willing to observe uncertainties. In this sense, his natural philosophy is a study intended to understand and elucidate the reality of nature in its whole form.

In Aristotle's time, when people had very limited knowledge of objects and phenomena, speculation was essential for a unified understanding of nature, even with advanced knowledge about concrete facts. Indeed, to those who understand present-day natural science, one troubling aspect of Aristotle's natural philosophy is its speculative character.

However, to gain a unified understanding, how much of nature can we scientifically recognize, even with our far more advanced present-day knowledge of natural objects and phenomena than in the time of ancient Greece? This question forms a challenge facing science that is presented throughout this book. I would like to argue this issue from the natural historian perspective. The intellectual curiosity of humans always anticipates the conclusions that science delivers, instead of scientific interim reports.

Theophrastus

Theophrastus (or Theophrastos in Latin notation of Greek, 371–287 BC), often considered the father of botany, was born in Eresos on the island of Lesbos. This name, as he became popularly known, was given to him by Aristotle. It means a person "to phrase" (phrastos) like "God" (Theos), indicating his eloquence.

Theophrastus studied in the Platonic Academy, as a member junior to Aristotle. Despite a considerable difference in age, they studied hard as friends. Aristotle bequeathed his writings to Theophrastus. In fact, the role of heading the Lyceum, founded and headed by Aristotle, was transferred to Theophrastus, who served in the position for 35 years. The Peripatetic school flourished during this period. According to records, Theophrastus left a vast number of works, as was the case of great people of letters at the time. Most of them have been scattered and lost. 'Historia Plantarum' is one of few surviving works of Theophrastus.

Aristotle is said to have left more than 500 writings, including those on plants. However, only a third, about a hundred and some works, are extant today, and none of his works on plants have survived in their entirety. While Aristotle is regarded as the founder of science, Theophrastus is known as the father of botany due to his extant detailed descriptions of plants.

Theophrastus lived to the great age of 85, until his death around 287 BC. Records say that a state funeral was held for him in Athens. The Athenian custom of the time was that Athenian citizens held these funerals on their own accord. That this event took place indicates that he was a great scholar respected by the citizens. This is understandable because he was the consistent head of the Lyceum for 35 years, in an era when people's reputations were quite volatile.

To write his primary work 'Historia Plantarum' (9 books), Theophrastus observed 500 hundred and some plant species known at the time, recording detailed information, including their relations with humans. Prior to this, he had already commenced his attempt to systematically catalog relationships between species included in this work.

Of the nine books in the original work, the first two provide a general description of botany. Book 3 begins with general explanations of wild trees, followed by descriptions of individual species. Throughout the nine books Theophrastus describes plants, including their relations with humans. However, due to his own beliefs, he strictly avoided supernatural explanations. In this sense, it is one of the first thoroughgoing, purely Western scientific works.

In documenting plant species, Theophrastus went beyond simply listing them, presented a classification method and introduced hierarchical classification. He assumed that several *eidos* (species) occur under one *genos* (genus).

The text of 'Historia Plantarum' seems to have functioned as working notes for lectures at the Lyceum. Rather than being written at once, they appear to be revised as he gave lectures. Hence it is natural that alternative versions would exist. While the title 'Enquiry into Plants' is used as an alternate name for the work, this book refers to it as 'Historia Plantarum.' The date of completion is not known with certainty. Some scholars estimate around 314 BC, the year Theophrastus began to teach at the Lyceum. This was probably the time when he prepared original material for his lectures.

(2) Understanding nature in medieval Europe

In the history of science, the Roman Empire, which had a succession of political conflicts, produced few scholarly works, compared to ancient Greece, which had the work of Aristotle and Theophrastus. When it comes to history, few history books describe the lives of the general public, although they constitute a large part of humanity. Historical texts are records of politics and wars, which are thrilling and exciting for some people. This may be a consequence in the interest of an era biased towards the tactics of people active in the arena of politics and towards war events.

Dioscorides' 'De Materia Medica' As an extension of Aristotle's natural philosophy, Pliny published a tome that described all objects and phenomena in nature, as mentioned earlier. In the history of botany, the first written work is Theophrastus's 'Historia Plantarum,' as described in Tea Time 1. Further, Pedanius Dioscorides (c. 40–90), a contemporary of Pliny, diligently surveyed and studied plants for their medicinal uses and published a cohesive five-volume work, 'De Materia Medica.' This mirrors the development of herbalism in the Orient.

Dioscorides, a native of Asia Minor, served as an army doctor for the Roman Empire in the reign of the emperor Nero. He traveled extensively throughout the empire to conduct field research on medical materials and sorted and classified them according to their pharmacological effects and functions, such as convergence, diuresis, and laxation. 'De Materia Medica' covers 600 medicines of plant origin, along with 80 animal and 50 mineral substances. In the sense that it encompasses nature's creations from a perspective of medicine, the work is a record of natural objects and organisms akin to medicine books of the Orient, rather than a book of medicinal plants. Of plants alone, Theophrastus's 'Historia Plantarum' covers 480 herbal medicines, while 'De Materia Medica' presents some 600 plant species and their uses, partly based on existing studies.

Even during the era of the Roman Empire, human knowledge of natural objects steadily accumulated, although this was inconspicuous in the history of the empire. Such knowledge was probably recorded by knowledgeable elderly people engaged in local medical services. Knowledge about medicines prevalent in everyday life preceded other genres of knowledge, which in a sense is a natural consequence of human history.

The records say that the enthusiasm for learning, which came into flower in ancient Greece and developed into contributions to society in the early Roman Empire, could not thrive after the fall of the Roman Empire, and before long, society experienced the medieval times known as the intellectual Dark Ages. In the field of botany, Dioscorides' 'De Materia Medica' played the role of an absolute reference book, because it was critical for people to look for medicinal plants directly relevant to their everyday life. Studying plants involved working hard to interpret the work and conduct historical investigations. To mention another noteworthy publication, Albertus Magnus (c. 1193–1280) published 'De vegetabilibus' in the 13th century. Writings that surpassed Dioscorides were not published until the appearance of illustrated manuals by O. Brunfels (1488–1534), L. Fuchs (1501–1566), and C. Clusius (1526–1609) as addenda and corrections to 'De Materia Medica' and later, the publication of 'De plantis libri' by Andrea Cesalpino (c. 1519–1603).

Referring to the medieval times generally as the Dark Ages is criticized by some researchers. Indeed, the Notre Dame Cathedral, built during the 12th and 13th centuries, is sometimes mentioned as a counterexample. However, in European society dominated by Christianity and Islam, science found no room to advance. It is likely that people with superb abilities occasionally appeared; however, they probably lacked the societal background needed to demonstrate their abilities. It should seem an undeniable fact that stagnant intellectual activities inhibited the growth of civilization during the medieval times, in contrast to the Greco-Roman periods. Nonetheless, one cannot overlook the historical fact that Aristotle's works were rediscovered in the medieval times.

(3) Development and growth of natural history outside Europe

Modern humans having emerged in Africa, moved to the Eurasian continent 50,000 to 100,000 years ago. Some of them proceeded to Europe, while others headed for Asia. Apparently, they did not part cleanly as soon as they reached Eurasia. They seem to have settled first in the present Middle East areas and become diversified into a number of groups, some lineages of which migrated to the east and to the west like ripples. It is said that the first group of people who arrived in Japan saw the Japanese archipelago some 40,000 years ago.

It is highly likely that, as they migrated, people carried their culture or civilization in some form. Moreover, individual groups must have built up something new of their own during the migration process or after settlement in a new region. Historically, each ethnic group that emerged in different regions developed respective indigenous cultures or civilizations adapted to the climate of the region they settled in. Of course, it is also true that they basically retained universal human aspects inherited from ancestors that originated from Africa.

Humans coming to Asia developed an advanced civilization, primarily in China. Sophisticated cultures developed in China have remained at the core of Asia to date. In China, the majority Han people ruled the territory for most of history, with land occasionally controlled politically by other ethnic groups. Although controlled by other ethnic groups, the Han population maintained their culture within a line of tradition, while slightly under the influence of other rulers.

Japan, an island country, is naturally highly independent, despite its small size. It appears that inhabitants of the islands accepted groups that came ashore the archipelago many times. Accordingly, Japan's people have never been a consanguineously homogenous group. The people on the archipelago smartly incorporated incoming peoples and cultures to form

Japanese people and culture, adapted to its climate. In Japan, natural history developed under the full influence of the neighboring superpower China, yet some noticeable distinct evolutions took on Japanese character.

According to some scholars, in the primitive age the Middle East was a verdant region covered with the Lebanon cedar. If it is true that the people who settled there cleared the forests of the Lebanon cedar and turned the region into desert, it may be safe to say that the culture that developed in the desert was not simply under the influence of nature but was an outcome of human activities that transformed nature.

China: Herbalism and Natural History Historically in China, the terms herbalism (*bencaoxue*) and natural history (*bowuxue/bowuzhi*) have been used interchangeably at times, and in different senses at other times.

The term natural history (*bowuzhi*) seems to have been used to cover extensive topics in an encyclopedic manner and beyond the study of natural objects and organisms. This was exemplified by Zhang Hua's 'Bowuzhi' and the works that followed as extensions. The development of natural history in China can be viewed in parallel with the origination of natural history in ancient Greece, as it developed in Europe.

In contrast, the term *bencao* (herbalism) was used in China to mean records of medical materials that described medicinal natural objects and their uses. The term *bencao* is also used to imply general plants. Incidentally, in the Japanese language, the term *honzō* (herbalism, the same characters as in Chinese but in different pronunciation) is used to specifically refer to books of medicine as well. Medicine books in China cover animals and minerals, although examples of these are fewer than those of plants.

In the West, the term herbalism is used as a generic term to refer to topics on plants that serve as medicines. The term once had a similar meaning to its Chinese counterpart, *bencao* (medicinal plants). In recent years, subjects of herbalism research include plants (and other natural objects) used as traditional folk medicines in different parts of the world, attracting great interest as gene resources for drug-discovery purposes. The term herbalism overlaps with its Japanese counterpart term in meaning. However, herbalism also appears to imply modern pharmacognosy.

The term *bencao* first appeared in the Treatise on Sacrifices in the records of the history of the Former Han dynasty, the 'Book of Han.' It is said that, although not surviving today, there were several books on medicinal plants around the first century BC. Nonetheless, while containing information on medicine, medicine books at this time, an era when the veneration of deities and immortals had an influence on medical practice, likely also contained descriptions relating to deities and immortals.

The oldest surviving medicine book is 'Shennong Ben Cao Jing.' Although the original is not extant, its revised editions with addenda have been widely used as a tutorial for medical practice. Tao Hongjing (456–536) revised and compiled pieces of text and alternative versions used discretely at the time and published 'Bencao jizhu.' Since its completion around 500, it has been used as a model book of medicine.

Around 1090 in the Song dynasty, existing knowledge, including additions to the above book of medicine was compiled into 'Jingshi zhenglei beiji bencao' (A book on materia medica). At this point in time, as before, 'Shennong Ben Cao Jing' was widely used and held in high esteem as a sacred writing on medicinal plants. However, some errors had been introduced and incorrect copies were in use. In the Ming dynasty, Li Shizhen (1518–1593) examined the details of 'Shennong Ben Cao Jing' and other books then in circulation, conducted additional field research, polished his knowledge of plants, and carried out the enormous task of editing 'Compendium of Materia Medica.' He completed the manuscript in 1578. However, it was a misfortune for him that the Jinling edition (the first edition of the book) was published after his death in 1596. In China, there is no surviving copy of the first edition.

'Compendium of Materia Medica' was introduced to Japan in 1607. HAYASHI Razan presented it to the Tokugawa shogunate. Hence, the book had a strong influence on herbalism in Japan. Moreover, it was also used in the practice of folk remedies. ONO Ranzan (1729–1810) compiled his working notes based on the book into 'Accounts of Compendium of Materia Medica.' He further edited it into 'Dictated Compendium of Materia Medica,' which had a colloquial flavor and was published in 1803. Ono was an outstanding herbalist. He revised 'Compendium of Materia Medica' for Japanese readers and added articles about Japanese materials and his perspective, publishing it as a book of natural history. Consequently, his efforts added value to the book for use as a textbook and substantially contributed to the growth of herbalism in Japan, although it remained a compilation of working notes, or a commentary to the original book.

The development of natural history and herbalism in China is regarded as a phenomenon occurring in parallel with the development of natural history and materia medica in Europe. Throughout history, the East and the West have had cultural exchanges in various forms. However, details regarding how the East and the West transmitted knowledge and influenced each other have not been ascertained. Substantial differences exist between the East and the West in natural objects, especially in the species present in the wild. Dioscorides' 'De Materia Medica' might have been almost of no use in China. If 'Shennong Ben Cao Jing' were introduced to Europe and translated into Latin, the cultural level in the region before the Renaissance was not sufficiently advanced to make use of it. When 'Bowuzhi' and other books of medicine were brought to Japan from China, Japanese people gradually developed abilities to use them, revising them into Japanized editions with addenda. Nonetheless, it is not surprising that the momentum to exchange knowledge in these fields of science did not exist between the East and the West.

India and Southeast Asia Cultures that emerged in India are notably characterized by thriving streams of philosophical thought, such as Hinduism and Buddhism. However, since ancient times, when culture flourished ahead of other regions, neither projects to compile natural history works faithful to facts, nor the practice of documenting nature's creations have developed.

Similarly, in Southeast Asia, people did not develop a custom of diligently collecting intellectual information that was driven by scientific curiosity. There was no known culturally advanced activity relating to natural objects in the region.

Natural History in the Middle East UMESAO Tadao, the foundation Director at the National Museum of Ethnology described the *Chuyo* (Mediant), or the Middle world, as lying between the Orient and the Occident (Umesao, 2003, original Japanese edition in 1957). The world's most ancient Mesopotamian civilization flourished in the Middle Eastern region. It played a major role in forming the skeletal outline of modern civilization, taking the lead in the world of science through the discovery of 0, the establishment of a mathematical system based on Arabic numerals, and the use of a solar calendar to name a few. Moreover, development of moral codes relied on the region, beginning with the Old Testament. Before long, Islam was founded, which led to the formation of powerful caliphates, in clear contrast to the dark Middle Ages in the West. However, the Middle Eastern region fell behind the rapid advances in science and technology of the material- and energy-oriented Western civilization. Nowadays, ordinary people in the region suffer from poverty.

When viewing the Middle East from a regional perspective, the Islamic world can be described from the standpoint of its philosophy. To examine the Renaissance that occurred in the Western Christian world and understand the spread of Greek culture in the West, one needs to remember the role played by the Islamic world, which resisted the early Christian world. After conquest by Christianity, the Roman Empire discarded Greek culture and most of the works by Aristotle and Archimedes were lost to society.

Nonetheless, the Renaissance occurred thanks to valuable Greek literature – which Christians had not been able to study – left in heaps of books discarded by Muslims on the run. These were found in Toledo on the Iberian Peninsula, under control of Islam in about the 12th century during the Reconquista. The literature in Arabic was conserved in the Eastern Roman Empire and the Islamic world and then translated into Latin in the Western Christian world, thus preparing the way for the Renaissance.

From the perspective of the natural history of living things, the Middle Eastern region has fallen behind other regions in compiling records of fauna and flora. One reason for this might be that people in the Middle East had little motivation for compiling natural history books due to the general lack of biodiversity in the wide expanse of arid lands in the region. Despite technological advances for urban development since ancient times, such as the construction of *karez* (underground channels), civilizations that emerged in the region were not strongly oriented towards a wealth of materials or energy. This may have led to the region's delayed advance in the field of natural science compared with developments in social sciences, such as religion, philosophy, and law.

Almost no records of natural history remain in this region from ancient times or the medieval times (Budge, 1978). People in the region must have had medical treatment and made effective use of medicinal plants and other natural substances learnt from Greek and Roman academic knowledge and art. However, no records indigenous to the region of the materials and their usage remain. It is thought that information on traditional medicines was collected and recorded by elderly people, who held the knowledge in their memory.

In the Islamic world, since theology consistently rules everything, preceding science, people may have had no chance to develop free scientific ideas. Moreover, the Islamic world expects scientists to be cultural generalists as well. As such, Muslim scientists might have failed to specialize and develop outstanding achievements in research in a specific field. In early periods, Muslim scientists translated and then studied the Greek literature in Arabic. Why is it that, before long, they became stuck to doctrine and made sparse contributions to natural science?

Natural History in Mayan and Other Cultures Humans are said to have moved to America some 15,000 years ago when the Bering Strait became dry during the glacial age. Other hypotheses say that humans migrated to

America far earlier, about 40,000 years ago. In either case, considering that the indigenous Mayan people are Mongoloids, it is likely that people of Mongolian race, probably moved eastward. It is speculated that after migration to America, ethnic fractionalization took place in some areas, while other ethnic groups arrived at different periods. Furthermore, some scholars hypothesize that Polynesian people also migrated to America. Whatever the case may be, evidence shows that modern humans were inhabiting the southern region of South America about 12,000 years ago. Incoming people consisted of many ethnic groups, with various civilizations.

The Mayan civilization, which was devastated by the invasion by Spain, formed cultures completely independent from the civilizations of the Old World. Mayan people were wholly defeated by the Spanish invaders because, as some say, they were innocent and deceived because of strategic failure. In addition, their technology was inferior to the Spanish people's. They certainly knew the principle of wheels. However, they did not put it into practical use, and instead relied on human power. Their farm work also relied on human power, as they had no cattle nor horses. Even in the 15th century, Mayan technology could not compare with that of Europe at that time, mainly operating slash-and-burn farming, terraced fields, and swamp farming.

Mayan people had superb architectural techniques, as seen with magnificent structures at their historical sites. In association with this, they had advanced knowledge of calendrical system and excellent skills of astronomical observation. They probably developed a civilization more oriented towards religious practice than towards the use of technology to pursue conveniences in everyday life.

It is certain that many ethnic groups of indigenous Americans developed diverse cultures, although neglect and destruction of their cultures continued during the invasion following the 'discovery' of the new continents by white people. They might have contributed little to the advance of modern science. However, it cannot be missed that the indigenous people's attitude towards nature provides direction for present-day ecological lifestyles.

(4) Natural history in Japan

Herbalism in the 10th Century The record of natural history in Japan began around 918 when FUKANE Sukehito (date of birth and death unknown) wrote 'Honzō Wamyō' (Japanese names of medicinal plants) by command of Emperor Daigo (885–930; Reign: 897–930). The work followed the style of 'Xinxiu Bencao' (newly revised Materia Medica) from the Tang dynasty, assigning Japanese names to medicines and verifying whether they were also produced in Japan. 'Honzō Wamyō' serves as literature that reflects the knowledge of the time, although it contains some errors regarding medicine identification. Moreover, its value lies in that it contains some text of 'Xinxiu Bencao,' which has been lost even in China.

'Honzō Wamyō' influenced subsequent Japanese medical and natural history knowledge. Although the original was lost, in the Edo period, TAKI Motoyasu (1755–1810), doctor for the Tokugawa shogunate, discovered an ancient duplicate copy of the book in the Momijiyama Library in Edo Castle. In 1796, Taki corrected errors and published the transcription, which survives today. The ancient transcription went missing again; however, photographs of it exist in Taiwan, which serve as valuable research material today.

Around the Engi era (early 10th century) in Japan, Japanese culture did not exist, as this was shortly after the formation of a state from what was known as the country of Wa, an assemblage of tribes. Nonetheless, around that time, a natural history book was compiled in Japan, although it was a Japanese version of a classical work originating from China. No cohesive work on science was created at the time (or at least does not survive) in either Mesopotamia or India, despite their cultural advances. The only exceptions at this point in time, were in the Mediterranean region and China.

Some scholars say that Japanese culture is a part of Chinese culture. Nevertheless, 'Honzō Wamyō' is not a simple copy of Chinese culture. It is a compilation of natural history that covers plants in Japan. Even in the Chinese cultural sphere, there was no local natural history book at this point. This notion shares commonality with the view of N.J.T.M. Needham (1900–1995) that modern science developed in Japan and Europe at the same time. It also could align with the understanding of UMESAO Tadao who, in 'Ecological Conception of the History of Civilizations,' grouped Western Europe and Japan into the first zone. (Umesao argues that it is incorrect to divide civilizations on the Eurasian continent into the Orient and the Occident, and that Europe and the Japanese archipelago should be grouped into the first zone characterized by advanced culture, with other vast areas into the second zone). Grouping the Mediterranean region and the Japanese archipelago is quite an idea. This topic may need consideration when examining science in Japan after the Meiji restoration. The following sections look briefly at tenth-century Japan.

Based on plant species, vegetation, and landscape names that appear in the 'Man'yoshu,' HATTOTRI Tamotsu et al. (2010) attempt to decipher nature of the Japanese archipelago at the time of the book. They conclude that people of the time had established clear divisions into inhabited, in-between, and deep mountain zones. The 'Man'yoshu' is said to have been compiled in the second half of the eighth century. The collection contains poetry composed during a period of more than one century, with the composition of the earliest verified to be in the first half of the seventh century. The anthology also has a distinctive feature, unique in the world, in that it contains poems of soldiers and anonymous ordinary people, as well as those composed by people of upper and educated classes, including royalty and titled nobility (although it is likely that the soldiers were probably of noncommissioned officer class and there is no positive proof that poets included people forced to live like slaves). The 'Man'yoshu' is also unparalleled in the world in that it contains poems referring to many wild plants, enabling scholars to trace plants and vegetation at the time. The compilation is an excellent record for botanists.

It is regrettable that subsequent poetry collections compiled by imperial command contain poems composed exclusively by aristocrat poets, often skilled to flirt with words, as seen when utilizing the technique of emulating works of great predecessors. Apart from artistic evaluation, they are not useful as straightforward records to indicate how the Japanese people connected with nature. In this regard, those imperial anthologies are in the same league as classical works of other countries.

Mainstream historians view tenth-century Japan as an underdeveloped country with no emergent indigenous culture. However, around that time, the Japanese had already begun to syncretize the traditional Shintoism concept with Buddhism, to borrow Chinese characters to express Japanese syllables, and created *kana* characters to represent the Japanese language. In this way, they began expressing the character of Japan culture. Culture unique to Japan, although not totally new, was beginning to emerge, modifying borrowed Buddhism and Chinese characters into a Japanese style. This trend also applied to natural history. A notable example is that 'Honzō Wamyō' was compiled in the style of 'Xinxiu Bencao,' which the Japanese people arranged into a Japanese style and began to use as Japanese literature.

For readers' information, Japan's first dictionary, 'Wamyō Ruijūshō' (931) compiled by MINAMOTO no Shitagō (911–983), contains plant names that appear in 'Honzō Wamyō.' In addition, another related book 'Ishinpō' (The essentials of medicine) (984) by TAMBA no Yasuyori (912–995) portrays the era of the tenth century. Following this, other books of medicine published in China were introduced successively to Japan.

In Japan, science in the proper form of modern natural science did not develop until very recently. Even in Europe, academia as proper science emerged only in the Renaissance period. In contrast, the Japanese people began to record the nature of the Japanese archipelago in text first in the form of literary work. Before conducting exploration based on scientific curiosity, they had captured impressive natural phenomena in an artistic manner.

For an example of a literary work themed on natural history, readers can refer to 'Tsutsumi chūnagon monogatari' (Tales of TSUTSUMI Chūnagon [middle counselor of the state] – year of creation unknown, though likely in the second half of the Heian period [794–1185] or later), which describes a princess who loved insects. This implies that there was a girl interested in insects as early as Heian-period Japan, although the story depicts her as an eccentric character.

Japanese Natural History Li Shizhen's 'Compendium of Materia Medica' was completed in 1578 and published in 1596. In Japan, this was immediately before the establishment of the Tokugawa shogunate in 1603. The Sengoku period (age of civil wars; mainly 16th century) had ended, followed by the dawn of the peaceful Edo period, which continued for 260 years. The records say that, in 1607, HAYASHI Razan (1583–1657) presented 'Compendium of Materia Medica,' obtained at Nagasaki, to TOKUGAWA Ieyasu, the first Shogun in the Tokugawa shogunate. It is said that health-conscious Ieyasu valued the book highly and studied it thoroughly. Japanese society also widely accepted it. Its Japanese editions also published repeatedly, like 'Tashikihen,' an edition of the compendium to which Razan added Japanese names. The work by Razan had a strong influence on herbalism in Japan and benefited medical care for the general public.

Using Razan's book as a guide, KAIBARA Ekiken (1630–1714) studied medicinal plants in Japan and published 'Yamato honzō' (Medicinal herbs of Japan) in 1709, a representative book of fauna and flora from the Edo period. Ekiken intended to accurately record the properties, rather than the efficacy, of living organisms. The result was a work that should be evaluated in terms of natural history or fauna and flora, instead of as a book of medicine. Ekiken was working on the study of fauna and flora beyond the perspective of efficacy, as early as when Linnaeus (1707–1778) was an infant.

In 'Yamato honzō,' Ekiken uses a term that means a study of diverse matters (*hakubutsu no gaku*). I speculate that his intention might have been to shift the focus from medicinal plants in the category of pharmaceutical literature to the study of diverse matters, as a shift towards natural history. This is probable despite the use, in the work title, of a term that also signifies medicinal plants. The use of the term meaning a study of diverse matters survived into the Meiji period, when natural history, as the foundation of science, was established as a study that recognizes and analyzes diverse matters in an impartial manner. At this point in time, herbalism fragmented into two parts: that specializing in pharmacy and that subsumed by natural history. Subsequently, both advanced greatly in their respective ways.

As the 18th century advanced, a growing trend emerged among general medical care insiders towards research and education concentrating on natural history. ONO Ranzan (1729–1810) devoted himself to studying plants, opting not to serve for the government, and instead opening a private school to make a livelihood. His disciples created a compilation of his lectures and other information, which Ranzan proofread and published as 'Honzō Kōmoku Keimō' (Dictated compendium of Materia Medica) (48 volumes), with the first edition released from 1803 to 1806. In addition, handwritten records of lectures given at his school survive as 'Honzō Kōmoku Kibun,' which neatly portrays how lectures were presented at the time. These serve as working notes and a commentary to the original work. However, they later contributed substantially to the development of herbalism in Japan, after revision by Ranzan, an excellent herbalist himself, for Japanese readers and after the addition of articles on Japanese materials and Ranzan's perspectives on them. These changes added value to the works as natural history textbooks.

Many works of literature were imported from Europe at this time, including superbly illustrated manuals. Learning from these existing studies, Japanese scholars published illustrated manuals of Japanese plants. In their 'Honzō zufu' (Illustrated manual of medicinal plants, completed in 1828) and 'Sōmoku-zusetsu' (An iconography of herbaceous and woody plants of Japan, 1856–1862), IWASAKI Kan'en (1786–1842) and IINUMA Yokusai (1782–1865), respectively, integrated botanical knowledge of the Edo period with pictures, signifying the high accuracy of studies conducted by Japanese scholars. With 'Sōmoku-zusetsu,' which follows the Linnaean taxonomy in the arrangement of plants, Yokusai cited, to the greatest extent possible, knowledge based on Chinese herbalism disseminated in Japan, yet he edited the book in line with Western biology.

The favorable influence of Dutch studies came to fruition in 'Kaitai Shinsho' (Japanese version of 'Anatomische Tabellen') (1774), which led to further developments. Prior to the influence of Dutch studies, medical practice had relied exclusively on Chinese medicine. With the Dutch introduction, the incorporation of anatomic techniques in such medical care allowed practitioners to diagnose patients' bodies simply by visual observation. Like the incorporation of Western medical techniques in Japanese medicine, Japanese scholars studying natural history under the influence of Chinese herbalism began to adopt more advanced Western techniques in natural history, as well. Nonetheless, despite E. Kämpfer (1651–1716) and, for botany, C.P. Thunberg (1743–1828) coming to Japan, Japanese scholars needed some more time to fully understand the advanced knowledge of the West.

Around that time, having learnt of Western science from various sources, UDAGAWA Yōan (1798–1846) wrote 'Botanika Kyō' (Sutra of botany) (1822), the three-volume 'Shokugaku Keigen' (Principles of botany) (1835), and other works. In these he coined a Japanese term for botany, informing Japanese people about botany in Europe at the time. In these introductory works, his understanding of natural history is pertinent even from a present-day perspective. This implies that Japan was smoothly absorbing advanced biological information transitioning into the Meiji period. In addition to botany, Yōan also introduced chemistry to Japan by writing 'Shemi Kaisō' (Introduction to chemistry).

In the second half of the Edo period, the level of education rose in Japan owing to the long years of peace. At the same time, remarkably advanced natural history studies from Europe reached the Japanese Archipelago, spurring the Japanese people to undertake similar studies in the field. Individuals such as Kämpfer, who authored 'Amoenitates Exoticae' (1714), Thunberg, and A. von Siebold (1795–1866) came to Japan successively. They helped develop natural history in Japan. Moreover, they published outstanding introductory books about Japan. They also compiled scientific books on Japanese flora to elucidate it from a current Western scientific point of view. Thunberg's 'Flora Japonica' (1784) was introduced to Japanese readers by ITO Keisuke (1803–1901), who obtained the book from Siebold and added Japanese plant names side by side with Western ones.

(5) The renaissance and scientific revolution

General historical commentaries explain that the dark medieval times broke with the advent of the Reformation and the Renaissance.

The Reformation was a movement that occurred in the Roman Catholic Church in the first half of the 16th century to reform the church system. The movement is symbolized by the Protestant schism initiated by Martin Luther (1483–1546). In parallel with Luther's movement, J. Calvin (1509–1564) led the English Reformation, which urged the Roman Catholic Church to transform. The movement in Germany was augmented by a return to the Bible. In fact, this was enabled by the dissemination of a German version of the Bible, on which the establishment of printing technology by J.G.L. Gutenberg (c. 1398–1468) had a great influence.

Scientific Revolution and the Rise of Modern Science The popular notion in current science historic thought is that the rise and success of the modern scientific view should focus on the scientific revolution in the 17th century following the Renaissance. The view of the world that bloomed in ancient Greece, based on a natural philosophy systematized by Aristotle according to a teleological view of nature, grew little in the Roman Empire, which would soon be dominated by monotheistic Christianity. This period is known as the dark Middle Ages.

After Christianity dominated the Roman Empire outstanding Greek works, such as those by Aristotle, were cut out from the Christian world and Greek- and Latin-speaking society. They were lost until those translated and conserved into Arabic in the Islamic world were Latinized to contribute to the Renaissance.

In Europe, control by monotheistic Christianity advanced, while the rise of Islam led to an expansion of acceptance of monotheism based on the Old Testament. That while all things are at God's will, humans are the noblest of all animals blessed by God and creations in the natural world are resources gifted by God for humans became a prevalent, common view.

In the mid-17th century, R. Descartes (1596–1650) in his series of works including 'Discourse on the Method' (1637) (many of them being published after his death) presented a mechanistic view of nature in contrast to existing scholasticism and established modern philosophy based on the proposition *Je pense, donc je suis* (I think; therefore, I am. In Latin, *Cogito ergo sum*). Descartes wrote his books in French, although it had been customary to write books in Latin at the time. He became interested in mathematics when he was a child and was the first person to use the concept of coordinates currently in common use. In 1610, when he was studying at the Jesuit college at La Flèche, he was delighted by the news of Galileo Galilei's (1564–1642) construction of a telescope and observation of the satellites of Jupiter. In 1633, the trial of Galileo, who had published his arguments on the heliocentric theory, began at the Inquisition. This was the historical background to the publication of 'Discourse on the Method.'

Transitions in the mid-17th century, such as towards the modern philosophy of Descartes and, at the same time, the use of modern scientific methods that incorporated hypothesis testing originating from Galileo, are sometimes referred to as the scientific revolution. The contributing factors to this revolution were, in addition to the departure from Scholasticism and the penetration of the mechanistic view of nature, the establishment of modern scientific methodology that existed, more specifically, to ascertain facts in the manner of hypothesis testing, while closely and accurately observing the universe with the use of an advanced telescope. Before long, the inductive reasoning of F. Bacon (1561–1626) came into focus in philosophy, and I. Newton (1642–1727) and other scholars contributed to the birth of modern physics.

Modernization of science progressed in the domain of medicine, as well. In 1628, W. Harvey (1578–1657), attacking Aristotle's view, published the theory of pulmonary circulation. In embryology, he inferred the life cycle as it is known today and suggested that all life, including mammals, come from the egg, although he was not able to provide evidence for this assertion.

In biology, the invention and advance of microscopy played a major role in scientific advances, like how the telescope facilitated the progress of astronomy and drove scholars to change their view of the universe. The first microscope is said to have been invented by Dutch spectacle makers Hans (father) and Zacharias (son, c. 1580–c. 1638) Janssen. Galileo also used the microscope. In the field of medicine, M. Malpighi (1628–1694) used the microscope to observe various objects. 'Micrographia' published in 1665 by R. Hooke (1635–1703) presented illustrated microstructures of living organisms. The term "cell" originated in his work. Hooke is known for Hooke's law of elasticity. He served as the President of the Royal Society as a physicist. However, because he did not get along well with his successor, Newton, Hooke disappeared for some time from science history. The microscope was improved by A. van Leeuwenhoek (1632–1723) and used for various observation purposes. He observed unicellular bodies, such as microorganisms and sperm cells, for the first time in history. Hooke appraised Leeuwenhoek's achievements and recommended him as a member of the Royal Society.

Through the scientific revolution, scientific studies grew as an intellectual activity to discover universal principles in line with the natural world, rather than through God's Eye. The obtained findings were soon appropriated to develop technology, and then in the mid-18th century, led to the Industrial Revolution. This technology enabled great accomplishments for those who used it.

Age of Exploration: European Christian World and the Globe The Age of Exploration refers to a historical period from the mid-15th century to about the 17th century during which Europeans explored various parts of the world. It was a social phenomenon that preceded the scientific revolution. Nations in southwestern Europe, such as Spain and Portugal, first made expeditions, which were then followed by England and the Netherlands. The term Age of Discovery is also used from a European-centered historical perspective. However, on the global scale, the discovery of the new world was made by the ancestors of the indigenous people when they migrated across the Bering Strait 15,000 years ago. Moreover, evidence also exists that northern European Vikings reached North America around the 10th century. When Columbus reached the new continent by sailing from southwestern Europe, rather than the discovery of the new world, it was the beginning of the subsequent history of hostile conquest by Europeans in pursuit of resources in America.

After experiencing the military failure of the Crusades (from about the end of the 11th century to the 13th century) and the extension of Muslim influences and the expansion of the Mongolian Empire's territory (around 13th to 14th century), Europeans increasingly sailed to Africa in the 15th century, with supporting factors including advances in navigation and the establishment of imperialism in Spain and Portugal. In 1488, they reached the Cape of Good Hope, the southernmost point of Africa. Vasco da Gama (c. 1460–1524), having left for his expedition in 1497, arrived in India in May the following year.

In 1492, C. Columbus (c. 1451–1506) became the first ever Christian to reach the new continents. Records say that the conquest of America by his crew and the succeeding Spanish army was a voyage of dreadful massacre of indigenous populations. With plagues and the far more advanced military equipment that they brought, Spanish people killed indigenous populations at will and obtained resources from the new continents.

The Folk Museum of Cordoba, sited in the place Columbus is said to have embarked on his voyage, exhibits dining tables that compare before and after the introduction of resources from the New World to Europe. The exhibit demonstrates how resources from the New World are important for the life of present-day people.

The Age of Exploration can be viewed as an era in which the European Christian world turned its eye towards the globe, imported resources outside of Europe, and began to overconcentrate wealth. In the history written by Europeans, this was known as the Age of Discovery. Thereafter, Europe became wealthy, facilitating people's cultural enrichment and the advance of natural science.

Age of Enlightenment Although the term Age of Enlightenment may not be in common use, from the late 17th century to the 18th century, when Enlightenment thought spread in Europe, Europeans, at least the upper reaches of society, enjoyed wealth resulting from the Age of Exploration. At this point in time, the scientific revolution was in progress. Around that time, there was a growing movement to spread this guiding thought among citizens. The scientific revolution was taking hold. Natural scientists observed and analyzed diverse natural phenomena. Scholars with intellectual curiosity in universal principles underlying the natural world were beginning to conduct analytical research in a concrete manner. However, their scientific studies still fell short of efforts to deepen analysis, with the extent of research specialization well below the present-day level. Despite this situation, natural history was making unique contributions, while being aware of general trends in the science of modernized surveys and observations.

Scientific studies of nature were receiving government subsidies granted in part with an aim to promote industry. Organizations such as the Royal Society and the French Academy of Sciences were successively founded. (The Royal Society, founded in 1660, is a voluntary body with no financial assistance from the royal family, although the word "Royal" is in its name. Nevertheless, the society plays the role of a national academy. The French Academy of Sciences was founded in 1666, funded by Louis XIV.) It was an era of absolute monarchy, and around that time, mercantilism dominated Europe.

Lastly, in that era, worthy of special note, the Enlightenment culminated, with the Encyclopédistes flourishing. The 'Encyclopédie' compiled by Enlightenment thinkers including D. Diderot (1713–1784) was published between 1751 and 1772. In addition to prominent figures, unrenowned writers also participated in this huge undertaking. It took more than 20 years to publish the work due to issues commensurate with a project of this size. Regarding the encyclopedia, many superb studies have been published.

It is said that as a backdrop to the compilation of the 'Encyclopédie' was a plan to translate 'Cyclopaedia' (1728) of E. Chambers (c. 1680–1740) into French. This is indicative of the fact that society at the time sought this type of undertaking. Indeed, when published first, the 'Encyclopédie' was welcomed by the wealthy class in France at the time, and the first edition numbered 4,250 copies, according to records. As history notes, the penetration of the influence of this work to everyday citizen was one factor that drove the French Revolution.

In the 18th century, literature that contained articles about diverse matters began to be referred to as an encyclopedia. From the perspective of the idea of showcasing diverse things, it has similarities with natural history in ancient Greece and China. However, natural history at the time was limited to articles about natural objects, as seen with Buffon's 'L'Histoire Naturelle' [see 1.1 (2)].

Subsequently in the age of the industrial revolution, enabled by the advance in technology based on science during the period from the mid-18th century to the 19th century, political events such as the independence of America (1776) and the French Revolution (1789) occurred successively in Western countries.

1.3 Natural History and Scientific Modernization

(1) Modern science and natural history

After Europe developed natural science, America took the lead from the mid-20th century, as people in America began to lead an affluent life with an advanced culture. This is demonstrated by looking at concrete numbers. Of 648 Nobel laureates awarded in three natural science categories by 2023, the number of North American winners (United States and Canada) was 298, accounting for more than 44%.

In scientific studies, particularly in the natural sciences, reductionist analysis ensures exquisite accuracy. In this way, subjects of analysis are fractionized, and emphasis is placed on close investigation of individual phenomenon. If a study has a flaw, however trivial it is, the study must be reworked from scratch, even after many hours have already been spent. To proceed, it becomes necessary to provide reliable evidence that eliminates all doubts. This is critical for science to grow.

Indeed, the correctness of scientific methodology used to verify hypotheses has become increasingly meticulous, which is regarded as an accomplishment of present-day science. Based on elucidated scientific knowledge, the techniques have become even more advanced and stronger, helping build affluent and safe human life. Like a two-edged blade, however, power buttressed by technology cuts both ways. If used incorrectly, the blade behaves like a devil. Advance in, and the unregulated use of, technology has caused the global environment to deteriorate. Since the 20th century, mass massacres wars that embroil citizens have become common. Accidents associated with technology are hideous both in quality and quantity, as expressed by the nauseating term 'traffic war.' These results might not be surprising with the use of technology by politico-economic experts who make light of, and are ignorant of, science.

This section is not intended to debate the virtues and vices of science. However, in the history of observing natural phenomena driven by scientific curiosity, humans have occasionally satisfied curiosity through partial answers. A likely natural consequence of a human society that has evolved as a group of intellectual creatures, is that humans pursue analysis as a means to fulfill scientific curiosity, as well as to make beneficial use of technology derived from science for society.

Scientific Revolution and Natural History While Aristotle's natural philosophy was based on a teleological view of nature, analytical methodology that has enabled present-day science emerged from the mechanistic view of nature initiated by Descartes. After the Renaissance and the scientific revolution, scholars, curious about the forms and phenomena living things exhibit, switched from methods oriented towards observation and interpretation to those oriented towards objectivization in the manner of hypothesis testing.

While looking at the growth of science in the above-described simplified manner, this section also briefly discusses how natural history has developed. It is speculated that at the time of Aristotle's natural philosophy, science was equivalent to natural history, with telescopes and microscopes being unavailable. Findings were compiled in the form known as natural history. Hence, at that time, a general view of scientific findings would have equaled natural history. This situation was similar in Greece and China.

Likewise, in Western Europe, practical aspects of natural philosophy grew through the study of medical materials, probably due to their relevance to medical care. Collecting and studying medicinal herbs led to the construction of botanical garden-like facilities at the medieval Vatican. Similarly, in China, compilations of herbalist knowledge grew on their own.

In medieval Europe, the pace of knowledge accumulation was neither smooth nor rapid. One probable reason for this is the concept of a monotheistic God's control affecting circles of intellectuals. The momentum of the Reformation guided scholars towards a path of understanding nature beyond teleology, in a mechanistic way. Scientific curiosity began to look for universal principles as a solution.

There was a time when Europeans' knowledge of the natural world was confined to the European world. Gradually, their view widened to a global scale through events such as the Crusades traveling through the Middle East and Asia, the Mongolian influence, and European inroads, in the Age of Exploration, into Africa, Asia, and the New World. As a result, compilations of natural observations, achieved through natural history methodology, amounted to tremendous volumes.

Fauna and Flora before Linnaeus The process of adding literature to supplement Dioscorides' 'De Materia Medica' progressed, although slowly, even during the period known as the dark Middle Ages. For medical care providers with a role of curing patients, it was probably imperative to strive to deepen their knowledge of medicinal materials and improve the quality of medicine not only by consulting classical works, but also by collecting or growing medicinal herbs themselves. Their efforts thrived even more as the Renaissance ended the Dark Ages.

From the first century, Dioscorides' 'De Materia Medica' was inherited through repeatedly transcribed copies at monasteries that provided medical care services. One result of repeated transcriptions was errors in and simplification of illustrations. The invention of printing arrived in the 15th century. There seem to have been few valuable medicine books produced in the early period of printing. It was likely not possible to achieve a Renaissance in science in short order. Leonardo da Vinci (1452–1519) had no interest in compiling a natural history book, although he left sketches of animals and plants. Showcasing diverse things seems to have been outside Leonardo's interest.

In the 16th century, scholars such as Brunfels, Fuchs, and Clusius printed and published books of medicine that contained beautiful illustrations. This was followed by the publication of 'Cruydeboeck' (1554) by R. Dodoens (1517–1585; his Latenized name is Dodonaeus). This book was brought into Japan and translated into Japanese. These represented the tradition of medicine books. Nonetheless, owing to the publication of these works, people began to have deeper basic knowledge of the forms of various plants.

Many of crude medicines were of plant origin. Animals were marginal in terms of medicine, despite their appearance in medicine books. Regarding animals, Aristotle described over 500 animal species in his 'History of Animals.' Meanwhile, in addition to Theophrastus, C. Gessner (1516–1565) was another father of zoology, who published 'Historia Animalium' (1551–1558). Furthermore, J. Jonston (1603–1675) published an illustrated manual (translated into Japanese by HIRAGA Gennai) as a copper-plate print, albeit with illustrations cited from previous publications. His work attracted people's attention to the diversity of animals, along with plants, from the perspective of natural history, irrespective of medicinal uses of materials. A Dutch edition of this book was introduced into Japan, which was translated into Japanese by NORO Genjō and published in 1742.

When looking from a broad perspective, various organism species are not randomly diverse. Their diversity is hierarchical. For example, works of Gessner and Cesalpino state that different species can form groups. J.P. de Tournefort (1656–1708) defined the group by the concept of genus. In his 'Éléments de botanique,' published in 1694, he classified 7,000 plant species into 700 genera. Moreover, Tournefort recognized order and class as ranks higher than genus.

John Ray (1627–1705) was the first scholar who defined species as a fundamental unit for recognizing diverse living things. In his three-volume 'Historia generalis plantarum,' published between 1686 and 1704, he presented a summary of objects he had collected during his expedition across Europe. Volume 1 defines animal and plant species. In response to the demand at the time for descriptions based on scientific empiricism that rejected scholastic rationalistic reasoning, he acknowledged the need to recognize listed species as the fundamental unit of diversity. Nonetheless, although definitions in Ray's style were established at the end of the 17th century, the issue of what defines a species remains an eternal discussion for those who study biological diversity today.

Linnaeus and Post-Linnaean Natural History Carl Linnaeus (Carl von Linné, 1707–1778) is known as the restorer of taxonomy. However, as briefly described above, natural history knowledge had already largely accumulated, showing a trend towards the recognition of diversity as hierarchical phenomenon. At the same time, the task of organizing and overviewing all animals and plants on the Earth in accordance with one set of standards requires an individual with a broad range of discerning abilities. Additionally, it is critical to have superb collaborators and create conditions for them to travel across the world, bringing back new knowledge. Linnaeus's outstanding talent was exactly what was in need at the time.

What Linnaeus pursued and what he achieved need not be described here. For people with interest in natural history, this discussion is like returning to the alphabet to discuss the English language. However, before looking at post-Linnaean natural history, let us examine two features of his achievements described below.

Binomial nomenclature became established after Linnaeus. In a sense, Linnaeus did not propose binomial nomenclature. However, he did propose that such a system would be applicable to the naming of all living things. Subsequent researchers found the method easy to follow and very superior.

An international consensus on nomenclature for plants was formally built at the International Botanical Congress II (Vienna) held in 1905, when it established Linnaeus's 'Species Plantarum' published in 1753 as the starting point of priority for plant names. However, the rules of nomenclature provide various addenda. For animals, the rules of nomenclature adopted at the Fifth International Congress of Zoology (Berlin) held in 1901 were the first international standards, which established the 10th edition (1758) of Linnaeus's 'Systema Naturae' as the priority for naming. Both rules of nomenclature have been repeatedly revised. Moreover, rules of microbial nomenclature have also been developed.

To recognize biodiversity objectively, it is critical to hold international meetings, adopt codes, and establish international nomenclature standards, although this is also a difficult challenge. For plants in recent years, nomenclature committee meetings have been held every six years during the International Botanical Congress. For several days, participants discuss whether to adopt proposals that are made in advance of the meeting and supported by preliminary voting. A final decision is decided by vote after the discussions. (When I attended, some speakers started a lengthy argument about the grammar of Latin and Greek, and I became bored more than a few times. Nevertheless, such arguments may be necessary to build an international consensus. There are some scholars who ignore the Congress, claiming that it is not a scientific conference. However, due to these efforts, an international consensus has built the rules of nomenclature for all organism groups. Consequently, as an information processing issue, it is an extremely important process of standardization.)

The other achievement worth mentioning here that Linnaeus made for later generations is his dispatch of collaborators to conduct surveys in different parts of the world. Linnaeus's own achievements in field research were limited. However, following his suggestions, superior collaborators conducted field research in many parts of the world. Thunberg, who visited Japan, was one of his most excellent successors. Thunberg's 'Flora Japonica' (1784) was an introduction of Japanese plants to the world. After returning home, Thunberg became the successor to Linnaeus's son, who had taken over the position of Linnaeus, and later took the post of the chancellor at Uppsala University.

After taking up the post of professor at Uppsala University, Linnaeus took over Rudbeck's old garden. He also raised animals there and lectured on zoology and mineralogy, as well as botany. (Rudbeck's old garden is the botanical garden of Uppsala University, established by Rudbeck the elder, Linnaeus's predecessor. After being promoted to the post of professor, Linnaeus lived in the garden. Presently, it is maintained under the name of the Linnaean Garden. The botanical garden of Uppsala University was expanded when Thunberg was working as a professor.) Thereafter, Linnaeus stayed in Sweden, while sending his excellent disciples to various parts of the world.

Explorers who were known as Linnaeus's apostles numbered 17. In the Age of Exploration, D. Solander (1733–1782) joined Captain Cook's first voyage, and A.E. Sparrman (1748–1820) joined the second. P. Osbeck (1723–1805), who conducted surveys in China and Java, and P. Forsskal (1732–1763), who did so on the Arabian Peninsula, are quite famous. In addition to Forsskal, C. Tarnstrom (1703–1746), F. Hasselquist (1722–1752), P. Lofling (1729–1756), C.F. Adler (1720–1761), J.P. Falk (1732–1774), and A. Berlin (1746–1773) died in their respective places of study. This death toll illustrates how challenging their surveys were at the time, although Thunberg and other explorers who visited Japan likely experienced the luxury of a feudal lord during their survey trips.

The mode that post-Linnaeus scholars adopted to study biological species diversity was to recognize living things, both living and extinct, on Earth by species and arrange them into a taxonomic hierarchy. This study is termed taxonomy. Taxonomists produced records for a tremendous number of plants and animals and traced relationships between species. Initially, their work was to develop a general view of living things on the Earth created by God. As their research studies advanced, they became confident that the diversity of organisms resulted from biological evolution. Consequently, the need arose for the taxonomic hierarchy to indicate evolutionary lineage. Among the strongest supporters of Darwin were Hooker, an important figure in the British botany community and Gray, who led the rise of botany in the United States, as emphasized later in this chapter (see Tea Time 2 on page 47).

(Dr. TAGAWA Motoji, one of my teachers at Kyoto University, wrote 'Colored illustrations of Japanese ferns' [Tagawa, 1959], which is his sole book written in Japanese. While this book provides a wealth of information, he authored no easy-to-read introductory guide to science. To be familiar with his entire perspective, the reader can consult his treatises. However, they are very difficult to understand unless the reader is an expert researcher with solid basic knowledge. Only a small number of scholars, including myself, could appreciate his profound knowledge. It is our mission to hand down and advance his knowledge. Nonetheless, my interpretation of his knowledge is, in a strict sense, of my own and not equal to his original thought. Culture is created by socially integrating the knowledge of those who have grown in their respective distinctive ways. That said, I wonder how much of the precious knowledge developed in individuals is stored in society. I often consider the significance of handing down academic achievements and feel my responsibility towards this.)

It is appropriate to provide a brief justificatory explanation of why it is necessary to consider the history of science as closely as attempted above. Descartes initiated the modernization of philosophy and paved the way for modern science. This does not mean there was absolute truth in his philosophy. For example, the philosopher Wittgenstein, who flourished in the early 20th century, said that all philosophy is a critique of language; however, cognitive science has grown into a completely different sphere.

That said, what knowledge is currently in our hands? Looking back at the knowledge of today's people, people 50 years from now will probably remark on the petty level of human knowledge of the early 21st century. Nonetheless, in every period of history, people are driven by scientific curiosity and intend to make the maximum use of achieved scientific findings to create affluence and ensure safety.

Accordingly, it is desirable to accurately know how human knowledge developed and contributed to society at various times in the past and how it became outdated. The history of science is expected to serve as literature for understanding the present, or knowing the present implications of individual past events, rather than simply looking at what occurred and what was correct. This is a matter of fact, although often forgotten. Although it may be needless to remark on this, one should be keenly aware of the history of science when contemplating the present situation of natural history.

(2) Science, religion and art as intellectual activities

Examining history is more than simply yearning for the good old days. There is good reason for summarizing a specific area of history, such as the historical growth of natural history. Historical development suggests much about the current state of the area of culture in question, and what potential it has for the future. Given this perspective, it is appropriate to contemplate the contemporary significance of the above discussion a bit further.

Culture as an Outcome of Intellectual Activities Science emerged, and has developed, as a means of fulfilling human intellectual curiosity.

Homo sapiens evolved among wild animals, began bipedal walking by migrating from forest to plain, developed language to efficiently communicate thoughts within a group, and succeeded in expanding information stored in society. They transmitted information efficiently from generation to generation, became more sophisticated in their activities based on this knowledge, and evolved into humans as intellectual animals. In biological terms, the genetic difference between *Homo sapiens* and chimpanzees is counted as only 2%. However, *Homo sapiens* transformed themselves into something that is completely different from other living things by developing culture founded on science and technology. *Homo sapiens* is an animal. However, humans have become isolated from wild organisms to such an extent that it is difficult to even determine where to place humans on a taxonomic tree of life, in terms of being in a domain delimited by an awareness of itself as existence being that exists and thinks. Humans regard their actions, or art, in opposition to nature (including all wild life) and understand the term 'artificial' as an antonym of nature.

As intellectual activities grew, human society elevated art motivated by beauty, produced religion attracted by mysteriousness, and developed science driven by a sense of wonder. There is a notion that since science, religion, and art came into existence, human intellectual activities have elevated humans to the noblest of all animals, differentiated from other living things. It seems that sophistication of intellectual activities lies behind the phrase "the noblest of all animals."

The backbone of art, for which people are moved by beauty, remains still today activities that see pure beauty and aim exclusively to create this heart-moving beauty. Because exquisite works of art bring about a small fortune, it is common to think that great artists realize economic affluence. Until very recently, however, many stories of those who strived for the creation of their ideal beauty remaining unappreciated still existed. Even nowadays, there are some artists who continue to struggle without recognition in the world, although not every struggling individual may be a great artist. Perhaps Romain Rolland said ironically that César Franck was a poor church organist precisely when, at a social occasion, a nicely dressed fine lady, having heard that Mozart had died in poverty, said that she would not have let the great artist have such a rough time if she were around at his time.

Excellent religious figures are said to attain great liberation and a power to save people by being free from all distracting thoughts and discarding ego. It is not certain that such an excellent religious figure is present today or not. Some people build wealth by viewing religion as a vocation, and religion has become an enterprise common to all countries. Nevertheless, many people today are vulnerable and opt to depend on religion. To be honest, I am not interested in immediately asking for religious instructions, although, personally, I have had some acquaintance with religious figures in the leadership of a religious sect.

That said, I have neither contemplated nor trained myself in art or religion as much as those who have devoted themselves in their respective fields. Therefore, I will not go further in this, as it is too much for me to discuss religion and art in this book based on my biased knowledge that formed through my own way of learning. However, as an individual who has dedicated his life to science, I must address current major issues in science.

Natural History as Mainstream Science The origin of science dates to when early people, beginning activities that can be described as intelligent, became curious. They started asking why when they experienced various phenomena from objects in the natural world.

No evidence indicating when *Homo sapiens* began asking why has yet been proposed. Since the acquisition of language by *Homo sapiens*, many objects and phenomena have been given names specific to them. If lining up objects, some have names, while others do not. Then, their similarity and dissimilarity draw one's curiosity. Regarding the phenomena they display, intellectual curiosity arises around the formation of their mutual relationships. Activities of humans who grew in intelligence resulted in an increase in recording objects and the dawn of the analysis of phenomena. This stage is considered as the origin of science.

In Aristotle's time, science had developed into a decent form, with thoughts being transferred into letters and compiled into books.

Consequently, it is currently possible to consult records and understand scientific knowledge of the time. According to the extant literature, achievements of science at the time were compilations that crystallized intellectual activities, fulfilling scientific curiosity.

Although science had developed into a decent form, people of the time simply observed objects and kept track of phenomena in the real world. They did not even think of analyzing causality between them. It goes without saying that they had not the slightest idea about conducting analytical research to discover the universal principles commonly underlying diverse objects and phenomena. Simply put, they watched phenomena, recognized motions in phenomena, and accurately described what they understood, with all their heart and soul. This is the origin of natural history. That said, Pliny compiled his 'Naturalis historia' four centuries after Aristotle worked on natural philosophy.

In the times of ancient Greece, discovery of the natural system meant arranging God's creations in accordance with divine providence, although a general view of diverse facts led to systematization for this arrangement. This arrangement technique was a method used in both the East and the West. However, although uniformly referenced, God was interpreted in varying ways by different regions and ethnic groups. In this interpretation, regional cultural characteristics were manifest. The characteristics of a culture, or the notion of God, resulted from interactions between natural features of the region and the people living there; however, in the process of the systematization of cultural characteristics into religion, the gift of unique individuals made special contributions, to some extent, to the process. In Japan before the arrival of Buddhism, the belief in myriads of gods residing in all things underlay its culture. Therefore, having a general view of diverse facts, or making a list of all things based on naturalist thought, was accessible to the Japanese people.

The character of the natural history of ancient Greece clearly contrasts with that of *bowuzhi* of ancient China. [The history of the introduction of *bowuzhi* to, and its growth in, Japan may be similar in style to natural history in Greece. The comparison was clearly made by P.L. Hearn's (1850–1904), also known by the Japanese name KOIZUMI Yakumo, unintentional declaration in his 'Insect-musicians' (P.R. Hearn, 1899).]

The volume of written knowledge steadily grew in the times of ancient Greece. By the time when Roman imperialism was interwoven with Christianity, a long period of languishing culture known as the Dark Ages began and continued until the Renaissance, a time when even recording new knowledge was confined to a limited range. Nevertheless, the long Dark

Ages came to an end, and the Renaissance commenced, probably due in part to the gradually increasing overall volume of knowledge in human society. Europe, ruled by the Roman empire, was under the political, as well as religious, control by the denomination that came to be known as Catholic. The glorious period of the Renaissance began, as seen in the field of art, in parallel with the end of autocratic control by the Catholic Church as brought about by the Reformation. That said, the Renaissance began in a region under the control of the Catholic Church.

The Renaissance, which denotes the great revival of art and literature, was a period of historical development characterized by remarkable advances in fields such as literature, art, music, and architecture. In the domain of science as well, which was pushed ahead by the Renaissance, people shifted from predominantly extensively reading literature to actually observing real objects to acquire knowledge. In this way, new findings were rapidly collected. Before long, this rapid expansion of understanding led directly to the progress of science. With a clear awareness that things present with their respective historical background, scholars began to recognize that interactive systems of things are related by lineage. They understood that to identify universal principles underlying diverse phenomena it was necessary not only to track phenomena, but also to analyze their causal factors. Subsequently, they introduced experimental analysis to scientific study.

In this way, scholars modernized science and attained an advanced understanding of nature. Thereafter, it did not take long for them to understand that the things that existed in nature and the phenomena displayed by them, which had all previously been explained as God's creations, were nature's evolutionary creations, rather than a result of divine providence. (That said, if one regards natural providence behind all creation as nothing but God, one will have a view that all creation in nature has been created by God. In this regard, in a sense, there is a fundamental commonality between Japan's traditional religion, *Shinto*, and the notions of natural science.)

Science and Technology Scientific knowledge may certainly be an outcome of cultural growth driven by scientific curiosity. That said, human knowledge has also grown from the perspective of how knowledge can be useful for human activities. Based on findings accumulated by humans, our species evolved from wild animals. While fear of death drove people directly to religious salvation, human knowledge stored in society helped in relieving the sick. Drugs used to alleviate suffering of the sick or pain of the injured, were particularly found in knowledge used and valued from the perspective of its utility for humans. This was seen with the very early faunas and floras, which were herbalism and medicine books.

Science is described as an intellectual, curiosity-driven activity devoted to knowledge. From its beginning, science was expected to make useful contributions to society by obtaining knowledge. Indeed, it did play the expected role. More accurately, rather than as an expectation, those who were engaged in science probably were oriented towards social utility as they obtained findings.

Nonetheless, the basic development of fundamental science has consistently been driven by intellectual curiosity. In Japan, the virtue of researchers was once considered to lie in being ignorant of the real world and living in honorable poverty. Utility-oriented knowledge was rather built through experience. Scholars engaged in science did not expect immediate utility from their studies. Moreover, society plainly assessed them as curiosity seekers. Until the late 20th century, scientists believed that scientific knowledge would grow only from the science-for-science perspective. Embracing this understanding, scientists made a collective effort to study basic science.

Science evolved and advanced steadily in human society. In parallel with this, highly intelligent individuals developed various technologies to ease life in society. Innovative hunting and fishing techniques and improving gathering methods were crucial for stably providing resources to sustain an increasing population. Additionally, cooking ingredients collected by foraging and the preparation of safe and delicious meals facilitated the creation of wealth that was most important for human society. Personal findings obtained through experience accumulated as knowledge in society. Technology advanced because of a long succession of learning through experience. Technicians acquired superb techniques through years of training. Such techniques were inherited from skilled individuals, called masters, to their apprentices through on-the-job training, or a handing down of experience.

Technology advances partly through sophistication based on human aesthetics, as seen with exquisite folk crafts. When useful everyday utensils developed, some were designed to appeal to human aesthetics. In that process, artistic excitement played a major role in producing useful items. Indeed, beautiful folk crafts suggest that items were made guided by an aesthetic sense that were also useful and convenient. This aesthetic is also an aspect of technology.

To advance technology, humans have used scientific knowledge to the greatest extent possible. Science and technology can be viewed as having evolved along different lines from the very beginning. Technology supported humans in their very fundamental activities to survive from the time when *Homo sapiens* looked like a species of wild animals and before they even began to call themselves humans.

However, activities that could be called science emerged and advanced along with the development of human intellectual activities. Before the acquisition of language, *Homo sapiens* were able, only to a quantitatively limited extent if any, to systematically conduct activities to satisfy their intellectual curiosity. Even if, at the times of *Homo sapiens*, humans acquired knowledge like science, it differed from a well-organized and arranged system of knowledge.

Since beginning to develop culturally, humans produced and developed intellectual activities driven by curiosity. In very early times, such intellectual activities were not useful in everyday life. Nevertheless, scientific thought, was not ignored, despite its lack of utility. It has had its place as one field of human culture, along with art and religion.

As science advanced, a variety of knowledge accumulated. Meanwhile, further innovation of advanced technology required knowing the true nature of materials. Thus, human society experienced the beneficial effects of knowing the principles behind things and events. It became common for scientific findings to make huge contributions to technical innovations. Before long, advancements based on scientific findings developed into the concept of science and technology.

To understand the relationships between science and technology and religious art described above, it is helpful to arrange these concepts in graphical representation by using natural history thought processes, although this may be a somewhat bold approach. I had an inspiration for this approach from "Octopoda, sparids, and *Homo sapiens*," (Hara, 1995), a small article that appeared in 'Biohistory,' published by JT Biohistory Research Hall. The article was written by the video artist HARA Tetsurō to present OKADA Tokindo's cherished theories in graphical form.

Imagine that you arrive at a fish shop. The question arises, whether to buy sea bream or octopus for your supper. First off, this question implies that you think that both sea bream and octopus are seafood. Within the domain of human intelligence occurs human contemplation of the fact that it eats. Additionally, in this paragraph, popular use terms are used for the living organisms in question.

Meanwhile, when thinking about a taxonomic chart, one follows the common practice of biology, that is, the use of biological scientific names. As an accomplishment of science, the structure of nature is depicted through understanding nature as objectively as possible. Consequently, scholars give names to things present in nature and use the name as a means of identification. Moreover, classifying and arranging objects at their own convenience (artificial classification) is avoided. Rather, scholars attempt to understand objects by arranging them according to their natural relationships (natural classification), if present.

A taxonomic hierarchy is created to understand relationships between a variety of living things. To view them as a subject of science, taxonomy aims at tracing the natural classification system along the biological evolutionary process. The relationships between the above three objects in question may also fit into the natural classification system. For instance, *Homo sapiens* are in the order Primates of Mammalia and sea bream vertebrate animals, while octopus is in Mollusca of invertebrates.

Both sea bream and octopus have nerve cells and a brain (group of nerve cells). However, they do not think of what they are eating or what mutual phylogenetic relationships their prey has. That said, for commercial transactions, seafood and humans are not in the same group.

Our common understanding is that humanities-oriented functions deal with discussions on art and religion, while science-oriented functions focus on science and technology. However, it is probable that science has been a purely intelligence-oriented human activity, along with religion and art, while technology, a result of advancing culture, arose for the purpose of making human life more convenient. In recent years, technology has been based on scientific findings. In contrast, up until recently technology developed among ordinary people relying on their senses. For this reason, it was placed in the category of art, which nowadays is sometimes known as folk art.

Present-Day Science In recent years, science has advanced very rapidly. Specifically, the domain of natural science has moved quite fast. The pace of advancement during the period of little more than half a century since 1957, when I began my career as a full-time researcher (my enrollment in a graduate school in 1957), until today is perfectly described as a succession of Copernican revolutions.

Natural science has recently made exceptionally noticeable advances in the field of biological science. One consequence of this advance, due to improved accuracy in the analysis of living things, is an increasing volume of related information. Naturally, this facilitates activities in a given area of specialization. Moreover, joint research projects involving many scholars can form due to the application of various analysis techniques. That said, it is not uncommon that joint research results in a simple collection of information, with individual informants not taking responsibility for the overall research results.

To figure out research fund allocations and carry out personnel selection, it is necessary to assess researchers' abilities. However, even within a paper, not every coauthor is familiar with and takes responsibility for the content. Considering this, even if limited to a specific domain, there is almost no individual who can know the all details of the many published papers. (One recent trend is that the name of a person appears as a coauthor because of the person's technical contributions to a very small part of evidence used to prove the published facts, and the person avoids taking overall responsibility for the paper. This touches on issues of responsibility and moral character of the researcher.) In such a situation, it is very difficult to assess the magnitude of contribution made by each coauthor. Broadening the range of comparable elements as wide as possible would make it difficult for the selectors to determine what are to be relied on.

A noticeable recent trend is that the quality of researchers is most reliably determined by the reputation of the journal that has published their papers. It is believed that journals are objectively rated by the quantity of frequently cited papers they have published. There are journals that are highly rated according to this criterion. Researchers with many papers appearing in such a journal are assessed highly.

Meanwhile, in recent years, there have been few researchers who, when involved in research fund allocation and personnel affairs, obtain applicants' papers and scrutinize the contents to help make a hiring decision. Given this perspective, an excellent clerk unfamiliar with research topics may be able to make a more objective and correct decision than a capable researcher. Referees for the relevant journal are the responsible parties for research evaluation. That said, when on duty, to what extent referees are aware of that responsibility is uncertain.

Research papers tend to be written based on specific data relating to a fragmented domain because they aim to increase their chances of acceptance. To precisely respond to arguments created in a fragmented domain, one needs little knowledge of studies outside the domain. Consequently, researchers attracted exclusively by immediate challenges in their own field have little interest in the current state of study in even closely related domains. Young researchers, as a matter of course, fall into a situation in which they are almost incompetent even in closely related domains, despite the superb abilities demonstrated in their own fragmented research domain. In actual reality, advanced science is evolving at this extreme level of fragmentation.

Not a few individuals are aware of the current situation of increasing fragmentation in science. These people quite often voice their concerns about this trend. The need to bridge humanities and science is stressed repeatedly in, for example, international reports issued by the Science Council of Japan. Nonetheless, no sign of improvement of the situation is in sight. This is perhaps because researchers, aiming to earn more research funds and a better post, would not dare to undertake an adventurous project without it promising a boost in reputation. Concerns are largely voiced, seemingly by individuals already in retirement who are not situated to take practical action themselves in line with their concerns. Most individuals engaged in current research are busy with immediate projects and have no leeway to be concerned about the future situation. It is regrettable that no clue is emerging as to how to break this vicious circle.

Transformation of Natural History Natural history is not recognized by some scholars as a domain of natural science. Among fast-moving biological sciences, natural history is sometimes regarded as a domain which has outlived its usefulness, although it was once the focus of attention. Some people even make little of it, considering it laypeople's amusement and not contributing to currently relevant studies. Perhaps against this backdrop, in Japanese, rephrased terms, such as study of natural history (*shizenshi gaku*) and natural history and science (*shizenshi kagaku*), are occasionally used in place of natural history (*shizenshi*) (see 2.1).

As far as biodiversity is concerned, natural historians half a century ago faced a situation in which, admittedly, natural history studies could be disregarded. At a certain point, as analysis techniques were inefficient in inquiring into the issues they were facing, they conversely and defiantly were insistent on using traditional techniques, claiming that current biological analysis methods were not applicable to solving taxonomic problems.

There was (and remains), of course, a vast number of challenges to be met by inquiry with traditional techniques. Moreover, the number of researchers involved at the time was limited. Therefore, they were fully occupied conducting studies using traditional techniques. That said, many scholars in natural history-related fields continued ignoring the arena of biological science, confining themselves within their circle. They were, instead, apt to turn their back to the rapid advances in biological science. Moreover, when biological science as a whole began rapidly making advances in analysis techniques and taking steady steps forward, it became worth attracting attention as a basic natural science.

For individual researchers, the need to introduce innovative analysis techniques was indeed recognized to some extent. However, when the overall state across the field of taxonomy was criticized or disregarded, they took the situation, in a paranoid manner, as an attack on their own study. Instead of advancing techniques, they set up an invisible obstinate barrier around them. As a result, they turned their back on the opportunity of healthy overall developments in the field. This remark might be a harsh criticism. Yet, looking back at our work at the time, as a fresh researcher in taxonomy, I am now filled with renewed self-admonishment.

Nevertheless, in those periods, as in other periods, some outstanding scientists were aware of how important fundamental fields, such as taxonomy, were for biological science. Those outstanding people were influential in science policy. Accordingly, they were, in my opinion, attentive to natural history studies more than commensurate with the performance of natural historians.

In 1992, the Convention on Biological Diversity was adopted to address the biodiversity crisis as a social issue. Sciences that dealt with species diversity became a renewed focus of attention. Scholars in the field of taxonomy could no longer stay inside their invisible obstinate barrier. This was precisely when advanced analysis techniques developed in biological science, particularly molecular biological analysis methods, which became applicable to the analysis of species diversity.

My personal experience reflects events around that time. In 1981, I left Kyoto where I had lived for nearly a total of 30 years since my undergraduate student days and moved to the University of Tokyo. The Botanical Gardens, Faculty of Science, the University of Tokyo at the time was under difficult management conditions. Nevertheless, I, in conjunction with excellent, but few, mid-career and young researchers boosted efforts to build taxonomy as a field that competed in the arena of biological science. An early example of such efforts occurred around the mid-20th century when Professor MAEKAWA Fumio's laboratory, at the same botanical garden of the University of Tokyo, began to explore broader perspectives beyond the framework of traditional taxonomy. However, most researchers who studied at the laboratory gave up studies in the narrowly defined field of taxonomy and became active as researchers in other domains. This is probably because of the historical stage at which biological analysis techniques had not yet advanced to such an extent to be suitable for the analysis of plant diversity. In contrast, in the early 1980s, researchers began to look for various analysis techniques to apply to species diversity analysis.

It is certain that since the turn of the century in Japan, biodiversity study has played its proper role, basically developing as a part of biological science. It is not only that studies of species diversity have been transformed. Because of such historical development, or advances, biology has become able to tackle diversity, which is a fundamental characteristic of organisms, as well as studies based on specific model organisms. However, this situation is not recognized by those for whom the term 'natural history' still has implications of being out-of-date or amateur (non-professional). Such individuals probably need to be aware that their outdated notion has led them to feel like a frog in a well, or a scholar in a fragmented domain of specialization. Of course, I do not mean all researchers in natural history are free from any thought that deserves criticism. What I anticipate is that from the perspective of natural history, outstanding studies will be conducted and lead the field biology.

The species diversity study that plays a role in biological science, in a sense, typically depicts the current-day state of natural history. In practice, researchers in plant species diversity conduct their studies based on traditional flora research studies, while adopting analysis techniques used in various biological fields formerly known as biochemistry, morphology, genetics, and ecology. Two examples clearly illustrate this trend: 1) In our research group, a researcher who had earned a degree in phytochemistry joined as a staff member and was active as a leading researcher; 2) Some graduates from our research group took a full-professor post and are active at a laboratory that was once regarded as different area than taxonomy such as an ecological lab. It seems safe to say that the field defined as taxonomy no longer forms a particular circle isolated from the arena of biological science. Indeed, species diversity study has become a core challenge in overall biology. It has become clear that biology can no longer meet research challenges if it relies solely on analyses conducted within the framework of traditional domains.

Expectations for Natural History Because science demands accurate and positive evidence for every phenomenon, a referee who reviews scientific journal papers evaluates the originality of the analysis conducted to verify the hypothesis and the soundness of argumentation based on the analysis results. On diverse phenomena in the natural world, scientific findings have steadily accumulated through research driven by intellectual curiosity, enabling scholars to have an advanced understanding of nature. Obtained findings have been effectively used to develop technologies, which in turn have made helpful contributions to the acquisition of further scientific findings.

Meanwhile, although science is said to have advanced greatly to reach today's level, scientists, when faced with questions such as what are humans and what is the meaning of life, say that these are not an issue of science and that, rather, religion or philosophy provides solutions. Some scientists who advocate agnosticism even declare that natural science can provide no answer to such questions, no matter how far analyses are made. This topic will be discussed later in more detail (Chapter 5). In either case, at present, natural science logic is not expected to answer whether a solution can be reached. Similarly, what the solution is also remains unknowable. Finally, whether a solution exists or not is a question that can be clarified only once science makes further advances.

Even if each individual phenomenon is elucidated through advanced analysis by means of reductive techniques as a principle of modern science, simply piling up elucidated phenomena cannot be expected to provide a satisfactory answer to the question: What is the meaning of life?

In the domain of natural history, scholars are expected to have an integrative understanding of all phenomena that occur in the natural world. Analyses conducted to look at the entire biosphere, such as taxonomic hierarchy, are meant to piece together all known analyzed facts into an integrated whole picture. Take a specific plant species as an example, one would attempt to obtain all kinds of information on the species, driven by one's scientific curiosity. Consequently, he or she will not only know all information on features intrinsic to the present individual living organism, but the researcher will also analyze its comprehensive lineage and biosphere information, such as its position in the species, the phylogenetic position of the species, and its role in the ecological system that it inhabits. After all, knowing the entire three- and four-dimensional background of the natural world translates into knowing specific individual species.

Meanwhile, it cannot be overlooked that in practice, researchers tend to look only at a specific aspect of their immediate challenge. I wonder how much they are, while analyzing individual phenomena, conscious of the principles that commonly underlie the whole, free from confusion by the excitement solving scientific challenge.

Recognition of biological species diversity depends on compilations of basic findings. Because such findings have yet been achieved only in a limited range, a huge amount of research study effort needs to be made in the field of biodiversity. In addition to having information that covers only a limited number of species, information about individual species is also limited to the range made available by the specimens themselves. That does not justify being engaged exclusively in enriching fundamental information; otherwise, researchers would simply aim for finding new species for all time.

Natural history study is vulnerable because it fails to make thoroughgoing individual analyses, as it is always particular about the whole. Natural history is occasionally made light of in the domain of science, probably due to individual analyses being less than rigorous. The results appear poorly scientifically grounded if, despite the collection of data from various sources, the individual analyses are insufficiently compiled instead of complete with data from rigorous analyses. Criticisms that such results cannot constitute a scientifically excellent study are certainly correct. However, from the perspective of advancement in science, it is true that no accurate conclusion can be reached at an intermediate stage. Yet, human scientific curiosity desires to know the final answer.

Researchers who desire to study and know the entity of a species should constantly have interest in differences and similarities between the species in question and similar species. In this way, they can express the scientific recognition of it as a species. One thing expected to substantially contribute to advances in this field is that researchers not only retain their findings but also make public to the extent they are elucidated, sharing the information in the world of science. It should be noted that in this paragraph, the term *species* is used casually, or in a vaguely understood sense without providing a scientific definition of a species (see Tea Time 5).

Natural history emerged as a way of observing nature around the time when science first arose. Subsequently, along with the expansion and deepening of human intelligence and, notably, with the emergence of the nominal designation, *modern science*, science grew to demand rigorous evidence. This included experimentally ascertained reproducibility based on reductive and analytical methods. Modern scientific techniques have enabled researchers to understand things only through simple observation and interpretation of nature. The findings have been appropriated to develop technology and greatly benefit people's lives.

Against this historical backdrop, the attitude of disregarding natural historians who only used observation and interpretation took shape in the world of science. However, in recent years, the spotlight has been on the revival of natural history. This is perhaps due to natural science legitimizing the domain of natural history, as hypothesis testing techniques have been gradually established, for example, in the study of species diversity.

It is not that the specific field of study known as natural history is set up as one of natural sciences. Natural history is that aspect of human intelligence that desires the elucidation of what nature is. Being driven by scientific curiosity that always expects an ultimate answer, human intelligence desires elucidation through natural history studies. However, to clarify the truth by modern scientific methods, it is imperative to conduct an analytical study. Studying an issue thoroughly in a fragmented field and elucidating the entity of each individual phenomenon certainly helps expand human intelligence. However, out of scientific curiosity, humans crave for an answer about the whole, such as life and the universe. Simply elucidating parts does not enable them to see the whole.

Natural history aims at drawing the whole picture from an integrative perspective. Some scientific domains, such as species diversity study, align with the way that natural history aims to depict life. This relationship was clarified by recent advances in scientific analysis techniques. Scholars in ancient Greece and ancient China in the times of Han and Wei pursued natural history out of curiosity. Modern-day scholars, out of the same curiosity, expect growth of comprehensive or integrative studies based on present-day sciences and technologies.

(3) Darwin and Mendel: Bird's eye view and analysis in biological science

In the 19th century, in the process of biology evolving as a science, Charles Robert Darwin (1809–1882) put forward the Theory of Evolution, and Gregor Johann Mendel (1822–1884) proposed genetics. Their theories symbolize the growth that biological science achieved in the 20th century and thereafter. Darwin approached understanding truth by using penetrating scientific logic, while having a bird's eye view of the biosphere. Mendel pursued the principles of universally underlying the biosphere using a modern scientific sense, including the use of quantitative analysis. The two geniuses are compared below, referring to background information on how they got along with or dissented from the academic world and society.

Theory of Evolution Prior to Darwin Erasmus Darwin (1731–1802), grandfather of Charles Darwin, used the term *evolution* to explain biological evolution. Because of his ideas being speculative with little emphasis on facts, he was criticized by Jean-Henri Casimir Fabre (1823–1915), who closely observed insect life. His grandson Charles was unwilling to follow his grandfather's concept and intentionally avoided the use of the term *evolution*, even up to the fifth edition of 'On the Origin of Species.' In biology at the time, the term *evolution* referred to development following a planned schedule, as exemplified by changes involved in ontogeny. It did not mean evolutionary processes, such as phylogeny.

Lamarck and His Theory of Evolution

Jean-Baptiste Pierre Antoine de Monet, Chevalier de Lamarck (1744–1829), a French naturalist, worked on the classification of invertebrates, which was then an untapped field of study. Although it was not his active choice, he made substantial contributions to the field. In the process, he recognized dynamic changes in biological phylogeny and tried to rely on an evolutionary theory. However, the use/disuse theory that he proposed as grounds for phylogenic changes was not persuasive and did not lead to natural laws.

While he recognized dynamic changes (evolution) in biological phylogeny as a hypothesis, Lamarck explained it as the spontaneous generation of living organisms and the inheritance of acquired characters. In addition, he applied the use/disuse theory, which was advocated elsewhere at the time, to demonstrate dynamic changes that living things exhibit. His greatest achievement as a scientist was 'Histoire naturelle des animaux sans vertèbres' (seven volumes) that he compiled through studies that continued even after he turned blind. He first used the term *biology* in 1802, which was subsequently established as a word to denote the science of living things. At the time, he also coined the term *invertebrate* and established the concept that distinguished it from vertebrate.

For the discussion of evolution, Lamarck's use/disuse theory is that of the past. He, of course, mentioned nothing about the mutation of genes because of his lack of knowledge about the existence of genes. If one thinks about biological inheritance as a phenomenon, putting aside the question of whether it is progressive and comparing the concepts of use/disuse and adaptation to environment, one would understand that, considering the knowledge at the time, Lamarck in fact pointed out an important aspect of these phenomena.

Starting his career in botany, Lamarck founded invertebrate zoology. He was also interested in paleontology and made moderate achievements in meteorology. In this way, he became experienced in diverse fields as if he were a natural historian. During these studies, he recognized the existence of biological evolution and attempted to explain its underlying principle. His explanation was speculative because sufficient grounds for argumentation of his theory were not available due to the small amount of scientific knowledge at the time.

Lamarck's evolutionary theory met strong criticism by Baron Georges Leopold Chretien Frederic Dagobert Cuvier (1769–1832). Throughout his career starting at the French National Museum of Natural History (as with Lamarck) and climbing to an academic authority, Cuvier made huge contributions to the fields of comparative anatomy of animals and to paleontology. Based on these studies and the results thereof, Cuvier advocated catastrophism. According to this theory, animals do not gradually change to evolve into various forms. Instead changes in the forms of animals in the geological age are attributed to extinctions of ancient animals, due to a rapid change in the Earth's environment, and subsequent fresh creations.

Cuvier was faithful to modern scientific methodology based on positivism. However, considering the historical backdrop, sole reliance on positivism resulted in a misunderstanding of the fundamental feature of living things, or evolution. His conclusion was inevitable because of the lack of empirical evidence. That said, his conclusion is still questionable because it emanated from the standpoint of a scientist who played a leading role in science policy.

Lamarck did not explain that a frequently used trait alters genes to strengthen the trait, because he failed to take genes into consideration. More specifically, it is true that frequently used traits will appear by selection as such traits become favorable. Recognizing the fact of evolution is not the same as demonstrating it with scientific and empirical evidence. It is incorrect to say that, because it cannot be proven, a fact, even if recognized, is not there or that recognizing it is not acceptable. Needless to say, it is also true that unless proven, a hypothetical theory is not science, it is merely a story.

Biological species diversity is not something that God made at the time of creation that has remained intact. Rather it has come as a developmentally created phenomenon by living organisms. This notion was widely shared among biologists as early as the first half of the 19th century. The important point is that they were searching for what phenomena were involved in diversification and how to explain the universal laws of diversification.

Darwin and the Theory of Evolution Under the above-described historical backdrop, Charles Darwin set out for a world voyage on the HMS *Beagle* to observe and compare the lives of living organisms in various parts of the globe. During this trip, he was convinced that species are not static; they alter in response to the environment and evolve over a long period of time. In search for facts that would serve as grounds for his conviction, he made comparative observations in the field. He also extensively read literature relating to the topic and was absorbed in contemplation. Without holding public office after returning home, he spent time deep in thought at his home. At the same time, he exchanged views with outstanding researchers, including the botanist Joseph Dalton Hooker (1817–1911) and the geologist Charles Lyell (1797–1875).

Darwin collected various corroborative evidence based on the theory of natural selection, nearly finalized his evolutionary theory to derive the conclusion of biological evolution, and almost completed writing 'On the Origin of Species.' That was when he received a letter from young Alfred Russel Wallace (1823–1913) who was working on field research in Southeast Asia. Wallace was one of Darwin's researcher friends. Darwin had some knowledge about his active field research. The letter explained a theory that perfectly aligned with the theory that Darwin was about to write. Upon reading the letter, Darwin was faced with the issue of priority disputes with his unpublished theory. In the end, advised by Lyell and others, Darwin read a paper on the evolutionary theory based on the theory of natural selection at the 1858 Linnean Society meeting as a joint presentation with Wallace. This is a very well-known episode in the history of biology, so this book refrains from describing it further. 'On the Origin of Species,' written by Darwin alone, was published in the following year (1859) in an abstract form of the huge work he had been preparing. Wallace was informed of the publication and opted not to claim his priority. Instead, as the most eminent supporter of Darwin's evolutionary theory, Wallace actively participated in the discussion.

The notion that biological evolution is a fact had become a mainstream at the time, at least among biologists. However, before Darwin, no one had been successful in systematizing the phenomenon into a single logic and making it persuasive to ordinary people as well. Since the beginning of the 19th century, despite Christian doctrines to strictly follow what the Bible said about God's creation, there had been an increasingly strong concept, along with a deepening understanding of living things, that the various forms of organisms resulted from evolution over time. Although it was not at all likely that species resulted from inheritance through generations intact from God's creation, in reality, proponents had difficulties building an objective and persuasive logic for evolution in the face of Church doctrines.

As anticipated, Darwin's theory of evolution met with harsh criticism from society at the time, especially from the Church and their admirers. Nevertheless, eminent biologists, with a perceptual understanding of evolution and who longed for the construction of a persuasive logic, supported his theory. Accordingly, it was rapidly established as a biological theory. Of course, Darwin's theory had many speculative aspects, which would later be met with a fatal criticism as biology advanced thereafter. However, Darwin's theory was truly established as a biological theory because it provided a basis for proving evolution, specifically by the concept of emphasizing the process of repeated transmutations and selection of adaptive forms. Considering the limited available biological information at the time, it was necessary to provide speculative explanations, in addition to evidence. Therefore, even with some aspects vulnerable to scientific evaluation, Darwin's achievement, as recognized by anyone today, is his affirmation of and construction of a theory for the phenomenon of evolution. (Nonetheless, I have heard that some groups in the United States, such as conservative religious circles, still refuse to accept the theory of evolution. It is hopelessly difficult to argue with these types of people. This is also partly due to evolution existing without scientifically conclusive evidence, although it is a theory accepted in the academic world.)

Erasmus Darwin and Lamarck had insight into biological evolution. However, they were unable to successfully provide arguable grounds and organize their insight into a theory. Additionally, even given that Wallace developed a logic for the theory of natural selection in the same way as Charles Darwin, he had not reached a point where various corroborative evidence was provided to build the theory in a persuasive way. By writing 'On the Origin of Species' as an abstract, Darwin became the first person to successfully convince the biological science community and enable the general public to recognize the biological truth to the extent influencing their thought. In this regard, the attitude of Darwin, who hesitated to publish the theory in consideration of Wallace's originality and priority in the theory of natural selection, was modest and faithful. As a result, the history of biology seems to have evaluated them fairly, understanding their contributions commensurate with their respective efforts.

During his round-the-world voyage, Darwin saw with his eyes and encountered diverse lives of many organisms living in various environments. He could elaborately observe the dynamic aspects of living species. Through field research, he recognized dynamic evolution of species and established his view of living organisms based on the theory of evolution. Wallace also discovered the theory of natural selection at the same time, which in a sense attests that their theories were a reasonable development in biology. Darwin was supported by Hooker (who in 1865 took up the post of director of the Royal Botanical Gardens, Kew, succeeding his father). Hooker had also researched plants in many parts of the world, and therefore consistently supported Darwin's theory of evolution, even undertaking a support campaign. Moreover, records say that Asa Gray (1810–1888), who established natural history studies of plants in the United States and became a professor at Harvard University, continued encouraging Darwin by sending letters expressing his views on 'On the Origin of Species.' These first-class researchers in natural history at the time eagerly awaited the completion of Darwin's theory of evolution, although they did not endeavor to prepare such a work on their own.

Darwin's arguments are subject to criticism in various ways, specifically from the standpoint of modern science. However, no one can deny the historical fact that, in the second half of the 19th century, the Darwinian logic helped biology recognize the truth of evolution and corrected the course of the field. For evolutionary biology to become a science beyond the Christian Charch's control, the recognition of individual facts of evolution was essential. Darwin's scientific understanding of the detailed facts he enumerated was substantially corrected in later years. Nonetheless, the significance of the role that the establishment of the concept of evolution played in biology and, additionally, in society's understanding of science should not be obscured.

Discovery of the Laws of Mendelian Inheritance Mendel discovered the laws of inheritance based on data from his carefully organized experiments. He was active mostly in Brno and its nearby areas in Czech (which was a part of the Austro-Hungarian Empire at the time), although he studied for some time in Vienna the then metropolis of Europe.

Acquiring empirical data through experiments to contemplate a subject by hypothesis testing was uncommon in biology at the time, or more specifically, in domains relating to natural history, the very basic biology. Naturally, Karl Wilhelm von Nägeli (1817–1891), an authority in botany at the time, who was giving advice to Mendel on experimental materials, did not respond to a reprint of Mendel's paper, although he had received it. It is said that for him, Mendel's quantitative analysis technique was probably beyond comprehension, and he was rather critical of Mendel's approach to studies. However, in other fields of biology, analysis incorporating experiments had already begun. For example, Louis Pasteur (1822–1895), praised as the originator of modern bacteriology, performed experiments intended to negate autogenesis of life. Incidentally, Pasteur and Mendel shared the same birth year.

Mendel, as a young man, taught mathematics at a gymnasium. However, he failed his qualification exams to become a certified teacher due to low scores in biology and geology. It is sometimes said that he failed because he insisted on his theory and opposed the examiner. When studying in Vienna, Mendel learned mathematics and physics under Johann Christian Doppler (1803–1853), known for the Doppler effect, who apparently had a profound influence on Mendel. The structure of the logic Mendel used to derive the laws of inheritance was not in line with the common study methodology of natural history at the time. Rather, it was based on analyses that followed a physicochemical way of thinking. There seems no doubt that Mendel had read Darwin's 'On the Origin of Species.' However, he was not influenced by Darwin's work.

The laws of Mendelian inheritance were read to the Natural History Society of Brno in 1865 and compiled into a treatise in 1866. (The latest Japanese translation of this treatise was published in 1999 by myself and a friend scientist. The translation includes a little lengthy commentary, written in consultation with 'Gregor Mendel' [Orel, 1996] and other works.) Immediately after the publication of the treatise, Mendel took up the post of abbot and became involved in disputes on tax reforms at the time, resulting in termination of his study in genetics. In his last years, he contributed to science as a meteorologist. The academic world continued ignoring the laws of inheritance he discovered, probably due to the laws being way ahead of the times. After his death, since rediscovery of his work in 1900, 20th-century biology grew around the laws of inheritance. However, some literature suggests that Mendel was confident in his theory and believed that his time would come before long, with his achievements being used at the core of biology.

Mendel's laws were rediscovered by three researchers separately and in parallel. This historical fact concretely indicates that in the domain of biology, there were multiple highly motivated researchers with the intention of furthering experimental results who received stimulus from Mendel's paper. This implies that at that time biology was in an era of yearning for analytical methods such as Mendel's.

Both Darwin and Mendel publicized achievements far ahead of their times. Although many scholars, even in the academic world, could not understand the significance of their achievements, Darwin had enthusiastic supporters, and Mendel had a small group of inconspicuous proponents. These supporters and proponents respected their respective achievements and helped them evolve, an interesting episode of the history of science. Although it took some time before rediscovery, Mendel, slightly earlier than the tide of history, provided a clue which, at the beginning of the 20th century, spurred the start of a so-called century of genetics. Darwin's theory of evolution concretized the independence of natural science in Europe from the Church.

From Evolutionary Theory to Evolutionary Biology The above description may appear to imply that Darwin and Mendel have helped biology grow in their respective fields, in completely different directions. On the contrary,

I would like to describe that with achievements being made in molecular biology in the second half of the 20th century, present-day natural history is evolving through the integration of Darwin's theory of evolution and Mendel's laws.

It is even said that Mendel's laws were at the heart of the development of biology in the 20th century. Researchers have verified that the diverse phenomena that living things exhibit are basically dominated by common universal principles. They have elucidated how apparently diverse life processes are passed from parents to children via DNA and how all traits develop under the control of the laws of manifestation, based on genetic information that the composition of four bases determines.

To elucidate universal principles of life, particularly in the second half of the 20th century, biology chose *Escherichia coli* as a model living organism for analysis. Through this, biology has accumulated results as if thoroughly knowing this model living organism equals clarifying what life is. Indeed, the study of life that has come to be known as life sciences has revealed many things by utilizing techniques that deserve to be called science.

Meanwhile, the diversity of living things has continuously been recorded in diverse manners. The origin of diversity, and the mechanism underlying the retention of diversity, have also been genetically analyzed based on Mendel's laws. Major advances in analysis techniques have enabled researchers to produce an enormous amount of literature on paleontological records that demonstrate the ancient history of diversification. In the late 20th century, it became possible to analyze species diversity with molecular biological techniques, along with advances in life science research methodology.

At the same time, theoretical studies advanced. In 1982, having conducted population genetic analyses, Dr. KIMURA Motoo at the National Institute of Genetics proposed the neutral theory of molecular evolution, which affirmed basics of the evolution of genes at the molecular level (Kimura, 1982). His theory led to the use of traits on the molecular level as an indicator in techniques for taxonomic hierarchy tracing and species identification.

A growing portion of the domain of natural history has become verifiable through analytical study. Natural history has become competitive in the center of the life science arena. This may be akin to the initial attitude towards challenges, in which Aristotle's natural philosophy gradually developed into natural science.

This book covers Darwin and Mendel as giants from the 19th century. To explain the intention behind this, I hope to clarify the present-day notions of natural history, bearing in mind the significance of compiling studies based on Mendel's scientific analytical techniques and using

Darwin's techniques of collecting corroborative evidence and making integrative theorization.

Impact on Society and the Academic World Mendel's genetics became mainstream in the world of biology in the 20th century. However, his achievements were not valued before his death. In terms of his contributions to society, he failed to make his valuable theory useful to society. Three researchers who recognized the significance of Mendel's experiments, Hugo Marie de Vries (1848–1935), Carl Erich Correns (1864–1933), and Erich von Tschermak-Seysenegg (1871–1962), reviewed his experiments, confirmed their scientific significance, and strived to make them common knowledge in biology. The world of biology recognized Mendel's work owing to the contributions made by the three scientists who were inspired by Mendel's experiments and, in part, due to the successful results of dissemination by William Bateson (1861–1926) and others after the turn of the century.

Darwin's theory of evolution began to attract the attention of the academic world when he read it to the Linnean Society and rather quickly attracted public attention. In this sense, Darwin's treatise and 'On the Origin of Species' can be recognized as a huge success contributing to society.

Of course, between the two theories, there were clear differences in the academic and social backdrop, that is how much advanced knowledge was in place to accept them. Nonetheless, they will serve as important examples when evaluating the contributions of scientific achievements to society, questioning the value of individual achievements if they are not recognized by the academic world or society, and debating which is desirable: science for the sake of science or for the sake of society (see 5.3 (2)).

It is said that Mendel was confident as a researcher, with certainty in his experimental results and speculation. However, he died without having an impact on the academic world or society. He strived to know the truth and produced successful results as a scientist. Nonetheless, a question remains, although its implication may differ between the beginning of the 20th century and today. That question is: What are the contributions that science makes to society in the present-day context?

Chapter
2

Contemporary Challenge of Natural History

While science has modernized, the field of natural history may give the impression that it remains a descriptive research field adhering to the tradition of compiling records on diverse matters, rather than a legitimate successor to the natural philosophy that began with Aristotle. Moreover, in the view of some scholars, natural history has metaphysical aspects, because naturalists attempt collaboration between analytical research methods and a comprehensive understanding to fulfill scientific curiosity. Natural historians are motivated to understand analyses in an integrative manner with the unified whole in mind. What are their objectives, and what are they trying to understand? The following sections cast light on contemporary challenges in natural history.

2.1 Biological Science and Natural History

(1) Japanese terms derived from the term *natural history*

The term *natural history* is contained in the title of the Natural History series issued from the University of Tokyo Press, as well as in the title of this book. This section first begins with investigating what the term *natural history* means. This term is open to various interpretations, depending on the user. Taking the magnitude of usage disparities into consideration, the following discussion expresses my understanding, even if vaguely, of what this book explores with the term *natural history*.

[Common definitions of the term in Japanese are summarized hereafter but this part is skipped in this English version, as the discussion is mostly on the Japanese expression and is unnecessary to be recognized by the English-speaking people.]

For context, the Graduate School of Science at Hokkaido University has a Department of Natural History Sciences. The phrases "natural history and science" and the "science of natural history" are used in Europe and

America. In comparison, the term *natural history science* is seldom heard. 'Natural History Sciences' is a long-established journal published in Milan since the mid-19th century. Tracing back to its predecessor, this title seems to mean a journal of natural science and natural history.

The Department of Zoology, the Division of Biological Sciences, and the Graduate School of Science at Kyoto University run courses in the study of natural history, including laboratories of Animal Ecology, Ethology, and Systematic Zoology.

Furthermore, in Japan, corporate bodies related to natural history, namely the Institute of Natural History and the Shikoku Institute of Natural History, also exist. Meanwhile, the Osaka Museum of Natural History contains "natural history" in the traditional form in its name. Established in 1952, this museum is a popular destination for Japanese citizens. In Japan, there are many institutions whose names contain "natural history." Needless to say, the term *natural history museum* is an internationally accepted common noun, which is not misleading. Natural history museums will be discussed in detail later. The Natural History series includes two natural history museum-related books, namely, 'Natural History Museums in Japan' (1993) and 'New Natural History Museums' (1999), both by ITOIGAWA Junji. The 1993 book begins with the following title: From Natural History to Study of Natural History.

Furthermore, the term *natural history* in medical context means the course of a disease. The term is also used to mean the recording of changes occurring over time in the life of a living thing, such as apparent seasonal or functional changes.

Usage of the term *history* has also changed. Originally, the term *life history* was used to refer to and describe changes with generational shifts in living organisms. However, at some point of time, the term *life history* began to be used increasingly to describe changes in the phenotype of a living thing, similar to the medical term *natural history*. In recent years, when explaining intergenerational structural changes, the term *life cycle*, in contrast with the term *cell cycle*, has taken hold. This might be intended to avoid conceptual confusion. However, until the mid-20th century, the term *life history* was commonly used to refer to phenomena nowadays explained by the term *life cycle*. Consequently, care is necessary to avoid misinterpretation when reading classical works.

A large number of plant and fungal types live in different forms between generations. Their life cycles should be carefully recognized. In contrast, in the metazoa, few types change form between generations. Scholars exclusively studying these animals need not be particular about differentiation between the terms *life cycle* and *life history*. The phenomenon known as *generationswechsel* (alternation of generations) in the life of jellyfish is actually a radical change (metamorphosis) in form observed in the growth process of a single generation. It is not a phenomenon observed at an alternation of generations, but a case in which the natural history of one generation is referred to as a life history. That phenomenon differs biologically from the alternation of generations exhibited by the life cycle of a plant and/or fungus.

The term *natural history* is used occasionally in tandem with the term *history*, signifying past events. When defining history using the term *historical science*, the term *history* may be combined with the term *science*. That said, the term *history* is used to indicate a domain that differs from the concept and techniques of natural science, which is regarded as a science itself. As the German philosopher W. Windelband (1848–1915) stated, historical science is: "a science that adopts the method of describing unique individual properties of events, in contrast to natural sciences which tend to establish general laws" (Windelband, 1903). Needless to say, the above discussion has nothing to do with the use of *history* in the phrase *history of natural sciences*.

The term *natural* also needs to be defined. However, this section refrains from examining it in detail. The usage of the term *natural* and its Japanese counterpart *shizen* will be compared more closely in section 2.3 (1) as well as in the postscript. In the Greco-Roman era, the term *history* was used to mean *search* or *writing*, as exemplified by the following book. It apparently began being used to mean *past events* in far later times. That said, the title of the primary work of the ancient Greek historian Herodotus (ca. 485–ca. 425 BC) is 'Historiai' (in Latin notation of Greek). It is true that the term *history* has been in use since the pioneer days of historical sciences. (Regarding history, natural history and civil history are occasionally contrasted.)

The Chinese character 史 (*shi*) was originally a ritual term. The character was later used to refer to registrars who compiled records of rituals. In the Former Han period, Sima Qian (145/135?–87/86? BC) wrote his work 'Shiji.' The term 歴史 (*lishi* in Chinese) apparently came into use by the end of the Ming dynasty. In Japan, since the Meiji Restoration, 歴史 (*rekishi*) has been used as a translation of *history*. This usage of 歴史 has apparently taken hold in China as well.

The term *natural history* once meant a description of natural phenomena present in the real world. In this sense it was used interchangeably with *records of diverse matters*. Since around the 18th century, when diverse phenomena and objects in the natural world were recognized as historical outcomes, the perspective of viewing nature as a historical entity clearly emerged, particularly in the West. The use of the term *natural history and science* differentiated from *historical sciences* probably suggests differences between the awareness of those engaged in natural sciences and those in the field of humanities. This is notwithstanding the historical development of the terms 自然誌 (*shizenshi*; literally, records of nature) and 自然史 (*shizenshi*; literally, natural history).

(2) Records of nature (自然誌 – *Shizen-shi*) and natural history (自然史 – *Shizen-shi*)

The term *historia naturalis*, in use since the Greco-Roman era, was first intended to mean an objective description of all things present in the natural world. (In fact, diverse phenomena and things observed in human society were also subjects of description. In this context, the term *naturalis* probably meant *all*.) Therefore, the notation 博物誌 (*hakubutsushi*, records of diverse matters) in Japanese is a reasonable counterpart of *historia naturalis* at the time. Indeed, Pliny's 'Naturalis historia' was translated into Japanese as 博物誌. Later in the mid-18th century, in line with the above-mentioned trend, 'L'Histoire Naturelle' by Buffon was also translated as 博物誌.

The Japanese word 博物 (*hakubutsu*) was borrowed from the Chinese language. In China, the term 博物誌 (*bowuzhi*), in no connection with the *historia naturalis*, titled a volume listing all creations present in the natural world (including the human world). By describing objects and events in the natural world, humanity formed the first domain of human activities that fulfilled intellectual curiosity commensurate with the state of human cultural growth, whether in the East or the West. In China, the study of taxonomy (in this context, it is rather closer to the tradition of herbalism than to natural history) has been steadily promoted since these times. While the study was in an early developmental stage, it had traditional associations with listing of natural assets for the country. The efforts continued even during the stormy period of the Great Cultural Revolution, despite difficulties encountered during this time.

In the 18th century, as a natural consequence of the development of scientific thought, all creations present in the natural world began to be recognized as four-dimensional entities. The historical background of individual objects began to be increasingly considered. Along with the emergence of the thought that all creations and events present in the natural world have a four-dimensional (historical) background and the rise of the intent to describe objects as historical entities, the term *historia naturalis* began to be understood in the context of natural history. In present-day Japan, most scholars use the term 自然史 to refer to natural history, because the term was introduced after the Meiji Restoration took place. Before then, the term 博物, originating in China, was in common use when referring to the recognition of objects in the natural world. In line with this tradition, the term 博物 was used in subjects taught at Japanese secondary schools up until the Second World War.

Meanwhile, botanist KIMURA Yojiro (1912–2006) was particular about the notation 自然誌, reasoning that none other but the character 誌 (*shi*) represents the correct understanding of the word *historia*. (Kimura, 1974). Incidentally, NAKAMURA Keiko, who is, although a biochemist, making continuous efforts to meet the challenge of observing living things from a comprehensive perspective, uses the notation 誌 (*shi*) in unfolding her bio-history (生命誌) concept.

Leaving aside the question of which Japanese notation to use, the study of the broad domain of natural history shaped the beginning of scientific thought and firmly secured its position in the development of science. Nevertheless, Japanese counterparts for the terms *study of natural history* and *natural history and science* were coined as late as the second half of the 20th century. Why is that so? This might result from the same mind-set using the term *biological science* instead of *biology* or in expanding the domain of biology to *life sciences*. More specifically, biology used to center on observation, description, and interpretation. However, physicochemical techniques brought about fruitful analytical results, and research conducted in a hypothesis testing manner produced substantial findings. Consequently, biologists sought to show that their biology differed from the ancient biology referred to as *natural history*, a term with a pre-Christian feel, by adding a modifier to the name of their field. Of course, attaching a new designation may have had the pragmatic intent of enticing young motivated researchers to join their circle.

(3) Coordinates of biodiversity study: Phylogeny (time-axis) and biota (horizontal axis)

Studies of Biological Species Diversity 1: Records of Nature Leaving aside the examination of variants of the term *natural history*, this section takes a step further in examining the growth of the study of biological species diversity and its implications, considering similarities and differences in simple notions about natural history.

The most fundamental and uniformly promoted aspect of natural history study is to sum up species diversity into biota and taxonomic hierarchy.

The study of species diversity likely originates from species identification, which was an integral part of life in the days before humans began intellectual activities. Wild animals keep a highly discernible eye to sense and discriminate similarities and differences between species closely linked in their lives, although, unlike humans, they did not give names to species. This is no different from *Homo sapiens* living as wild animals before they evolved into humans. This fact suggests that humanity's study of species diversity is instinctively inherited from the pre-scientific activities of living things.

In Aristotle's time, human knowledge was progressively compiled using scientific methods. For species diversity, a comprehensive description of nature purposed listing as many biological species as were recognizable. This was known later as natural history. Aristotle and Theophrastus devised measures for easy-to-understand arrangements of living things, instead of random lists of a large number of animals and plants. However, as expected, their classification methods lacked the concept of evolution. Consequently, although dealing with the natural system, these classifications were artificial, with an emphasis on intelligibleness. When understanding the interrelationships of all creations in the natural world, they were not aware of the primacy of the historical background of objects. That said, they seem to have intuitively recognized inherent relationships between living things.

The spirit of showcasing diverse things was respected in China, a common feature for all countries in a developmental stage of human knowledge. Similarly in the West, after having undergone the Renaissance, wider knowledge was compiled. However, advances in natural sciences resulted in natural consequences. In the 18th century, scholars understood that the presence of things in the natural world was not chaotic (even if they were creations present as an act of divine providence) and that all objects and events had a historical background. In fact, the idea that they were outcomes of historical development gained importance. Before long, this concept would grow into an evolutionary theory that would gradually secure its position in the world of science.

In the compilation of biota by Linnaeus (first edition; Linne, 1735), the organization of knowledge was based on the concept of a system, as indicated by the book title 'Systema Naturae.' He had the notion that the system of nature underlying the diversity of living things was, at least partly, due to historical outcomes, although later evolutionists criticized his artificial taxonomic hierarchy. His recognition of Rosaceae and Liliaceae (Linne, 1751, 1753, 1754) suggests the budding of early thought about natural classification in the natural world. Of course, Linnaeus fell short of scientifically recognizing evolution, a fact for which there is no need to reread the history of biology.

With the introduction of the system of nature when identifying and classifying similar or different species, the resulting system led to evolution.

This is a natural consequence of advances in natural sciences. In Japan, one conceptual consequence of viewing the system of nature as a prerequisite for natural history was the replacement of the term 自然誌 (records of nature) with the term 自然史 (natural history). In Latin, however, the notation *historia naturalis* remained intact. As a progression of the history of science in Japan, it may be easier to regard this as a transition from 誌 (records) to 史 (history). (I do not mean that those who are particular about the use of the character 誌 are not aware of the fact that real-world diversity is a historical phenomenon.)

Study of spatial species diversity starts by making a list of living organisms in the researcher's region, or in concrete terms, from the compilation of a regional biota (flora and/or fauna). Specialists arrange lists of individual organism groups (such as plants, insects, and birds), rather than compiling lists of all living organisms at once. Numerous compilations, such as books on Japanese flora and lists of birds on the Ogasawara Islands, have been produced for individual regions and taxonomic groups, gradually casting light on the actual state of biological species diversity on Earth.

Even with a specific organism group within a certain region, it is difficult to ascertain the diversity of many biological species, clarify the characteristics of individual species in writing, or compile findings into identification criteria that can be useful for anyone. Indeed, saying is one thing and doing is another. Many books on fauna and flora have been published. While a large number of them are excellent, some, unfortunately, seem worthless. However, due to the accumulation of superb achievements, the compilation of regional biota information has expanded from confined localities to wider regions. Findings about diverse taxonomic groups have been summed at higher taxonomic group hierarchy levels.

For the study of plants, researchers across the world collaborated at the end of the 20th century to publish books on global flora. However, there still remain wide regions for which detailed flora information is not clarified. For taxonomic groups with advanced results at the rank of family, researchers are now able to take a broad view of all regions of the earth. Nevertheless, it will take far more time for researchers to compile results for all families. 'Flora of the World' (IOPI ed., 1999–2005) is an international collaboration project inaugurated by voluntary members as a collection of flora books on a global scale. It is regrettable that, partly due to funding issues, the project was discontinued after the compilation of several families was published as a model. At the present-day level of science, it may take a considerable number of years before the aforementioned project is completed.

Incidentally, the work of Global Biodiversity Information Facility (GBIF), inaugurated in response to the Organization for Economic

Co-operation and Development (OECD) proposition to form a unified network and promote the construction of a biodiversity-related database, has also proceeded slower than expected.

Studies of Biological Species Diversity 2: Natural History Studies that tracked the degree of similarity and difference between species (affinity), conducted in parallel with projects describing and making lists of all diverse regional species as biota, led to evolutionary biology, which explores the historical implications of species. In sum, studies of biota widely present in the biosphere, spatial research, analyses of phyletic evolution tracing affinity among diverse biological species, and temporal research begin to be conducted concurrently.

From the perspective of the history of evolutionary biology, this is a natural consequence. Lamarck, in charge of classification of invertebrate animals, a field in which, at the end of the 18th century, few biologists were interested, took note of evolution during his studies. His use/disuse theory was designed to explain evolutionary processes, but it presented an incorrect explanation. However, his notion that living things diversify through historical development and that their forms are not created and fixed by God was correct. His idea substantially contributed to the establishment of the concept of evolution.

The evolutionary theory that Darwin attempted to explain in 'On the Origin of Species' became so influential that it forced not only biology, but also social thought, to undergo a radical transformation. Joseph Dalton Hooker (1817–1911) continuously supported Darwin most strongly during his development of the theory. Hooker was a son of William Jackson Hooker (1785–1865), who was the restorer of the Royal Botanical Gardens, Kew and made a substantial contribution to plant taxonomy in the United Kingdom. Succeeding the post of director of Kew Gardens after his father, J.D. Hooker led global plant taxonomy efforts in the second half of the 19th century by implementing the compilation of the 'Index Kewensis' and the 'Index Londinensis,' a foundation for projects that showcased living things in two dimensions.

As 'On the Origin of Species' went through several editions, Darwin appended the book with a note revealing that Gray sent him an enthusiastic letter from America, which at the time was falling behind Europe in the field of science. Gray was the main figure among pioneers in plant species diversity research in the Americas that was conducted at Harvard University. He was one of the excellent researchers in *records of nature* who became interested in the historical background of species diversity. (Incidentally, Charles

Wright (1811–1885), who took part in the Rodgers-Ringgold North Pacific Exploring and Surveying Expedition that shocked Japan at the end of the Tokugawa shogunate, brought back plants from various places, including Ogasawara, Okinawa, and Shimoda on the Idzu Peninsula in Japan. Gray was the leader of the research team which identified Wright's collection and discovered a number of new species.)

It is an objective fact that in the recognition and popularization phase of evolutionary theory, natural historians played important roles by substantially contributing to fauna and flora research. Notwithstanding this, recognizing the veracity of evolution does not mean that the evolutionary historical backgrounds of diverse organisms were all elucidated. While lineage analysis investigates the evolutionary historical background of species diversity, it is necessary to develop new techniques, pursue further analysis, and accumulate research results for clarification of phyletic evolution.

Presently, biological species diversity counts some 1.8 million recognized species. However, many organism groups, such as invertebrate animals, fungi, and microorganisms, are not yet fully studied. The actual number of biological species recognized on Earth is estimated to increase substantially with further research. Estimates of the number of biological species range widely, from millions to tens of millions, because nobody knows the real number. Some scholar estimates exceed even one hundred million.

Moreover, with microorganisms, progress even in basic research, such as what species are found in what habitats, is very slow. This is because even the notion of species defined provisionally for mammals, birds, and vascular plants (Tea Time 5) is not always applicable to microorganisms, adding to the general difficulties also found in biota research.

Even with the study of vascular plants, deemed to be at an advanced level with slightly over 200,000 currently known species, the recognition of species on the list is provisional. The common notion is that there are actually 300,000 to 500,000 living species of vascular plants. The reality that biological efforts have not yet provided a scientific definition of the concept of species becomes manifest when one looks at the entire Earth's surface, where some areas remain to be fully surveyed. Leaving aside new discoveries, records of new species are unending due to changes in the concepts of known species. In addition, research results that clarify that known established species are a type of another species are occasionally published.

In light of the above-mentioned reality, further basic research and studies on species diversity must still be vigorously conducted. Work to produce records of fauna and flora, or in other words to recognize similarities and differences in individual species and descriptions of which species are found in which habitats, are still needed.

Nonetheless, clarification of species lineages through analysis should not wait until species diversity has become completely clear. There is no reason that species differentiation (speciation) occurs only when complete information on fauna and flora is available. To recognize biological evolution, it is sufficient to compare three organism species. Nearness and remoteness, or similarities and differences, are naturally recognized in the affinity of these species. Tracing species affinity is an interesting and vital scientific challenge. When a species related to the three species is discovered, the relationship between the discovered species and the already known species may be adequately recognized. Rather, advances in speciation analysis will lead to the correct recognition and sound identification of species, which records of fauna and flora use as basic units. The challenge is that, despite having limited scope of empirical verifiability, the analysis of four-dimensional relationships should explore the extent of its capability.

Studies of Biological Species Diversity 3: Phylogeny While biological evolution was acknowledged as a concept, until the 19th century the only viable method of tracing lineages was to compare phenotypic characters and make inferences based on similarities and differences emerging through comparative morphology. As the similarities of embryology and morphogenesis processes to trace morphologic homogeny and fossils became subjects of comparative studies, lineage tracking was regarded as grounds for verifying the truth of evolution. In the 20th century, encouraged by the rediscovery of Mendel's laws, genetic analysis of traits was established as a biologically viable analysis technique. As this progressed, analysis techniques developed that enabled improved verifiability in the analysis of inter-species affinity and lineage.

Subsequently, in the first half of the 20th century, cytotaxonomic analysis techniques were established by benefiting from advances in genetics. Genes were presumed to be present on chromosomes. The numbers and forms of chromosomes and dynamics observed during division were certainly known to be unique to individual species. Consequently, these chromosomal characteristics were thought to indicate species. Therefore, using similarity of these as a measure, efforts were directed towards the exploration of inter-species phyletic nearness and remoteness. The ratios determined by crossbreeding individuals from different populations were occasionally used as an indicator of affinity. Analysis that used chromosomes as a key, which would be known as the cytogenetic technique, is today an optimal analysis technique for the analysis of, among other efforts, species differentiation originating from chromosomal mutation.

Chromosomal mutation plays a critically important role in the species formation of vascular plants. As the ploidy of chromosomes was known to drive species differentiation in vascular plants, Professor KIHARA Hitoshi (1893–1986) at Kyoto Imperial University (present-day Kyoto University) and collaborating researchers ascertained lineage diversification by the combination of polyploidization and crossbreeding to trace the history of cultivated lineages developed from wild wheat. Their research produced very early cytogenetic analysis results that successfully analyzed lineages and, along with the subsequent growth of analysis techniques, developed into a study model. Facts ascertained through karyotype analysis in the time of Dr. Kihara were verified by research using deciphered DNA, later enabled by advances in analysis technology (Tsunewaki, 1993).

Triploid hybrid lines produced from frequently occurring natural crosses are sterile. However, if polyploidization occurs, such a line may be able to form successive generations. Through this process, species formation occurs originating from natural hybridization. Evolution that forms netlike inter-species relations (reticulate evolution) also exists, with several related lines serving as common maternal types. Additionally, it seems that new types may form through the derivation of an agamosporous triploid, likely due to gene mutation, as described in 2.3.

Similarly, during the mid-to-late 20th century, molecular-level traits attracted scholars' attention and identification of primarily secondary metabolic products was actively conducted. The technique that used chemical relationships between metabolic products as an indicator for tracing the species formation sequence was called "chemotaxonomy." Studies using this indicator were vigorously conducted.

Meanwhile, the technique of numerical taxonomy (Sokal & Sneath, 1963) improved. This technique attempts to deduce purely mathematic approximate relations by avoiding intuitive evaluations of traits and employing as many taxonomic characters as practicable.

For handling traits with which objectivity was occasionally an issue, the cladistic method (Hennig, 1965) was applied. Hennig proposed a method of recognizing and organizing evolutionary processes as branching patterns. The cladistic method attempts to ensure objectivity of evaluations of classified traits. When a large volume of molecular trait information became available, it grew into a method of organizing lineages based on gene indicators and has helped provide theoretical grounds for ascertaining the basic course of biological evolution.

In parallel with cytotaxonomic analysis and other studies that required live materials, experimental analyses were also tested using comparative cultivation of related groups adapted to similar conditions, such as on high mountains or seashores, and presumed maternal types growing in more stable environments. Some scholars used the term *biosystematics* to refer to these analyses using live materials, in contrast to traditional taxonomic studies that mostly relied on comparative studies of dead natural history specimens. They claimed that theirs was true biological analysis.

Of course, biological analyses should be performed based on living things. However, information about living phenomena in large part cannot be easily acquired from living organisms. As such, depending on analytical methods, information provided by dead natural history specimens can serve as an important foundation for revealing the state of living organisms. A biased viewpoint criticized that this remoteness from biology should not presume to analyze facts about living things. Some scholars attacked studies that analyzed dead substances extracted from living bodies, comparing them to studies of squeezed juice. Others pointed a finger at observations performed with a traditional electronic microscope, claiming that they were just observations of dead materials. The point, however, lies simply in differences in perspective towards analysis methodology, namely, what methods are appropriate to use to observe the state of living.

By the end of the 20th century, it became possible to view nucleic acids as taxonomic characters, more due to advances in scientific analysis techniques than to the field of biology. Molecular phylogeny, along with advances in the concept of cladistics, began to provide important evidence for improving lineage tracing by scientific techniques. Although not principally intended to describe the current practice of biodiversity analysis, this book introduces the reader to the fact that the application of analyses using DNA as a key helped natural history advance dramatically in its present-day development process.

In parallel with studies of fauna and flora, studies based on the natural history perspective rapidly advanced in methodology and resulted in the accumulation of solid results. In the study of species diversity, analyses over the spatial expanse of the biosphere and along the time-axis of lineage have been deployed in every direction, supported by rapid advances in analysis methodology and strengthened by information processing technology designed for material organization. Consequently, the accumulation of findings obtained through these analyses is properly under way.

Nevertheless, understanding species diversity involves an extremely large amount of information. Accordingly, the provision of basic information is still slow and inadequate for bioinformatics processing. Regrettably, the number of scientific findings about the historical aspects of life still falls far short of what is needed. While GBIF (3.3) has been active since 2000, I am acutely aware of the processing limitations defined by the organizational size of GBIF, considering the vastness of the estimated amount of information needed. In this field characterized by the handling of an enormous amount of information, mathematical analysis of database information is essential for future developments.

(4) Understanding biodiversity

The term *biological diversity* is commonly defined at three levels, namely, diversity within species (genetic diversity), between species (species diversity), and of ecosystems (ecosystem diversity). Biodiversity is normally understood at these three levels precisely when it is discussed from the perspective of sustainable use, according to the definition in the Convention on Biological Diversity. Nonetheless, this definition fails to refer to the whole picture, although it represents some aspects of biodiversity.

Phenomena that represent the diversity of living things, if examined in biology, go beyond the scope of understanding defined by these three levels. Diversity within species refers to diverse genetic variation that brings about somewhat different features within the population of the same species. That said, one can also look at constituent elements that make up individual bodies of, for example, multicellular organisms, including *Homo sapiens.* A single fertilized egg differentiates and grows into a multicellular body. In this process of change or diversification into cells, tissues, and organs, the manifestation of various genes takes place with time. Take, for example, an individual *Homo sapiens* body. Some thirty-seven trillion differentiated cells form skin, muscles, nerves and other tissues, making up diverse structures such as the hands, feet and head. Truly, diversity is expressed even in an individual body.

Diverse genes, species, genera, and ecosystems are the result of evolution throughout the history of life. Likewise, the diversity of cells and tissues of a multicellular body emerges through cellular differentiation and growth. In this process, ontogeny follows a similar developmental process to phylogeny. In either case, biodiversity is an outcome of four-dimensional phenomena. Biodiversity comprehended at the level of individual bodies has developed through the history of evolution. Likewise, diversity understood at the level of cells and diversity at the level of tissues and organs developing through the ontogenetic process have been framed through the history of evolution.

The above resembles variations of slightly modified expressions of the term *biodiversity*, such as "diversity of organisms," "biological diversity," and "diversity of living matter."

Many biologists today view living organisms as masses of matter or vehicles of life. In this case, a living thing is more like living matter. The expression "diversity of living matter," from the perspective of developmental biology, is occasionally used on the basis that traits of living matter are diversely differentiated. This might be the reality of modern biology. In this domain, a natural history that grows into evo-devo may be pursued, as pointed out later in this book (3.3).

Journals published by the Linnean Society include 'Biology,' along with the 'Botany' and 'Zoology' series. Of course, biology in this case refers neither to botany, zoology, nor molecular biology. The journal is more likely to feature papers in domains close to the field of ecology. In the context of biological diversity, as well, biology does not imply life sciences.

The technical term *biodiversity* acquired meaning of its own on the grounds of the Convention on Biological Diversity (CBD). However, the implications of this term are diverse and not easy to understand. Accordingly, its visibility remains consistently low. The United Nations Framework Convention on Climate Change, signed at the same time as the CBD and ratified later, used the term *global warming* to express the subject phenomenon. The visibility of this term rapidly increased. In this respect, a better way to gain an understanding from the general public may be to deal with the *spherophylon* (5.1) as the reality of living things and to discuss the life of the *spherophylon* alive on Earth, instead of using the abstract expression "biodiversity."

(5) Science and natural sciences in the broad sense

Positioning natural history methods in the science of living things inevitably requires referencing the very methods of science. However, this book cannot devote pages to a full-scale argument about science. The following is a minimal discussion on a range of topics relating to science.

On Science The term *science* has acquired a substantially broad range of meaning. As such, it is difficult to present a simple definition that encompasses all the implications of the term. A Japanese dictionary 'Kōjien' defines *science*: "1. Nomothetic or empirical knowledge proven through an experimental procedure, such as observation or testing; the generic name for studies conducted in individually specialized fields. While typical sciences are normally natural sciences, such as physics, chemistry, and biology, social sciences, such as economics and jurisprudence, and humanities, such as psychology and linguistics, are also pursued. 2. Natural science, in the narrow sense." In the first part of definition 1, science is knowledge, while in the last part, science is explained as the generic name for specialized study fields.

Furthermore, the dictionary explains the term *natural science*, which is the narrow definition of science, as follows: "Studies that deal with, and elucidate the rules concerning, various phenomena occurring in the natural world. Natural sciences generally cover fields such as astronomy, physics, chemistry, earth science, and biology. Natural sciences are also grouped into basic and applied sciences, according to whether focus is on application or not. → social science/humanities." The explanation emphasizes occasionally specialized disciplines.

The Japanese counterpart (科学) for the term *science* was coined in the last days of the Tokugawa shogunate, abbreviating the Chinese term 科挙之学 (studies for imperial examinations). The term originally meant individual disciplines. Around the time when the University of Tokyo was established, the term 科学 was used as a translation for the term *science*, which was later reimported to China (in a different pronunciation).

The term *science* originated from the Latin word *scientia* (from the word stem *scio*, which means "to know"), meaning knowledge in general. Consequently, the above explanation is consistent with the etymology of the term *science*. However, it should be noted that although Latin *scientia* refers to general knowledge, scientific knowledge today is limited to proven nomothetic and empirical knowledge, as defined in 'Kōjien.'

In the course of the meaning of Latin *scientia* settling to the above-explained present-day definitions of science, the process of human knowledge development has been intricately involved. This is an interesting topic in the history of science. However, this book refrains from delving into the history of the conceptual changes related to this term.

In the present-day definition of science, which refers to proven nomothetic and empirical knowledge, proof is vitally significant. It is crucial for scientific research to explore how to empirically track phenomena. Simple indifferent observation and recognition of plain phenomena will not reveal rules inherently involved in the phenomena. Reductionist analysis of phenomena is expected to lead to universal principles. In the domain of science, analysis is grounded in reductionism, which assumes that higher-level laws and concepts are replaced with lower-level laws and concepts.

Consequently, in natural science research circles, outstanding results are expected to be found in papers published by high impact factor journals, such as 'Nature' and 'Science.' The reason for this is that papers that pass review by these first-class journals are believed, as a prerequisite, to contain logically-derived universal laws, based on facts revealed from phenomena analyzed through high-quality reductionist observation and experiments.

Expectations for Science When humanity began to have knowledge that constituted culture, its responses to the true, the good, and the beautiful began to grow and be embodied in science, religion, and art. In this sense, humanity promoted science due to the rise and growth of its intellectual curiosity towards the truth present in the nature.

In the modern era, science served as a foundation for technology; mainstream technology fortified by science became powerful; and science evolved further by contributing to technology. Even an ironical comment, such as that war causes science to advance, is occasionally made. For the competitive creation of life-threatening technology, foundational science will necessarily advance to beat the competitor.

That said, science and technology are two different things. In Western languages, the connecting phrase *science and technology* is in use, while the modified form *scientific technology* is not valid. This may imply that technology subsists by itself whether being based on science or not. (In contrast, in Japanese, technology based on science is understood distinct from such techniques as used in folk art. The term *folk art* is indicative of somewhat artistic elements. The Japanese counterpart for the term *science and technology* is in practice often used to denote technology based on science.)

Since the growth of human technology based on scientific knowledge, humanity has achieved remarkable growth in civilizations. The result is successful construction and, at the same time, increasing destruction. The brutality of war increased and the pressure of human action (technology) on the global environment has become outrageous. At the price of these negative effects, humanity has come to enjoy material affluence and safety, the evaluation of which is closely related to the very evaluation of science and technology. On one hand, science and technology are evaluated by the level of technological quality; and on the other hand, most importantly, science and technology are assessed according to the significance of the roles they play for society.

Having come to manipulate technology for maintaining a life of safety and affluence, humanity expects further growth of science and technology to lead to a life of even greater affluence and safety. Nowadays, humans are demanding that science provide knowledge as a foundation for technology. Everyone prays for cures for diseases, a stable supply of resources, and ensured safety from disasters. Therefore, investments in the growth of science are tolerated, as science serves as the foundation for achieving the above goals. In this instance, applied sciences that offer immediate applications in technology, readily understood as useful, are distinguished from basic sciences that are expected to contribute someday to such fields.

Meanwhile, science and technology have brought about another set of realities. The human environment has actually degraded due to artificial causes. Unhealthy lifestyles affect people's health. Imbalance in the consumption of Earth's resources has led to an unbalanced supply of resources. A small incident can cause substantial damage to people's lives. The growth of science and technology does not mean that humanity uses advanced science and technology in a proper manner. Superb technology can be a devastating blow to human society, depending on how it is used.

I have greatly benefited personally from advances in medicine. Deteriorated parts of my body have been repaired. I have been allowed to continue to live, despite reaching old age, thanks to several kinds of drugs that I take on a daily basis. I can maintain a reasonable level of life, free from starvation, although not particularly rich. My life has withstood repeated natural disasters and evaded dangers encountered on a daily basis. For this reality, naturally, I am grateful for advances in science and technology. I think this way because I view my life objectively as an individual living in this world, which depends on the advance of science, rather than as an individual whose profession is to contribute to science.

While the benefits and problems that science has brought to Earth and human society need to be assessed elsewhere, the question arises as to whether humanity expects science alone to fulfill physiological needs, and whether science should be considered essential in carrying out a life of safety and affluence.

Art and religion have helped people enrich their minds, and they still play this role, although there are some who are completely indifferent to art and religion. Similarly, the intellectual curiosity of an individual who desires to know the truth gives raison d'être to joy of knowledge. Science has undoubtedly contributed to the advance of technology. However, there may be no individual who believes that the joy brought about by science is confined to direct contributions to affluent and safe lives.

As a scientist, I feel that I am leading a worthwhile life when I deal with living things, tackle the fundamental question of what it is to be alive, and find myself contributing to solving the question. When I began my career as a scientist, the general impression that outsiders had about researchers in plant taxonomy was that they were self-indulgent fellows who deviated from the trend of the time, turned their backs on and made no contribution to society, and plunged into what they loved. Although this went against the belief of plant taxonomists themselves, the contributions of this discipline to society were not highly rated. They failed to be considered highly in public esteem.

In reality, during my days in active service, I encountered an era in which challenges collectively referred to as biodiversity came to the limelight, with the significance of my research becoming understood by the general public. Nevertheless, not a few scientists sincerely working on their challenges are regarded as eccentric virtuosos plunging a lifetime into what they love, despite the actuality of their superb achievements and contributions to society.

When it comes to science's contribution to society, Mendel offers an accessible example. In preparation for the creation of new cultivated forms of grape, Mendel assumed that fundamental laws useful for breeding were universally present among living organisms and endeavored to discover them. This was one motivation for him to work on the study of genetics. He took the time to clarify the laws of heredity and it was successful. However, neither were the laws recognized by the academic world while he was alive, nor could he contribute to wine production in Brno by creating superb forms of grapes. Nevertheless, no one says that Mendel's study has contributed nothing to society.

Mendel's science might have made no contribution to the society of his generation. However, outstanding researchers contributing to the genetics emerged in later years owing to his research. With time, Mendel's laws proved to be extraordinarily beneficial to human society, a fact now known universally. Specifically, when the term *bio-industry* was coined, biology could not stand without Mendel.

However, despite their understanding of Mendel and claiming their involvement in biology, those who trumpeted the prefix *bio-* were almost largely ignorant of biodiversity, roughly until the end of the 20th century. This is a fact that exquisitely depicts the attitude of many scientists who cast a cold eye on domains not mainstream in the academic world.

Additionally, regarding Mendel's scientific contribution, one must not ignore the fact that, in addition to his significant contribution to enabling bio-industry and material-/energy-oriented lifestyles, he provided basic knowledge helpful in solving the fundamental question of biology, that is, what it is to be alive.

Returning to my personal experiences, I have successfully made a livelihood as a scientist and have gained reasonable visibility in society. Nonetheless, my treatises are not immediately useful for making people's lives more materially affluent or safe. Thus, I am unable to be completely free from doubt as to whether I have done the right thing or not. Science is originally meant to contribute to human society. However, the research that I elucidated will never find application in technology to build a safer and more affluent human society during my lifetime, at least not from the material- and energy-oriented perspective.

Notwithstanding the above, I persuade myself to work in this domain of science because I am aware that my pursuit of solving challenges driven by my scientific curiosity will ultimately clarify the truth. I also know that my endeavor based on a sense of purpose will be a humble step towards solving humanity's fundamental question of what it is to be alive. The good intentions of those who have been involved in natural history since the time of ancient Greece have principally focused on clarifying universally applicable truths.

However, if the above is the right thinking, it is necessary to produce proof for its significance in a persuasive manner. If one only says that he or she thinks so and plunges into one's own world, hobby-like from the perspective of outsiders, one is probably making oneself a hermit isolated from society.

Scientific Awareness It is difficult to sum up the general awareness of present-day scientists in an objective manner. To understand the thinking of scientists with raised awareness about the social significance of science, the report of the 1999 World Conference on Science held jointly by the International Council for Science (ICSU) and UNESCO in Budapest, Hungary provides useful hints. I am carefully disseminating the idea that in the 21st century, scientists should strengthen their mind-set towards science for society and depart from the traditional attitude of science for science.

At the World Conference on Science, participants had discussion on the responsibility, the challenges facing, and the obligations of scientists concerning the use of science and scientific knowledge. The Budapest Declaration was adopted at the conference, regarding four aspects of science, namely, science for the purpose of peace, science for the purpose of development, science intended for the growth of science, and science for society.

Regarding science for the benefit of society, the participants discussed the necessity of integrating, for a specific learning purpose, knowledge that had deepened in sophistication within individual specialized disciplines. The Science Council of Japan was poised to name this scientific methodology "design science."

For the integration of sciences, various arguments have also been presented at academic meetings and other occasions. This shows that outstanding scientists have thought deeply about the theoretical integration of sciences. However, no methodology for that purpose has been proposed, as repeatedly described in this book. The World Conference on Science has remained a one-off. Scientists with the Budapest Declaration in their mind number few. The majority opinion of scientists may be that it is more necessary to carry out many experiments and write quality papers than to afford spare time for those sorts of conference discussions. The reality of the present-day scientific community may be that the minority of scientists, with raised awareness, participate in these discussions at scientific conferences.

There has been no tangible methodology developed for design science. One probable reason for this is that, as pointed out occasionally in this book, scientists who are aware of this need are mostly those held in high esteem by and who have achieved outstanding results in sciences that employ the currently fashionable analytical reductionist techniques. They are not expected to come up with any excellent idea for an integrative methodology. It is for this very reason that expectations are high for researchers in the domain of natural history, predicated on an integrative perspective, to construct and propose a solid and persuasive methodology for integrative science.

Investigative Sciences and Integrative Sciences UMESAO Tadao used the expression "investigative sciences" to refer to currently mainstream sciences based on reductionism. In contrast, domains such as cultural anthropology, in which reductionist analyses grounded in mathematical logic alone are incapable of providing solutions, approach the truth by using techniques that accumulate observed and ascertained facts to the greatest extent possible. He called this approach "integrative science." [Originally, Dr. Umesao used the Japanese terms *tsuranuku* and *tsuraneru* for investigative and integrative, respectively. While *tsuranuku* may be translated as investigate, penetrate, or go through, and *tsuraneru* as combine, join, or integrate, the translation of this book employs *investigate* or *investigative* and *integrate* or *integrative* to convey the phonetic similarity of the original words.]

The non-natural-science domains of social sciences and humanities have difficulties in thoroughly analyzing events applying solely a reductionist manner and providing logical evidence on the reality of phenomena, at least within the current state of science. Considering the volume of information involved, it is impossible for these disciplines to provide perfect solutions under the current state of human knowledge. Even more, there are some who think that it is difficult for these disciplines to perform reductionist analyses due to their academic nature.

One thing that I learned through conversation with Dr. Umesao and have explored by my own efforts, is that investigative and integrative sciences do not refer to ways of thinking that should be applied in different domains. Regardless of domains, unless both are adopted, the integrity of scientific pursuits is not assured. Naturally, however, the proportions of these approaches differ substantially with domains. For instance, for reductionist analyses to hold, suitable basic information must be reasonably provided. That said, according to Dr. Umesao, the two are different research techniques. To promote an integrative science, analyzing specific phenomena exclusively within a given domain is meaningless.

Since the mid-20th century, biologists have conducted analyses using DNA as a keyword. By making optimal use of these reductionist techniques to describe phenomena exhibited by individual bodies from the subordinate structural units of cells and DNA, biology has grown remarkably as a science. Its results have been applied to medical and agricultural technologies, and have, as recognized by everyone, contributed significantly to the safety and material affluence of humanity. Further advancement of this type of research and its reliable application to technology is expected to proceed without problem, raising hopes for the future prosperity of humanity.

Until the mid-20th century, biology was pursued primarily by means of observation. Thereafter, the discipline began to produce successful reductionist analyses by physicochemical means. This shift clearly demonstrated actual improvements in scientific analysis techniques. For example, in the second half of the 1980s, my laboratory undertook a DNA mapping project to promote molecular phylogeny. At that time, it took about a week to map a single strand through laborious manual operation. In comparison, it has now become possible to obtain the required results instantly, owing to improvements in the accuracy of the sequencer. Dramatic technological advances enable mapped data to be evaluated directly by computer.

Until the mid-20th century, Japanese middle schools treated biological topics together with earth science or geoscience topics in one subject (*natural history*), distinguishing it from the other subjects that covered physics and chemistry topics (*physical phenomena*). In these subjects, *natural history* once was studied by memorizing written objects and phenomena, in contrast to *physical phenomena*, based on scientific hypothesis-testing analyses. Compared with that time, biological research based on reductionist methods that evolved in the second half of the 20th century has attained truly dramatic advances.

Nonetheless, in the early developmental phase of molecular biology, for example, arguments exploring possibilities such as “beyond molecular biology” were already present (e.g. F. Yoshimura, 1987). Doubts about achieving rapid forward steps solely through reductionist analysis were receiving attention. Again, it should be noted that the doubt was about rapid advance by **only** reductionist analysis. The rapid advance itself was, and still is, very important for the growth of science.

The issue here is whether, in the domain of bioscience (perhaps considering only life sciences in a narrower sense), the recent seemingly noticeable view that **only** investigative science can achieve rapid successful results is acceptable or not. Meanwhile, whether the use of integrative methods in biology is necessary requires additional discussion. However, there is no established integrative research method that sounds reasonable for everyone, which may be the true nature of the problem. This challenge may lead to contemporary questions about natural science and science as a whole. This book examines the issue from various perspectives in a constructive manner.

In the relationship between humanity and other living organisms, the focus of bioscience progress has been medical care and food production directly related to people's safe and affluent lives. Along with this, since the mid-20th century, expectations have been high for biology to play a role in solving environmental problems. In the fields known collectively as environmental science, those relating to living organisms are still lagging behind in scientific analysis. Therefore, it is difficult to find scientific solutions in these fields. Consequently, they have received the least attention in all environmental science. In Japan, particularly, physicochemical, i.e., engineering, fields have been at the core of environmental science. The human environment in connection with other living things has been given a low priority.

However, the degradation of the biological environment has raised a serious problem. Again, societal demand is affecting the direction of progress in bioscience, similarly to how war strongly drives the progress of certain fields of science for technical advances. Indeed, it is well known that differences in scientific and technological abilities manifest through differences in the growth of technologies designed for warfare.

Integrative Science and the Human Environment Developments and transitions of the universe or Earth have unfolded as an aggregation of natural phenomena. With remarkable intellectual development, *Homo sapiens* became humans and strengthened their technological abilities through science created by intellectual ability. Consequently, humans have become aware of the environmental issues currently facing humanity, which are due to the tremendous distortions in the natural environment caused by the application of artificial impacts. Phenomena noticeable to everyone are common discussion topics. Producing records of these phenomena is not difficult. However, in investigating their causes, the current ability of science is limited.

Observing abnormalities in biodiversity as phenomena is not difficult. However, when one attempts to depict the integrated whole persuasively, one will find that the issue is too complex to scientifically process. The issue of endangered species was picked up as a model for tracking the dynamic state of all of biodiversity, because it was viewed as a task that would enable scholars to scientifically comprehend and delineate artificially induced distortions in the global environment. Indeed, reports with evidence of some popularly known species being on the verge of extinction gave a strong blow to society and played a certain role in attracting the attention of the public to the sustainability of biodiversity.

Producing records of phenomena is relatively easy. In addition, it is possible to create a model to predict their future dynamic states. However, to analyze the causes endangering the species in question, to extend the phenomena of endangered species to the dynamic state of all of biodiversity, and to accurately track them to discover a universal principle of the world of living organisms are virtually impossible tasks considering the present-day capability of science. Individual species exhibit species-specific phenomena. They have adapted themselves to the environment through their respective evolutionary histories. It requires tremendous labor to individually analyze natural transitions in and artificial impacts applied to the environment. Provision of background information is also necessary. Consequently, it is difficult to prepare reports with technically guaranteed integrity. Moreover, it is also difficult to deduce how a species-specific phenomenon is exhibited by closely- or remotely-related species. For this purpose, currently known and collected data alone are absolutely deficient. Furthermore, this task requires analyzing data on all other individual species. To explore the universal principle underlying the above challenge, a vast amount of additional information should be provided.

Unless individual phenomena are analyzed to provide reliable basic data, nothing becomes clear. For this very reason, individual studies that produce data with scientific evidence about certain aspects of a species are all very valuable. Nonetheless, however scientifically useful data resulting from analyses pursued to fulfill scientific curiosity, it is rare that individual data alone contributes directly to providing solutions to environmental issues. Analyses that track the dynamic state of endangered species serve as a model for depicting the dynamic state of biodiversity. Based on this understanding, the dynamic state of endangered species must be analyzed to provide a bird's eye view of biodiversity. This allows determining of what is happening in the human environment and, if any dangers become apparent, devising a solution for course correction. This model is not meant to analyze a single phenomenon from a particular perspective. Rather, it is a model designed to depict the dynamic state of a whole biodiversity.

Consequently, it is essential for biodiversity studies to facilitate the analysis of individual elements. However, to meet the expectations of contributions to society, biodiversity studies need to make optimal use of all currently available data; infer, as accurately as possible, the current state of the human environment emerging from the data; and reveal what is needed for future study at any point of time. It is necessary to view the environment as an integrated whole. The required perspective differs from reductionist analyses conducted from a scientific investigative point of view. Moreover, these intended studies do not have full scientifically-guaranteed integrity. Perhaps, at least under the current state of science, economic and social science studies are also required to pursue the above-described analysis and integration. It is certain that these disciplines are developing differently from natural sciences, founded on strictly physicochemical methodology.

If biodiversity is analyzed as an aspect of environmental issues, the role biodiversity plays in the integrated whole of the human environment will become the focus of investigation.

Integrative Sciences The current trend is that achievements are not assessed to have scientific value if not based on a reductionist technology, at least in the domain of natural sciences. Therefore, observation of diverse phenomena and subsequent inference are held in low regard, criticized as being intuitively captured, lacking evidence, and presenting no scientific deduction or conclusion. In a similar way, the accumulation of diagnosis examples alone does not lead to describing the true nature of diseases, although the diagnosis made by a good doctor is valid for improving pathologic conditions. Plant taxonomy, which used to place priority on producing records of phenomena, was regarded as simply collecting diverse facts, lagging behind in scientific analysis. Certainly, doing nothing but recording mutually unrelated phenomena, selecting heterogeneous types through comparison of phenomena, and noting new species could be reasonably criticized as being an art of enumeration and not a science. Indeed, those who were involved in plant taxonomy were not completely free from the tendency to aim at noting as many new species as possible. That said, the above tendency is also present in more than a few present-day scientists, who use all their efforts *exclusively* to produce objective and impeccable records of specific facts and phenomena. Furthermore, these very researchers in question are quite not aware of this fact.

The species diversity of living organisms on the earth is tremendously vast in terms of the number of species. Due partly to that vastness, humanity has not yet achieved accurate knowledge about the state of species diversity of common animals and plants, despite the efforts of many people. As described earlier, the actual number of terrestrial plant species, of which over 200,000 have been identified, is estimated at between 300,000 and 500,000. This does not mean that all unlisted species are rare or unknown to humanity. Some are common but roughly grouped into a single species because of the incomplete identification or accuracy at the species level. As in-depth studies are conducted, one species frequently turns out to be a complex of several species. More concretely, many instances are present in which species already known and described are subdivided by further research. These include those regarded as cryptic species.

Humanity does not accurately know the number of species. Needless to say, estimates are poorly grounded. Various views presently exist regarding estimating the number of species. If particular about finely-defined species, a researcher will estimate a large number of species. If content with broadly-defined species, a researcher will naturally estimate a smaller number of species. Studies that aim to track the real status of species are conducted concurrently with basic research on diversity. In this process, estimating the number of species using ambiguous definitions is not a very scientific task. Nevertheless, without attempting the task, science at its current capability cannot draw the whole picture of biological species diversity.

The task of recording biological species diversity is not satisfactory if the results are simply numerous, indiscriminate records. Some might think that taxonomic classification is interchangeable with subdivision. However, producing records of species diversity does not equal creating random records of diverse types. It is necessary to clarify interrelationships between discriminated diverse types and classify them according to the clarified relationships.

Verification exists that living things on the earth originate from a single type. The biological species diversity observed on earth results from more than three billion years of evolution. To understand this diversity, the phylogenetic affinity among extant biological species must be clarified to satisfy intellectual curiosity.

If of course, as the number of earth's extant species reaches millions, tens of millions, or even hundreds of millions, newly recognized individual species add to the currently known species that number slightly less than two million. This is a challenge that requires an immeasurable amount of labor. Scholars devoting themselves to this documentation should not be underestimated. Their work is not wasteful. It is fundamental to understand that intellectual curiosity is not fulfilled simply by producing records. Records of phenomena alone cannot satisfy human intellectual curiosity. People must be highly curious about causal association and relationships among diverse phenomena. Science has clearly indicated that biodiversity results from the history of evolution, which started from a single type that emerged on earth more than three billion years ago.

Needless to say, exploration of species diversity, based on the process of biological evolution, cannot be completed simply by analyzing diverse individual species. The solution will not be reached without inquiry into the mutual affinity between diverse identified species, through which process knowledge about diverse species is integrated. The question of whether it is scientifically possible to carry out this task of integration by current biological analysis techniques should be addressed elsewhere.

Foundation of Integrative Sciences For reductionist analysis, or the method of investigative sciences, the fundamental methodology used by most scientists developed alongside the dramatic growth of natural sciences. Evaluation of reductionist results should certainly be understandable to any scientist in any field, whatever the subject of analysis, although the extent to which available analysis techniques are effective for problem solving depends on the competence of the researchers involved.

That said, let us look at the methodology of integrative sciences. The necessity of integration in science is enthusiastically discussed. Outstanding scientists repeatedly emphasize the importance of bridging humanities and science. Notwithstanding this, in reality there are many who are restless that there have been no achievements that meet the expectations in line with the above goal.

Nowadays, so-called outstanding scientists are researchers who have made superb achievements, specifically by means of reductionist analysis. Nevertheless, outstanding researchers clearly know the extent that their achievements have contributed to science because they commonly have broad and profound knowledge in their respective specialized domains. Accordingly, they continually self-evaluate their individual achievements relevant to the degree of their contribution to the system of science, while remaining confident in their achievements. In addition, they attempt to find the position of forthcoming achievements in the system of science. It is quite normal that individual research results are grossly inadequate to provide an answer as to their exact meaning in the whole of science, even when individual results make contributions to society, as in application to technology. Because they precisely know this fact, many excellent researchers mention the necessity of integrative sciences in bridging humanities and science.

All that said, researchers who have achieved excellent results in reductionist analysis often make light of, and have poor knowledge of, integrative sciences. Doubting the methodology of integrative science, they naturally opt to not seriously pursue this methodology. Consequently, even if they self-rate themselves accurately in investigative science, they are not good at positioning their achievements in the overall coordinates of science. Without possessing an established integration methodology, it is difficult, even for competent individuals, to take part in integrative science, unless they have attempted integration or have had experience in integration methods.

Individuals who have seriously worked on studies in the domain of natural history have contributed to integrative sciences. Of course, they are aware that their approaches towards compiling individual events into treatises lack thorough hypothesis testing, are not empirically ascertained, and involve many incomplete parts from the perspective of investigative science. However, they are conversant with the extent of their findings and the problems they solve in the pursuit of knowledge of the natural world. As such, they can evaluate the results of integrative sciences and know what problems need to be solved in the future.

(6) Methods of natural history

Understanding Diversity Natural history studies originally employed the methodology of integrative science. The accurate description of a species does not mean that the study is complete. Comparative studies with closely related species must be conducted. It is essential for the researcher to conduct integrative studies on related species in parallel with investigative studies of the attributes of the species in question. To know the mechanisms of various phenomena and interrelationships of diverse phenomena in the natural world, it is necessary to sort out and integrate the origin and the differentiation process of seemingly random diversity. How presently diverse forms have developed and what principles underlie the diversification must be clarified. In this respect, the fact cannot be tracked unless diverse realities are integrated.

In the study of monograph, which sorts out the whole picture of a particular group, the researcher must have an overview of all species within the group. To research the biota of an area, the researcher should direct attention towards all living organisms in the subject area. With this research challenge, looking exclusively at a single example will not lead to any concrete results, so the researcher must examine all examples brought together.

The current form of the natural world has developed through the orderly aggregation of phenomena and events since the birth of the universe, not through a recent sudden development. The present level of evolution is not conclusive. It is still on the way of an evolutionary process. History is always in the middle of a process. It develops into a new phase under the application of additional pressure. Every part of the earth has also undergone constant changes over geological time. Currently, the rate of change in biodiversity is astonishingly rapid.

In this context, the compilation of books on fauna and flora does not refer to simply listing species, although it may integrate described species. Essentially, it must track the historical background existing between the species of interest and construct a system representing their interrelationships (expressed in the form of classified system). Integration necessitates analyses to understand inter-relationship. While engaged in the task of integrating diverse species, some natural historians may only satisfy themselves by materials of interest, without contemplating the significance of the integration or being aware of the challenges facing them.

Homo sapiens is created as a phenomenon by nature. Hence, things described as artificial, or human-made, are also created by nature, in its principle. However, since human intellectual activity began to self-propagate, it has regarded human practices as differing from natural evolution, as exemplified by the definition of the word *artificial* as opposed to the word *natural*. Consequently, humans regard artificial things not as natural phenomena, but as non-natural practices affecting nature. What is not artificial is defined as being natural. This is pointed out in the context of this section, although it is a topic that should be contemplated elsewhere.

In an attempt to track biological species diversity from the perspectives of lineage and differentiation along a time axis, simply producing random records of diverse types will never achieve a solution. If viewed as the *spherophylon*, biodiversity is a large entity deployed on the entire surface of the earth (= biosphere) against the background of more than three billion years of evolutionary history. Unless one sees the whole as a single entity, only a partial understanding is available.

Integrating the Results of Investigation: Bridging Records of Nature and Natural History Dr. Umesao attached more importance to integrative science than investigative science, as in the methodology of cultural anthropology. He said that it is meaningful to have an insight into those findings that emerge from the integration of various bits of information and to deepen one's knowledge in domains that cannot be elucidated by the reductionist way. He stressed the significance of fieldwork conducted to obtain correct basic information for that purpose.

Nowadays, however, in most domains of natural sciences, scholars recognize that only results verified by reductionist methodology constitute correct knowledge. Therefore, facts intended for integration must be those verified by investigation. In this context, integrative science must not only bring observed facts, but also facts ascertained by investigation, into the causal reasons of phenomena; otherwise, this inquiry is meaningless. That said, one should know that while some facts can be currently investigated, other facts cannot be analyzed with current techniques or methods. (This section examines the significance of integrative research and studies in terms of natural history methodology. From other perspectives, integrated facts pose a scientifically insoluble challenge unless they are investigated down to individual parts.)

Inquiries into the implication of integrative science expect accurate facts to be integrated. Cultural anthropology should examine facts observed in vigorous field research and verify them by collecting specimens and other demonstrative materials to the greatest extent possible. Cultural anthropologists do conduct their research in this way. If integrated facts have a blemish, even slightly, the insight that emerges would be unclear.

In the domain of natural sciences, integration encounters the criticism that no picture can emerge from facts not yet proven through any reductionist analysis. Indeed, in more than a few instances, review papers, deemed to draw a systematic conclusion in an integrated manner, provide a list of unverified materials as basic facts.

Facts recorded in books of fauna and flora constitute information on living things existing on earth's surface. Diverse extant living things are identified and recorded based on reliable specimens and literature. Unfortunately, most records accumulated by hard work throughout history were produced by classical techniques. Driven by an international consensus, an attempt to reliably provide this information through a networked database has begun. However, the progress has been behind schedule [3.3 (2)]. Attempts to trace the natural history of biodiversity by applying bioinformatics methods to collected materials and information has progressed to some degree. However, the development of measures for the effective use of the data is slow.

Natural history should inevitably combine investigative science with integrative science. However, the actual situation actually is less than ideal. One reason for this may be that few researchers have a clear understanding of research in this domain.

There are a significant number of calls to address unsolved challenges, such as bridging humanities and natural science, or the integration of whole sciences. Such expectations are considered valuable for the development of science. Nonetheless, if the process is an incongruous aggregation of facts, bridging humanities and science is not likely to produce expected results. Simple addition of humanities and scientific research results will not lead to successful bridging. That said, the distinction between humanities and science is quite unique to Japan. I myself have questioned said distinction in the concept of science.

I once joined a meeting on academic curricula at a university. Afterwards, during casual conversation, a senior professor mentioned that he would like to deliver lectures on criminal law using a framework of mathematics. I was strongly impressed by this statement. It will take centuries before the methodology of natural sciences is applicable for clarifying social phenomena using robust data. Moreover, whether such a development will lead to the anticipated bridging of humanities and natural science will require additional examination.

In the process of maturity from the compilation of records of fauna and flora to natural history, integrative science is necessarily constructed through biodiversity studies that integrate investigative studies. It is desirable that this mechanism is presented as a model. Regrettably however, despite its past position on the high road in biology, present-day natural history lacks the necessary drive. Perhaps most researchers who should be conversant with natural history have lost track of how to address their primary challenges, have followed the trend of modern science easily and blindly, and have become particular about the analysis of a specific challenge.

The significance of integrative science lies in the inquiry into the implication of outcomes that emerge from sincere integration of observed facts, independent from investigation. However, the question arises as to whether integration is meaningful without knowledge gained by natural science methods. I would like to respond to this question by explaining that the actual methodology for integration substantially differs from the above by integrating facts determined by investigation. Some may argue that this is precisely the fundamental concept of science. Then, the question as to whether the present-day natural sciences are truly practiced in this way arises. By thoroughly adhering to this methodology, natural history should be approaching the truth.

It is true that in the domain of natural sciences, strict adherence to investigative methodology will produce outstanding achievements, and by considering integration, the results might be distorted. In contrast, in the domain of natural history, integration, which may produce a paper or two, will not truly solve problems. Integration combined with investigation, if predicated as a methodology, should establish a method for integrating sciences.

Natural History Falling Short of Expectations Aristotle, who created the system of science and is known as the first scientist, viewed 'Historia Animalium' as part of *physica*, or natural philosophy. When natural science began to take its very early, yet proper, form, each individual scientist produced records of individual objects and events, while contemplating the system comprising those objects and events.

Since Aristotle, many outstanding researchers in the history of science have had an interest and been active in natural history research. Advances in scientific methods, particularly analysis techniques, enabled the systematization of species. Progress in this research produced an accumulation of information, and before long, biodiversity became correctly recognized by researchers.

Meanwhile, an increasing number of researchers began to identify and record an enormous number of species. Many researchers with interest in species diversity began to embark on the recognition of unknown types, with little interest in tracking course of phylogeny. As it became easy for people to move around the globe, the profiles of living things in uncivilized regions, as well as the biota in culturally advanced Europe, were gradually brought to the forefront of the academic community. Records of hitherto unknown types were thus accumulated.

Notwithstanding the above history, identified species still fall short of two million. Researchers are unavoidably busy recognizing diverse types. As a result, even when I started my career, a researcher who produced more species records was considered more outstanding in the domain of taxonomy.

Consequently, researchers would strive more to quickly identify species than to explore species affinity, lineages, or the systematization of diversity, i.e., integration. Indeed, discovering, describing, and writing treatises on unknown types is an exciting task on its own, as doing so elucidates new facts and fulfills scientific curiosity.

If taxonomic researchers are content with this sense of fulfillment, the inevitable course for the domain of taxonomy is to leave the field of biology. Indeed, the Japanese biological community of the mid-20th century, when I started a career as a researcher, seemed to have this perspective. The consequence was, regrettably, that natural history was isolated from mainstream science. Moreover, as molecular biology began to produce brilliant results, records-oriented researchers were alienated as disagreeable fellows. They, in turn, reacted defensively to protect their fields of study, which increasingly deprived the field of natural history.

Producing records of as many new types as possible still remains an urgent challenge for the growth of biology. At the same time, it is a historical fact that outstanding researchers have continued to contribute to integrative science, consistently adhering to the high road of natural history. Those who contributed to the study of species diversity as an integrated whole constituted a minority in the circle of biological researchers. Regrettably, they lagged behind in making meaningful contributions, from taxonomy to biology, in both record documentation and systematization studies. This was also very unfortunate for biology. More than a few outstanding biologists, who had achieved successful results in non-taxonomic domains, recognized the above conditions and expressed their expectations for a thriving field of taxonomy. Through my experience, I have been continually aware of this.

Basic Information for Understanding Nature Notwithstanding the remarkable advances in science today, the enrichment of basic information is still very limited for correctly understanding nature. Approaches to environmental issues remain in a state of partial understanding, as society is acutely aware. In the domain of natural history, the enrichment of basic information tends to produce individual records of attractive, interesting phenomena and has been, in methods of conducting studies, even apt to deviate from the framework of natural science. More than a few researchers, failing to be aware of this fact, have been enthusiastically driven by their curiosity about diversity. Ironically, it is true that in hindsight, their research activities greatly helped to enrich information about nature.

That said, many scientists often fail to remember that to analyze biodiversity, diverse phenomena must be observed and their corresponding information well organized.

For biodiversity, to which the three diversity levels are frequently referred, it is inappropriate to treat ecosystem diversity as equal in rank to other diversity levels, partially due to the greater focus placed on the functional aspects of ecosystem diversity.

Species diversity studies involve tracking similarities and differences between species and species hierarchy. In this way, they help direct scientific curiosity towards lineages and affinity. Diversity does not present randomly. Species diversity studies elucidate that living things emerged originally as a single type, underwent the evolutionary process spanning more than three billion years, and have been diversely differentiated.

Diversity within species refers to genetic variation between individual bodies that constitute a species (genetic diversity). A species is not a set of uniform individual bodies that resemble industrial products. It is rather a group of individual bodies that feature diverse genetic configurations that are readily transformable for speciation. Genetic diversity is a phenomenon not limited to populations within a species. It is contiguous with species diversity and expresses the hierarchical structure of biodiversity beyond the rank of species.

To understand the above biodiversity structure correctly, it is necessary to completely depict diverse genetic structures at every rank and phenotypic character of individual bodies (ultimately, the structure and function of adults) created by such genetic structures. For this purpose, all traits ranging from the rank of molecules to individual bodies should be analyzed. In this respect, relevant information will be inevitably enriched through analysis of specific model organisms. In actuality, all individual bodies must be observed throughout their lifetime to produce comprehensive trait information. In this sense, the basic information brought about by modern science is still very limited.

Biodiversity holds the universal scientific principle that there are no two identical individual bodies. Gene variation constitutes a phenomenon present in individual bodies that may differ in every trait. By knowing what traits are shared between bodies, it is possible to describe the hierarchy of diversity.

For individual facts, analysis aligned with the essence of modern science is expected to produce valid results and is required to produce general information. If a specific model organism is the subject of analysis, analysis results will vary depending on model selection. Facts elucidated through the analysis will also vary in efficiency when universalized.

Simply put, notwithstanding that molecular-level information is essential for knowing lineages, observations that acquire all information from all individual bodies within a species are inappropriate for obtaining information on the species. The scientific methodology infers the whole picture by comparing particular sets of information for particular individual bodies. Thus, understanding the individual model body's position in its lineage by using all available information is essential. Such information cannot be acquired solely with advanced technology.

When selecting a particular species as the subject of study, a researcher puts full confidence in the existing species classification studies. Criticism of existing findings is predicated on the presence of existing findings. If no such findings exist, no model selection study will be viable.

A huge amount of biodiversity information has been accumulated and compiled into a taxonomic hierarchy through diligent efforts throughout history. Study subjects are selected, information produced, and new findings obtained based on the taxonomic hierarchy, no matter which analytical study a researcher undertakes. Problems in the existing hierarchy may be identified as a result. However, this does not mean that the existing hierarchy is unnecessary. (Specialists with knowledge of the existing hierarchy do not even always have a correct understanding.) New findings cannot be obtained without organizing the existing hierarchy, which must always be ready for service. Some researchers complacently say that if a study does not incorporate the most advanced techniques, it is not a study. Nevertheless, what is called for is the enrichment of the most advanced information. For this purpose, it is vitally important to prepare necessary data in every aspect. In this sense now, more than ever in the history of science, is the time to enrich the correctness of natural history data to lay a foundation for producing the most advanced information.

2.2 Natural History as a Means of Exploring 'What Is It to Be Alive?'

(1) Examining living things scientifically

As a foundation for technology, biological sciences must meet many challenges to ensure the safety and affluence of humanity. The study of living things currently pursued in the world seems to focus *exclusively* on the above aspect, from a material- and energy-oriented perspective.

If challenges that drive the curiosity of the intellectual human animal form the basis of scientific pursuit, biology must be the science that employs natural science methodology to solve the question of what it is to be alive. This question has intrigued human curiosity since humanity began to conduct intellectual activities. Of course, in this context, the phrase "what it is" implies a fear of death, which everyone must face. Moreover, this question is occasionally labeled as philosophical, rather than being scientific (Tea Time 10).

Needless to say, primitive people had significantly different scientific curiosity than modern people. Perhaps people thousands of years ago might have been more aware of life when they recalled the dead than when they contemplated the reality of their own life. It is inferred that it was their awe at death that made them think about life, with their awe at death being greater than their intellectual curiosity about life.

The death of a family member seems a serious issue for primates. Records document behaviors of a mother monkey carrying her dead child for days. Questions arise as to how this maternal love links with the recognition of life, and how the behavior of a boss monkey or an old elephant leaving its herd for a solitary death connects the recognition of life. Various explanations have been offered about what awareness living things other than *Homo sapiens* have about life. However, no existing scientific studies have provided evidence. Other questions arise from this, such as whether or not a sense of fear is involved when animals avoid death instinctively, and whether, when encountering a detrimental change in the environment, plants exhibit various responses simply as a physiological action because they lack nerves and the perceptual reactions caused by nerves.

The reason a person has a sense of being alive may not be exclusively because a person thinks. While some philosophers state that an object presents in the real world when the subject (i.e., humans) recognizes its presence, natural sciences attempt to prove the presence of an object or phenomenon not as a matter recognized by humans, but as an objective fact.

Questions arise, separate from the topics of cognitive science, as to how modern people analyze living things that even *Homo sapiens* knew and what modern people have come to know from these analyses. For humans, natural science poses the most relevant and eternal question of what it is to be alive, as separate from the philosophical challenge solved by the saying "I think; therefore, I am."

(2) Biology, biological sciences, and life sciences: disciplines

The term *biology* emerged to refer to the basic science of analyzing phenomena exhibited by living things to explore the question of what it is to be alive. As mentioned earlier in this book, Lamarck was the first person to begin using the term (Tea Time 2). *Bio* means life. *Logy* originated from *logos* and means logical analysis, or academic pursuits. Consequently, biology is a branch of science that deals with life, or in other words, a science that analyzes what it is to be alive.

Natural sciences basically look for universal principles underlying diverse phenomena, as analyzed by materials science. After all, living things are made up of materials and are alive due to physicochemical reactions. Physics and chemistry that deal with living things are known as biophysics and biochemistry (or biological chemistry).

Of the scientific subjects taught at Japanese pre-Second World War middle schools (five-year middle schools under the old education system), *diverse matters* was a subject that combined the study of *living organisms* and *earth science*. This subject was intended for pupils to learn the diverse phenomena exhibited by living things, the earth, and the universe. While physics and chemistry analyze phenomena exhibited by all matter present in the universe, the study of *living organisms* examines the characteristics of living things and *earth science* explores the characteristics of the universe and one of its constituents, the earth. *Diverse matters* was intended to explore the reality of a particular subject matter. (Incidentally, Japanese four-year girls' high schools taught all scientific fields under one subject, *science*.)

Physical phenomena was oriented towards the universal principles underlying the phenomena exhibited by all objects. *Diverse matters* observed the universe and living things closely to track their characteristics. In other words, while the course *physical phenomena* was meant to analyze phenomena in a reductionist manner and clarify causal principles, *diverse matters* attempted to gain insight into the reality of the universe, the earth, and living things on earth by bringing together findings about objects and phenomena.

In materials science, alongside the growth of natural science, findings were applied to technology. In turn, radical advances in technology have allowed people to lead secure and affluent lives. The accumulation of research in the field of *diverse matters* has provided noticeably advanced findings related to the safety of the earth and human health, greatly contributing to the health and safety of humans and the environment.

When high schools were established under the new education system (1947), *diverse matters* was divided into two subjects, *living organisms* and *earth science*. At colleges, these domains are covered by separate disciplines, such as biology, geophysics, space physics, geology, and mineralogy.

Actual Entities and Their Names: Evolution of Species and Changes in Species' Name

As the basic unit used to recognize biodiversity, species are dynamic and constantly evolving, even today. A species in this instant is not identical in material substance to the same in a previous instant, because diversity within species, or genetic diversity, is never even momentarily static.

However, with sexually reproducing metazoa and terrestrial plants, the formation of a new species through mutation is calculated normally to take millions of years. Even under extraordinary conditions, such as on an isolated oceanic island, it is estimated to take hundreds of thousands of years (Ito, 2013).

Speciation in the context of this section includes somewhat ordinary speciation, once explained as species differentiation, with originally one type differentiating into two types. However, the phenomenon of speciation is more comprehensive. For example, speciation can occur as changes in the genetic configurations of all individuals over time to the extent that they are interpreted as a new species, having undergone a substantial change in comparison with the species structure millions of years ago. In this case there is no change in the number of species, so the term *species differentiation* does not apply.

When a set of species undergoes the former speciation, a doubling (possibly more in some cases) of the number of the species takes place. The latter case will not result in any change in the number of species. Moreover, with the latter case, if the currently living species is identified to be another species compared to fossil specimens from millions of years ago, and if consecutive fossils of that type are available from every era, it is not possible to identify any specific point of time at which a species difference was established. No leap indicative of a species difference can be found at any specific point of time throughout geological era, although changes occur at every instant.

This is like the differentiation of a species into two types. Even if it were possible to observe every individual body at every point of time, it is not possible to indicate when the differentiation was established. This means that some species are near completion of the process of differentiation. With such species, researchers recognize species configurations, despite uncertainty about the present grounds for determining whether variations between populations indicate a definite differentiation or a course towards the elimination of differences between populations.

In practice, if the change is not diversification, but a species turning into another species that differs from the type revealed by fossils, then another species name is given. Of course, in most cases, currently living species that have names are compared to fossils for determination of species differences. In actuality, a new name is then given to fossil specimens if they are determined to be another species.

When a species differentiates and diversifies, vicarious species may be derived from the original type (mother species). In this case, the mother species is referred to by the original species name and new names are given to the derived types. If species differentiation results in two species, both of whose types differ from the original type, new names are prepared for both types.

New types of currently living species are given names in accordance with the international code of nomenclature. This nomenclature provides for naming in such cases that a currently living species, originally regarded as one species, turns out to actually be composed of two species. Of the two types, the type specimens used to name the species are called by the existing species name and a new name is given to the other species. This is also applicable to diversification through the time process known as speciation. However, there may be cases where further research reveals that fossil specimens and currently living species with different names turn out to be the same species. In such cases, regardless of priority, the name given to the currently living species will be formally employed in accordance with the nomenclature.

For dawn redwood, Dr. MIKI Shigeru (1901–1974), when he was a lecturer at Kyoto University, first set up the genus *Metasequoia* and named *Metasequoia disticha* (Heer) Miki based on fossil materials from the Quaternary (Miki, 1941). Meanwhile, the species locally known as *shui-shan* in Sichuan and Hubei provinces, China, was determined to be of the same genus as *Metasequoia*. In 1948, Dr. Hu Xiansu (1894–1968) at the Fan Memorial Institute of Biology and others named the plant *Metasequoia glyptostroboides* Hu & W.C. Cheng based on living strains. Since then, *shui-shan* has been identified as the same species as the fossil plants recognized by Dr. Miki. Currently, the fossils are also called by the specific epithet given to the currently living species (in accordance with the code of nomenclature).

Naming species is expected to go smoothly under the above-described rules. However, when expressing a concept, it is often difficult to select a name that will communicate the right idea, because the concept may change over time.

It is said that a word is equivalent to a concept. If a word exists, a corresponding concept exists. If a word is not present, the corresponding concept is missing. There is no Western word corresponding to the Japanese word *kyousei* (literally, living together). This means that the Western world lacks the corresponding concept. The biological sciences and life sciences have taken the place of academic pursuits encompassed by biology. This may result from a desire to define disciplines narrowly. The outcome today is that the term *biology* is understood as implying basic biology, although it once meant all branches of science that pursued the study of living things. Instead, biology is even occasionally regarded as a part of life sciences. This is an example of a transition in the use of words that denote concepts.

People change the use of words and communicate concepts by language flexibly. It is unconceivable that the interpretation of a concept can change definitively at a specific point of time. Similarly, the species differentiation of living things cannot occur abruptly at a certain moment.

---

(3) Analyzing life in a reductionist way

Sciences Investigating Living Things In Japan, *living organisms* [the subject on curriculum is written in italic], once a branch of *diverse matters* and later a subject on its own, was once considered as a memory subject in the public mind. This was true even in the early 1950s when I was studying hard to take college entrance exams. This was immediately before Crick and Watson published their treatise on a model for the structure of DNA.

Twentieth-century biology started with the rediscovery of the Laws of Mendelian Inheritance. This symbolizes biology's endeavors to analyze living matter with physicochemical methodology, aiming to elucidate life phenomena by modern scientific methods. Research investigating and clarifying life phenomena began to advance rapidly in the 20th century. Of course, at the same time, research in *diverse matters* in biology was also promoted. In the field of taxonomy, research expeditions were actively organized to survey many parts of the globe. In the tropics and other colonial regions of powerful countries, researchers from suzerains led vigorous research activities. Research of biota in undeveloped regions was driven by a growing intellectual curiosity, and by the intent to discover and develop unknown resources.

After the commencement of modernization following the Meiji Restoration, contemporary with this early period of biology, Japanese biologists started survey of the biota of the Japanese archipelago. Before long, they conducted research in pre-World War II Japanese territories, including Taiwan, Sakhalin, Kuril, the Korean Peninsula, and Micronesia. Research subject areas expanded even to regions that Japan temporarily occupied during the Second World War.

In the history of biology in Japan, the first international contribution that Japan made in botany was the discovery of the sperm of the seed plants, gingko (*Ginkgo biloba*) and cycad (*Cycas revoluta*). This achievement was also highly regarded in Japan and awarded an Imperial Prize of 1912 Japan Academy Prizes. This observation, which provided substantial evidence for the taxonomic hierarchy of plants, was truly when botany in Japan made its global debut in natural sciences.

It was a healthy development for 20th-century biology to delve into the physicochemical analysis of phenomena exhibited by life and, at the same time, to work actively on discovering and recording species diversity. On one hand, biology based on the analysis methods of modern technology grew under the name of molecular biology; on the other hand, species diversity research, the single-minded production of records of phenomena, noticeably became the mainstream. Of course, some researchers attempted to prove evolution by trial and error using challenging methodologies. However, it took a long time before these efforts were rewarded. Consequently, biology polarized into reductionist studies and diversity research, to such an extent that it became impossible for them to act in concert with each other.

In the second half of the 20th century, biology in which DNA was the keyword steadily developed. Species diversity analysis also began to incorporate DNA-based research techniques. This was, in a sense, a reunion between the polarized biological sciences and natural history. DNA barcoding advanced rapidly as a key problem-solving technology. Inference methods using data obtained for lineage analysis also steadily improved. In this situation, natural history began to be regarded as joining the useful scientific fields. This may be construed as the establishment of investigative techniques in natural history.

Although it may be needless to remark on this, but to ensure no room for misunderstanding, natural historians have always conducted investigative analyses in line with the times. They began to observe microstructures following the development of microscopes. Along with the establishment of genetics, they conducted analyses using chromosomes as an indicator. Although species studies appeared to depart from mainstream biology that began to center around molecular biology, communication between these fields became possible again in the second half of the 20th century due to, among others, the establishment of molecular phylogeny techniques.

Aspects of Life Investigated and Revealed Using advanced scientific analysis techniques, scientists have elucidated many facts about life. Names for biophysics and biochemistry have been coined by combining the scientific disciplines of physics and chemistry with the study of living matter. In fact, advances in biology (or maybe biological sciences or life sciences) have even elucidated aspects of what it is to be alive.

Along with advances in biology, many clarifications about various aspects of individual life phenomena have been discovered. This is also true for the current state of biodiversity. Advanced in-depth surveys have been conducted on the actual state of living things on Earth, including those in previously difficult-to-research tropical zones and deep-sea areas (although studies of deep-sea creatures proceed more slowly). Information organized about the current state of species diversity has included the latest data.

In biodiversity research, however, researchers are slow to conduct investigative studies of individual phenomena, busy producing information on what species are present and where they are located. It is currently an urgent and essential challenge for biology to promote analytical studies clarifying biodiversity, as emphasized repeatedly in this book.

Notwithstanding the progress of the current era, no satisfactory answer to the fundamental question of what it is to be alive has been achieved. Reductionist studies have elucidated facts about various objects and phenomena, providing empirical explanations for particular aspects of life. However, they do not provide an answer regarding the implication of the subject matter, namely, what it is to be alive.

There is a constant demand for academic analysis of political and economic affairs, even though in this field, it is not possible to prove everything scientifically. Indeed, humans have always had feelings of anxiety and awe towards unknown events, while at the same time developing curiosity towards various phenomena. The meaning of human existence lies in exploring all unknown matters through intellectual effort driven by scientific curiosity. Scientists know that it is not possible to immediately solve the question of life by a series of analytical studies alone. To fulfill scientific curiosity, it is necessary to explore an additional elucidative perspective. As one possibility, integrative perspectives should be promoted. This is why the natural history perspective deserves additional attention.

Analysis of Phenomena and Elucidation of Reality Natural historians are successfully incorporating the methodology of investigative science into their analysis process.

It is mysterious that when science in Japan was divided into the two categories of *physical phenomena* and *diverse matters*, why life, the earth, and the universe were not included in materials sciences (*physical phenomena* = physics and chemistry). Since ancient times, scientists have studied the constituent substances of living matter. It has been known that living matter contains organic matter not found in non-living matter. However, chemistry that dealt with organic matter (organic chemistry) was included in *physical phenomena* and was not a field of *diverse matters.* Because that which is specific to living matter was a subject of *physical phenomena*, the distinction between *physical phenomena* and *diverse matters* should have not been made on the grounds of methodology.

I have not examined why *physical phenomena* and *diverse matters* were developed as two branches of science. Nevertheless, it is certain that the basic goal of *physical phenomena* was to discover universal principles underlying the properties of matter present in this world, including living things and the phenomena exhibited by them. In comparison, *diverse matters* dealt with living things and the real existence of the universe and the earth. Biology examines living things that are actually alive, as well as what, exactly, it is to be alive. To answer these questions, not only the principles underlying the matter present in this world, but also the lives specific to living things that have evolved over time must be understood.

The above may be clarified by contemplating explorations in earth science. The field of earth science does not include substances or phenomena specific to the earth or the Milky Way Galaxy, unlike organic matter in living organisms. (In Japan, because of the use of fossils as an indicator of stratigraphic succession, the study of extinct organisms, falls in the domain of geology. In contrast, in Western countries, paleontology is commonly studied in biological laboratories, in the same way that biophysics and biochemistry are pursued within the disciplines of physics and chemistry.) Substances that make up the earth and phenomena exhibited on the planet are challenges that should be elucidated individually by physicochemical techniques.

However, individual in-depth studies of the properties of earth substances and their phenomena, and exploring underlying universal principles alone, do not sufficiently profile the universe, meteorological phenomena, and natural disasters important to human life. Other programs should also develop that are not bound solely by universal physicochemical laws and can closely observe and record various situations taking place on the earth to predict possible future situations.

Towards this end, subject matter studies should go beyond simply stating that the subject is present by chance. It is necessary to elucidate the changes over time spanning five billion years since the formation of the primitive earth and how the Milky Way Galaxy allows for the existence of this planet. This cannot be studied solely based on the constituent substances and phenomena individually exhibited by earth and the Milky Way. Because the earth and the universe are analyzed as four-dimensional phenomena present over a long period of time, findings from the perspective of natural history must be integrated.

Given that its subject of analysis are substances, earth science could be part of physics or chemistry, rather than the field of *diverse matters* paired with *physical phenomena*. This proposition seems reasonable. However, the fundamental difference is that earth science looks at concrete subject matter, such as the earth and the universe, and studies tangible phenomena, such as weather and earthquakes, to identify the raison d'être of the subject matter. This is why earth science can be differentiated from the fields of physics and chemistry, which aim to discover universal principles through the objective analysis of information. These different objectives lead to different perspectives for these fields.

The remarkable advances in science and technology that support human life today are made possible by physicochemical studies of materials. Similarly, analysis of material properties and resultant phenomena of the earth and the universe obtained from the perspective of earth science has demonstrated how to manage our environment, greatly benefiting the development of a safe and affluent human environment. These benefits for human life are driven by the intrinsic scientific curiosity of humans towards the universe and the earth.

The question as to the nature of the universe and the earth is one that most stimulates scientific curiosity. At the same time, phenomena in the universe and on earth deeply affect the safety and affluence of people living today. No one would say that people should wait until humanity knows everything about the earth to utilize information about earthquakes and volcanic eruptions. It is natural that people hope to know as much as possible about the earth from the present scientific information, even if small. Challenges that should be addressed by science include those for which relevant answers are demanded and ultimate solutions are anticipated by society. In fact, efforts for solutions to these challenges are in progress, deeply interacting with each other.

Evolution from *Diverse Matters* to *Living Organisms* and *Earth Science*

FROM *DIVERSE MATTERS* TO *LIVING ORGANISMS* Japanese middle schools under the old education system before the Second World War taught science first under the subject name *physics & chemistry*. This subject was composed of two categories, one of which was materials science, later termed *physical phenomena*, and the other *diverse matters*. *Diverse matters* included *plants*, *animals*, *minerals*, and *physiology & hygiene*. High schools under the old education system at the time taught science in four subjects: *physics*, *chemistry*, *living organisms*, and *minerals*. The curricula for these were more specialized than in middle schools. Although the comprehensive subject of *physical phenomena* was introduced in middle schools, teachers in charge of physics were often not the same as those in charge of chemistry [as their courses in the universities are usually different].

After the Second World War, the curricula were reorganized during the educational system reform. As part of this modernization, high schools under the new education system introduced physics and chemistry in place of *physical phenomena*. *Diverse matters* was divided into the two categories of *living organisms* and *earth science*. Consequently, four subjects comprise science at high schools. The subject name of *earth science* came into use in place of *minerals*, which was used under the old education system. This resulted from the introduction of an American system during the education system reform in Japan. It seems that no Japanese counterpart term had been established for *earth science*. Incidentally, within the fields included in *diverse matters*, in high school *physiology & hygiene* moved to *domestic science* under the new education system.

Subsequent developments included setting up subjects under the titles of *science I*, *science II*, and the introduction of *basic science*. This book refrains from discussing these nominal changes in detail.

Living organisms after the Second World War, in the second half of the 20th century, underwent a transitional period during which the subject incorporated more scientific analysis and distanced itself from *diverse matters* centering around observation and documentation. A common notion holds that this was because biology was developing by drawing on the laws of inheritance. Conversely, it can be said that biology's development was facilitated the establishment of the DNA model. Symbolically, Crick and Watson proposed a DNA model in 1953, followed by advances in biology using DNA as a keyword.

Nevertheless, in the 20th century, *living organisms* taught in high school was generally understood as a memory subject. As such, people developing this subject were constantly facing the question of how to develop textbooks and entrance exam questions that would help pupils learn and assess scientific thought without placing exclusive importance on memory.

I have had the experience of writing questions for college entrance exams. It was requested that exam questions for *living organisms* avoid testing pure knowledge gained through rote learning, testing well-thought-out answers instead. At the time, there was a strong demand for questions that could be solved with knowledge contained in the question to the greatest extent possible, although these questions required a minimum level of basic biological knowledge and not just like questions in a quiz show. Needless to say, simple questions that test memory do not assess the capacity to think. In this respect, to my mind, essay exams in earlier times were superior to questions only testing the amount of knowledge. The extremely difficult challenge constantly imposed on question creators is to develop non-essay questions that avoid solely testing memory but that can still be rated impartially.

The field of *earth science* that spun off from *diverse matters* at first encountered resistance, partly due to the title. For a long time it remained ineffective, failing to provide balanced education. The reason for this situation was that training in fields such as astronomy, meteorology, geology, and mineralogy was traditionally provided separately, according to college faculty and course composition. This resulted in a lack of curricula integrity. Because of this lack of integrity, the quality of high school teachers was also affected.

While in *diverse matters* emphasis was placed on the observation and documentation of phenomena, the study of biology moved towards using physicochemical techniques to analyze living bodies and elucidate phenomena exhibited by living things through hypothesis-testing and reductionist methods. This was intended to raise unscientific observation and documentation to the level of science underpinned by dialectical explanations.

Rapid growth in science and technology characterized this era.

Facts about living things and their phenomena obtained through analysis were applied to various technologies, leading to substantial contributions for the safety and affluence of human life. On one hand, humans utilized physics and chemistry for the benefit of society; on the other hand, in wartime, they utilized these subjects for mass murder. In contrast, biology did not exert such evil effects, with few exceptions such as bacterial weapons. For some time, however, humans were unaware that advances in science and technology were endangering the sustainability of the earth.

Notwithstanding the above-described growth of biology, the development of materials science, which examined living bodies as substances, was construed as an advance in the exploration of life. Therefore, biology continued to be a large, independent domain of science, along with physics and chemistry, despite the remarkable progress in biophysics and the accumulation of great achievements in biochemistry.

***Diverse Matters* and Biology** Biology remained as a domain independent from the materials sciences of physics and chemistry because living things were viewed to have a special existence, although, in common with non-living things, they are both aggregates of elements. In this context, viewing living things as having a special existence raises a question. The question as to whether there is a critical difference between living and non-living things stirs controversy in every era, unique to the understanding of the times. Even today, no conclusion has established the presence of any difference. The special research domain, biology, exists for this unanswered question. While a science that aims to know the true nature of this special existence, biology is also a science designed to ascertain whether living things actually have a special existence.

In *diverse matters*, scientists pursued scientific curiosity to know what vaguely understood objects, such as living organisms, the earth, and the universe, really are. To know the true profile of the object of interest, detailed observations were conducted about the properties and motion of the object. As scientific analysis methods advanced, studies were conducted based on the most advanced techniques, as it was recognized that observation and documentation alone were not enough to identify reality.

Aside from the continued exploration of the fundamental question of what it is to be alive, scientists also discovered the characteristics and dynamic states of living things. Their findings presented important information for increasing the safety of living bodies, understanding the richness of living things as resources, and helping to advance essential technologies for the good of humanity.

Consequently, aside from scientific curiosity, for those seeking a more fulfilling and safe life, detailed findings about living things presented highly valuable resources for material- and energy-oriented lifestyles. People began to look forward to more and more findings. In an extreme sense, they did not care about pursuing scientific curiosity that did not fill their belly, such as questioning what it is to be alive. Their principal goals were to ensure their immediate safety and to immediately fill their bellies with an adequate amount of tasty food. Accordingly, these people began to strongly demand research that gave a technological foundation for attaining these goals.

Science elucidates the universal principles that underlie diverse objects and phenomena in the natural world, providing a useful foundation for technology. Promoting research in line with this aspect of science is important to allow science to advance, as expected and supported by society. For this reason, some people's hopes for the future rest on the advancement of materials science. Society provides a large amount of funding for scientific research, not only for intellectual curiosity, but in anticipation of material- and energy-oriented returns.

However, the question arises as to whether science is then fulfilling its full role with this goal. If yes, then the analysis of living bodies as substances might be more effectively accomplished in the fields of biophysics and biochemistry (biological chemistry).

Furthermore, questions arise as to what biology, continuing to exclusively research, observe, and study living things, really is. Also, considering both are studies of structures composed of substances, what are the differences between studies of the earth or the universe and studies of physics and chemistry? In science teaching, neither the exclusive study area of water nor air has been an independent subject, unlike biology, at junior high or high schools, although water and air are individual matters of study. This indicates how human scientific curiosity has been focused on life, the earth, and the universe.

Moreover, the very starting point of biology was driven by curiosity about what it is to be alive. This question has remained active in studies. The research and education fields of *living organisms* and *earth science* were collected under *diverse matters* in the past. The challenge posed to these fields is how much can they fulfill scientific curiosity about the reality of objects such as living things, the earth, and the universe?

Having emerged from biology, studies in biological sciences and life sciences were promoted as the study of the physical chemistry of living bodies in a reductionist manner using hypothesis-testing methods. Information acquired through these studies was utilized to help augment the safety and wealth of human society. This part of biology, focused on phenomena at the rank of individual bodies and higher, was mainstream when the term *diverse matters* was used. However, the raison d'être of this part of biology was downplayed and criticized for not contributing to useful production in society.

In the 20th century, humanity made significant mistakes in the manipulation of technology, which resulted in environmental disruption. The raison d'être of biology pursued at the rank of individual bodies and higher is occasionally acknowledged as the science of environmental conservation.

When I began my career as a researcher, taxonomy was already criticized as an outdated domain. However, in a very short time, taxonomy became known as the science of biological species diversity. Since it warned about the existence of endangered species, taxonomy has attracted public attention as a domain deeply involved in sustainable use of biodiversity for the good of society. Although regarded as a specialist in that domain, I did not feel at ease with the abovementioned reputation. Since retiring from a full-time position and returning to the analysis of the species diversity of a particular plant group of interest, I have been able to continue fulfilling my scientific curiosity, while achieving results suitable for publication.

If not a biological science or life science, does biology have any raison d'être in our present-day context? People with different beliefs will have different answers to this question. Some say that science should contribute to the material- and energy-oriented growth of human society, while scientific curiosity should remain in realm of hobby, pursued by wealthy people with free time. Others think that the purpose of science is to definitively produce studies based on human intellectual curiosity, the results of which may eventually allow people to lead a life of safety and affluence.

Regrettably, the domain of science defined as natural history is not fully fleshed out and unable to directly address these questions. In fact, the powerful domain of science known as life sciences has elucidated universal principles and phenomena in the natural world, which have been applied to technology and made substantial contributions to a safe and fulfilling human life. Compared with this, natural history has been unsuccessful in achieving the level of esteem that society has for life sciences. However, this should not be construed as a death sentence for natural history. Natural historians should pause to explore their own challenges and what they should examine at this point in time, to fully carry out their responsibilities.

---

(4) Human efforts in biology

Biology explores the features of living things in comparison with non-living things to know the characteristics of life, or the state of being alive. Biologists follow a similar thinking process when ascertaining whether the species *Homo sapiens*, among all living things, is special or not.

Philosophers state that man is a thinking reed (Pascal) in the context of "I think; therefore, I am" (Descartes). The ability to think makes man the noblest of all animals, because the act of thinking is a priori ranked highest.

Natural science attempts to define the characteristics of the species *Homo sapiens*, using indicators such as bipedal walking, cerebral development, the creation of cultures, and the use of language. It is said that the genetic difference between modern humans and chimpanzees is only 2%. In actuality, without a philosophical definition, it is clear that the intellectual animal *Homo sapiens* has evolved to an extraordinary level even among the Primates. Despite a small difference in DNA, there are minutely documented morphologic differences. At the same time, the concept that *Homo sapiens* as the naked ape has also emerged, which stresses similarity. Incidentally, although described as "naked," the entire surface of the human body is covered with hair, the sole difference with other primates being in the traits of hair.

Along with advances in natural sciences, *Homo sapiens* characteristics have been closely analyzed by physicochemical techniques. Diseases and other abnormalities are handled in an exquisite manner. The survival of *Homo sapiens* is artificially controlled and maintained. The advance of science contributes to the safe maintenance of human life. In these processes, to what extent *Homo sapiens* shares commonalities with other living things, and to what extent it is species-specific, has been clarified. This should help elucidate what *Homo sapiens* is among all living things.

Needless to say, analysis of living things leads to the accumulation of basic facts useful for clarifying what it is to be alive. In this sense, the physical chemistry of living bodies is precisely biology. Hence, analyses of phenomena of living bodies in comparison with non-living things may constitute biology, and physicochemical analyses of said phenomena, regardless of whether the subject matter is alive or not, may be considered biophysics or biochemistry.

Because intellectual activity is specific to humans and not a universal characteristic in the biosphere, it may not be a basic topic for contemplating what it is to be alive. More specifically, the ability to create diversity is one basic characteristic of being alive. However, lists of characteristics that are universal to all living organisms alone will not lead to a solution to the question of what it is to be alive.

Natural sciences endeavor to analyze universal principles by means of reductionist techniques. As part of this endeavor, biology conducts physicochemical analysis of life phenomena with successful results. Hence, the field of biology has become committed to analyzing individual phenomena in a reductionist manner. As a result, biologists place importance on looking at differences between phenomena exhibited by living and non-living things. They tend to omit tracking the phenomena of living as an integrated whole. Consequently, they begin to neglect fulfilling the fundamental scientific curiosity about what it is to be alive.

Those who have vaguely felt the above-described historical reality may be increasingly hoping for the reinstatement of natural history. What they hope for is what present-day science should provide. As society obsessively pursues material- and energy-oriented affluence, rather than cultural richness, scientists increasingly consider that contributing towards this goal will yield successful results.

Human intelligence can be obsessively material- and energy-oriented. Perhaps wild animals are free of such bias. For instance, lions prey on zebras. However, under steady-state conditions, they do not demolish zebras. It is only the intellectual organisms, or humans, that negotiate a discount on environmental rehabilitation, despite being responsible for the environmental degradation of their own habitat.

It is certain that people's aspiration for a life of safety and affluence is limitless. They expect to indefinitely achieve greater safety and affluence. However, it is also true that individuals who are not simply satisfied with such a life can explore phenomena in the natural world, driven by scientific curiosity. In the science of living things, natural history is one domain that is suitable for such individuals. Nevertheless, I cannot deny that it is disappointing that natural history's capacity is still very underdeveloped in the world of science.

When it comes to the biology of *Homo sapiens*, the reader might anticipate reading the latest scientific information. However, this book is not intended to provide that information. What humans have begun to know about *Homo sapiens* are basic facts, such as the origin of *Homo sapiens*; its evolution, including diversification into present-day ethnic groups; and their distribution. Additionally, life sciences for human health and safety, social sciences, humanities, and all aspects of religion and art have also been elucidated.

2.3 Bird's Eye View of Biodiversity in 'Diversification of Plants under the Impact of Civilization'

A Slightly Lengthy Preface for This Example The previous sections have provided a solely abstract discussion of approaches that can be taken by biology to elucidate what it is to be alive. Notwithstanding the phenomenal achievements made by present-day materials science in analyzing living bodies as aggregates of substances, biology has been clouded in obscurity, despite its intended objective to fulfill humanity's scientific curiosity of what it is to be alive. I am aware that in expressing this, I may sound like I am complaining.

In this section, it is then appropriate to describe, based on the above theoretical grounds, what present-day biology can do as a branch of science to elucidate what it is to be alive and to what extent present-day people can explore this question by natural sciences methods, along with philosophy and religion.

Humanity's natural intellectual curiosity asks what it is to be alive. Innocently saying that little is known about this brings dishonor to humans as intellectual animals. Therefore, I would like to present an example to describe to what extent scientific achievements can approach this question and what one can do to know life using natural science techniques.

I am aware that, as a researcher, I am responsible, not only for abstractly describing what must be done, but also for presenting an example of how to theoretically approach the problem. I am slightly ashamed to only describe the efforts of my group as a small example, despite the existence of demonstrably successful examples in present-day natural history, as exemplified by primatological research in Japan.

Primatology is accessible to everyone due to a great number of both technical and introductory guides. There are numerous related books published by the University of Tokyo Press (e.g. Yamagiwa, 2012 and 2015, Izawa, 2014), even though the Natural History series includes no works in this field. That said, I know no work surpassing the discussion of civilization by UMESAO Tadao, who reported on primatological studies as early as in

1960 (Umesao, 1960). It is not easy to determine when these studies started. The study of semi-wild horses (*misakiuma*) in Cape Toi in Miyazaki Prefecture, which subsequently led to primatology studies, can be a clue. Within a little more than a decade from the emergence of primatology studies by a group of researchers including IMANISHI Kinji (1902–1992), the creator of primatology, Umesao clearly saw the potential of the results they achieved. Umesao's insight into the group's research results is splendid. Umesao was the only person capable of foreseeing the subsequent growth of primatology because he was alongside the primate research group and could discuss its study as an outsider, while cheering for them. As even more of an outsider, I view myself as one of the people who could understand their results because I witnessed the growth of primatology from a somewhat related point of view, receiving significant influence from the core researchers in that group.

In the aforementioned work, Umesao asks ITANI Junichiro (1926–2001), who led the group's activity, why the research was successful. Itani's answer was that they conducted their research scientifically. This answer may concisely express what they had been aiming at since the beginning of primatology. The use of the alternative term *natural history and science* does not lead to a more scientific approach towards natural history. Tangible scientific research results must be presented. However, this was not straightforward for mainstream biologists on the high road of modern science, who were producing significant analytic and reductionist results in understanding primatology in its early phase. The phrase *scientific research* has a broad and deep meaning.

Imanishi foretold that it would become clear that animals have cultures, as well as humans. Very early in the development of primatology, Dr. KAWAI Masao (1924–2021) of Imanishi's group wrote a paper to demonstrate the existence of a pre-culture based on close observations carried out on Kōjima Island, Miyazaki Prefecture, where wild monkeys were observed to wash sweet potatoes (Kawai, 1965). It took nearly a half century before his theory became an accepted notion in the academic community world-wide. This is one example of demonstration taking a long period of time before becoming persuasive.

Japan led the advance of this field of study in the world. This is because Japanese researchers tried to understand the society of monkeys by befriending them from the standpoint of living in harmony with nature. In contrast, most of western researchers looked at nature from an elevated standpoint as the noblest of all animals. This difference in perspective deserves attention in the context of natural history. That said, I am hesitant to expand the topic further in this book. I was an outside observer but was closely acquainted with the outstanding researchers who started up the research. From this standpoint, I am somewhat concerned about what Umesao or Itani would have to say if they witnessed current developments in this field.

In the research example described below, researchers infer what would emerge from an integrated set of diverse example species and findings obtained through individual investigations. Individual observations must be exhaustive and accurate. In addition, scientific methodology must be used to arrive at a sound conclusion through induction of the observation results to depict the deductive conclusion. This research will lose ground if partial, analytic, and reductive analysis methods fail. Moreover, the resultant inferences will be meaningless if the integration technique is deficient.

The following example illustrates developing research on a specific organism group. The theme is the phenomenon known as propagation that takes place for generational change. At the cell level, variations occur during the process of cell division. In comparison, variations at the individual level (intraspecific variations), known as genetic diversity, manifest through generational change. In this sense, propagation is a phenomenon that should be noted to approach the reality of biodiversity.

(1) Science pursued in the Botanical Gardens, the University of Tokyo in the late 20th century

Mode of Research Activities I began to work at the Botanical Gardens of the University of Tokyo in the early 1980s. At this time, this Botanical Gardens had a research organization slightly larger than one small unit [usually formed by 4 research staffs] at the time. The laboratories were located separately in the main garden (commonly known as the Koishikawa Botanical Garden), somewhat off the Hongō university campus and in the branch garden (Nikko Botanical Garden) in Nikko, Tochigi Prefecture. At the university, other researchers in the natural history of plants belonged to the University Museum on the Hongō campus and to the College of Arts and Sciences (present Graduate School of Arts and Sciences) in Komaba. They took part in research and education at these organizations. (The Botanical Gardens did not serve as a national joint research center [formally admitted by the government] at the time. However, researchers commuting to the botanical gardens for a considerable period to learn about and use the equipment, constantly joined the botanical gardens' research activities.)

Research groups in these organizations, which individually participated in research and education, collaborated flexibly with research groups at the main botanical garden. In one period, related researchers joined weekly research seminars held at the main botanical garden, even though they were not directly affiliated with the botanical garden. They made combined research efforts, additionally involving graduate students and, occasionally, alumni, as if there were no disparities arising from affiliations with different departments. They showed abilities comparable to those of a large organization, more than expected by their size, which was slightly larger than the small unit size plus several individual research groups.

It was a loosely unified body of research groups crossing organizational boundaries. Their activities were very extensive, and several individual groups formed. Although small, groups of researchers belonging to individual organizations naturally followed their unique courses of research. Some researchers summarized local biota in a traditional way and analyzed specific taxonomic groups of plants using the traditional methodology of comparative morphology as a fundamental means of analysis. An increasing number of other researchers used traits at cellular and molecular levels, as well as indicator forms, to analyze plant groups of their interest. Other researchers (on an individual or group basis), placed importance on an ecological perspective to track differentiation at the rank of species. Researchers used various techniques to analyze lineages, not only at the species level but also at higher ranks. While traditional methods were applied, advanced means were also aggressively incorporated in research, including molecular phylogenetics, which was rapidly evolving technically at the time. For these activities, the necessary devices were steadily provided over time.

These research activities were characteristic in that almost all researchers understood the details of their colleagues' research, in addition to their own specific research projects. They intimately exchanged insightful views in discussion with each other. In these, they shared frank views from various perspectives, on topics ranging from the collection of materials to analysis processes to how to process data and how to derive conclusions. Although individual challenges were focused on analyzing specific plant groups by specific methodologies, in actuality, researchers were maintaining a bird's eye perspective to determine the position of each research challenge in the overall plant kingdom.

(2) Diversification of plants under the impact of civilizations – Locations of problems

The results of individual research projects are published as treatises in relevant journals. Review papers summarize the results of efforts made on specialized subjects that take a bird's eye view of problems. However, even review papers do not go farther than making inferences based on scientifically verified information within a certain limited range. In domains with only limited fundamental information available, as in diversity studies, hopes rest with additional efforts to build knowledge. Meanwhile, bold inferences based on limited information are expected to provide new perspectives on the analysis of problems, although existing scientific findings alone may not provide sufficient evidence.

Such attempts may, in some cases, be subject to risks in logical reasoning when compiled as a general theory. That said, research compiled in 'Plants Grown by Civilizations' (Iwatsuki, 1997) may serve as an example of drastic inferences made with these said risks in mind. This example is predicated on the evolution of agamosporous ferns, including *Asplenium hondoense*. 'Natural History of Ferns' (Iwatsuki, 1996) in the Natural History series contains one chapter devoted to describing this evolution.

Based on the results of individual scientific analyses, these books boldly discuss the influences of human civilizations on the evolution of biological species in the natural world. If things under an artificial influence are considered not to be in a purely natural environment, then it may be impossible to report on the development of biological species in an artificial environment as evolution in the natural world. However, events presently taking place on earth are all more or less under an artificial influence. Hence, when discussing the evolution of living wild organisms in recent days, one cannot ignore the influences of civilization.

Again, the reader is expected to have an acute sense of discerning the real from the false when examining the extent of scientific proof or the sound grounds for the demonstration of any assertion. Furthermore, the reader is encouraged to be aware that this discussion is held solely based on the results of analysis that are driven by purely scientific curiosity.

At the Botanical Gardens of the University of Tokyo, through the second half of the 1980s and the 1990s, I tracked the implications of parthenogenesis in the speciation process of plants. My purpose was to approach the reality of a particular species evolution by noting the phenomenon of parthenogenesis in vascular plants in the process of diversification. This was a bypass, rather than the main path, of species differentiation study.

There was a reason that this phenomenon was interesting. It was discovered that 13% of Japanese fern species were agamosporous (Takamiya, 1996). (Agamospory is a type of alternation of generations from gametophyte to sporophyte without gametogamy. This type of generational change occurs always by way of spore formation, without following the commonly observed process of chromosome number halving that results from meiosis.) The rate of agamosporous fern species in Japan was slightly higher than global estimates of approximately 10% on average of all species (Lovis, 1978).

I embarked on this study driven by pure biological intellectual curiosity about the species differentiation by agamospory. As I proceeded with the study, I inferred that the particular phenomenon of this type of species differentiation was linked with environmental development by *Homo sapiens.* Thus, the interesting challenge of examining the relationship between humanity and nature emerged. While analyzing species evolution as a scientific study in botany and presenting the results to the academic community in scientific literature, I also attempted to track the environmental history of the development of human civilization as a backdrop to speciation.

In the first phase of this study, I aimed to assess the relationship between biodiversity and humanity in connection with the Convention on Biological Diversity. However, even the meaning of the term *biodiversity* was poorly understood by the general public. Therefore, by presenting the thought underlying the above-described scientific pursuit, I hoped for a clue that allowed the general public to think about biodiversity.

On its very first day of emergence on the earth, life was bound for variation and diversification. Interesting questions as to what flexible adaptation was involved in diversification and how life responded to artificial changes introduced by particular organisms, i.e., *Homo sapiens*, to earth's environment arise from this. I was motivated to explore the question of what it is to be alive based on the results of scientific exploration.

(3) Agamospory in ferns

The life cycle of ferns shows that the ferns commonly recognized by people are in the sporophyte generation. When mature, they form sporangia and produce spores that have a haploid nuclear phase as a result of meiosis (sporogenesis). Spores leave the mother stock. When favorable conditions are met, spores germinate independently and form gametophytes known as prothallia. The prothallium of many species is an inconspicuous structure, which is heart-shaped and several millimeters wide, composed of single-layer of cells. Some prothallia are shaped in a thread-like structure.

Gametangia are formed on the prothallium. The archegonium produces an egg cell, while the antheridium produces sperm. Mature sperm swims through a trace amount of water, such as a dew drop on the prothallium, to the archegonium. An inseminated egg cell becomes a fertile egg (zygote). The fertile egg undergoes repeated cycles of cell division, known as cleavage, to develop and form a new sporophyte. In a fertile egg, nuclei of two gametes are united, with the nuclear phase being diploid. Naturally, the nuclear phase of the sporophyte is diploid. Following this process, ferns alternate generations, involving the alternation of nuclear phases in which sporophytes of diploid nuclear phase and gametophytes of haploid nuclear phase are alternately exhibited.

The life cycle of bryophytes is same as the above-described life cycle (in the case of bryophytes, the plant body commonly known as a moss is a gametophyte, and the sporophyte has a simple structure, which is parasitic on the gametophyte). The life cycle of spermatophytes is also basically the same, although their sporophytes only are apparent and gametophytes are simplified to part of the flower structure. Some algae and fungi also exhibit similar life cycles.

The life cycle of common ferns is as described above. Textbooks relay this information as if it is the life cycle of all ferns. However, as mentioned above, 13% of fern species in Japan exhibit different life cycles.

Agamospory is also more simply termed apogamy, which means the omission of gametogamy. More specifically, cells of gametophytes form sporophytes through a similar developmental process to that of a fertile egg, without egg cell fertilization or any nuclear phase change. Therefore, the sporophytes that are produced have the same nuclear phase as that of gametophytes. Obviously, if a spore case is formed, spore formation involves no meiosis, and haploid spores are formed from haploid sporocytes. (Apogamy always entails apospory.) In sum, ferns that undergo this type of life cycle apparently exhibit the same alternation of generations between sporophyte and gametophyte as that of the other 87% of ferns, but their reproduction entails no alternation of nuclear phases.

(4) Locations of agamosporous species: Classification

Our research group analyzed the speciation process for the above-described agamosporous ferns, obtaining various interesting findings.

Asexual reproduction through vegetative propagation of plants occurs in diverse ways. In plant bodies, vegetative cells retain totipotency. They are commonly capable of cloning to form new bodies. More than a few fern species live life like deciduous trees, the rhizomes of which survive even as leaves on the aerial parts of the fern wither. Like bamboo, cutting a rhizome leads to an increase in the number of individual bodies. The oriental chain fern (*Woodwardia orientalis*) produces gemmas. Of the *Hymenasplenium*, specific examples of inducing gemmas have been observed in a form on Ceram, Indonesia (Kato & Iwatsuki, 1985).

However, this section will not discuss vegetative propagation, but a form of reproduction that does not involve change in the nuclear phases in the absence of meiosis or union, despite following an apparent normal life cycle. This form of reproduction occurs predominantly in agamosporous triploid species. Ferns of this type differ from sexually reproducing ferns in spore formation. Common leptosporangiate ferns produce 64 spores in each sporangium, while agamosporous species produce 32 spores.

This difference was inferred to be highly useful for distinguishing reproduction types. Hence, using this indicator, I observed *Crepidomanes minutum* to preliminarily identify the reproduction type, since its area of distribution was wide and it lacked an established species classification (Yoroi & Iwatsuki, 1977; 3. 2). Using herbarium specimens of various types distributed widely in the tropical zone, I distinguished reproduction types and collated them with various types of phenotypic characters. I reported on this study at an international symposium held in Aarhus, Denmark (Iwatsuki, 1979), which drew a lot of attention from participants. This technique of reproduction type identification thereafter came into wider use (although there are few researchers who have the careful manual dexterity needed for this observation at the level of the Japanese).

The *Hymenasplenium* was included as subjects in the first research projects identifying agamosporous species by using a vast number of materials and the spore number determination technique. By the definition at the time, *Asplenium hondoense* in the broad sense was regarded as a complex species as widely distributed as *Crepidomanes minutum*. [For a long period of time, the scientific name proposed by Lamarck was given to the Japanese type of *Asplenium hondoense* (3.2).] For *Asplenium hondoense* in Japan, our observations served to verify that the reproduction type difference between sexually reproducing diploids and agamosporous triploids serves as an indicator for distinguishing the morphologically identifiable two species (Murakami & Iwatsuki, 1983).

The number of spores in each sporangium serves as a convenient indicator suitable for identifying the fern reproduction type. It eliminates the need to observe the reproduction process of living plants. This handy technique enables the researcher to carry out comparative observations with relative ease on a large quantity of materials using dried specimens from various areas. Unfortunately, it is not an observation of the actual process of reproduction. Although this technique enables the researcher to infer reproduction type using circumstantial evidence, it does not provide definitive evidence. In this respect, there is a growing awareness of the need to improve the accuracy of observations, as a result of the diverse variations in the agamosporous groups of *Dryopteris erythrosora* and *Dryopteris varia*. That said, spore counts in sporangia are still in use even today in an auxiliary manner, as a convenient indicator to locate problems for knowledge building.

Using this technique and other means, analytic studies of species differentiation in the *Hymenasplenium* were further conducted. When the technique of analyzing genetic diversity using isoenzyme as an indicator became available, we conducted comparative studies of various types on the Japanese archipelago, drawing on support from the members of the Nippon Fernist Club for the collection of materials (Watano & Iwatsuki, 1988). Subsequently, along with the establishment of molecular phylogeny, N. Murakami led detailed studies on the lineage and classification of the *Hymenasplenium*. At the time, it became possible to conduct field research in various tropical areas, which also served as a driver for the study.

Murakami published a monograph on all species world-wide in *Hymenasplenium*. He used molecular phylogeny results produced in the very early phases of the method by manual work without a sequencer (Murakami & Hatanaka, 1988; Murakami, 1995). As a result, it was inferred on solid grounds that in *Hymenasplenium*, the agamosporous triploid species evolved not merely once (monophyletic), but it has concurrently undergone geneses at least three times. In other words, these results suggest that agamosporous triploids are derived through the simpler mechanism of mutation, rather than the normal course of speciation, i.e., evolution through repeated gene mutations along with phylogenetic differentiation. Moreover, this particular type of evolution has been known to have frequently occurred concurrently in other taxonomic groups than *Hymenasplenium*. Although developments of interest in this topic appear in the analysis of *Hymenasplenium* species, this book cannot devote pages to a detailed discussion on it.

The implications of detailed species studies of the *Hymenasplenium* were not confined within *Hymenasplenium*. They were greatly significant in contributing to overall studies in the family Aspleniaceae, which had previously not received much study. HAYATA Bunzō (1874–1934), third director of the Botanical Gardens of the University of Tokyo, proposed recognizing an independent genus for *Hymenasplenium* (Hayata, 1927). However, few botanists agreed to Hayata's treatise. (At the time, Hayata proposed a dynamic classification theory. Due to criticism of this theory, researchers were unwilling to take notice of the traits he used as grounds for his argument.) Sometime later, to ascertain the group's monophyly, our research group reexamined the dorsiventral structure of the rhizomes to which he attached importance. This research preceded independent of Murakami's in-depth research.

In morphology, it was common practice to interpret the internal structure of a rhizome by laying images of sliced specimens on top of each other; however, Hayata manually removed steles to illustrate their three-dimensional structures. This technique requires a manual dexterity that seems difficult for non-Japanese people. Tardieu-Blot (Tardieu-Blot, Marie-Laure, 1902–1998) of the National Museum of Natural History of France conducted morphologic studies of Indo-Chinese Aspleniaceae without making use of Hayata's method. I and others also conducted studies to verify the monophyly of *Hymenasplenium*, by utilizing Hayata's method to observe the internal structures of *Hymenasplenium* rhizomes (Mitsuta, Kato & Iwatsuki, 1980).

The genus *Boniniella*, native to the Bonin Islands, was recognized as a monophyletic genus independent of *Hymenasplenium* by Hayata (1927). I and others determined not only the internal structures of rhizomes, but also other characteristics of *Asplenium cardiophyllum* to the greatest extent possible. Consequently, monophyly again became the subject of examination. In reality, the species (ascertained to be monotypic, although multiple species had been noted) distributed in Bonin, Kita-Daitō, and Hainan (China) were included in *Hymenasplenium* (Kato & al., 1990). *Asplenium cardiophyllum* is a simple leaf species. The above conclusion proved that Aspleniaceae has been evolving concurrently in diversification of foliage. This finding provides another piece of auxiliary evidence that *Asplenium antiquum* and its associates with lateral veins joining at the peripheries of simple leaves and *Asplenium scolopendrium* and other simple-leaved species with complete free venation have undergone parallel evolution.

(5) Various agamosporous forms

One additional challenge was to find out how agamosporous types were involved in species differentiation of *Hymenasplenium* in comparison with other examples.

Apogamy in ferns was first observed with *Pteris cretica* L. (Farlow, 1874). Research on apogamy in ferns progressed along with advances in in-depth studies of apomixis in plants. Nevertheless, despite some findings about these phenomena, unsolved questions about the reality of apogamy in *Pteris cretica* L. remained.

In the relationship between this species and its relative species, interesting examples of the types that grow in Japan and its neighboring areas present. T. Suzuki conducted an analysis in cooperation with botanists at the Botanical Gardens of the University of Tokyo. Although the analysis is still in process, the results obtained so far reveal that, while diploid and triploid apogamy is known to occur in *Pteris cretica* L., there are two types of triploid apogamy, which can be distinguished clearly by nuclear form. It has been ascertained that these two types have differentiated to such a degree that they can be distinguished from each other, even in phenotypes.

Additional examination of these type variants based on isozyme and other indicators reveals that, in diploid apogamy types, four clone types equivalent to those identified in triploid apogamy types, and an additional clone, have been recognized. *Pteris kidoi* is a relative species, which is a sexual diploid. A similar examination was also conducted on *P. kidoi*, which verified that, of the variants of *P. cretica* L., three of four triploid clones were derived by crossing the diploid apogamy type and *P. kidoi*. These findings revealed that *P. cretica* L., characterized by high intraspecific variation rates despite being apogamic, not only presents diversity in reproductive and nuclear forms, but also has developed intraspecific variation through repeated crossing with *P. kidoi* (Suzuki & Iwatsuki, 1990).

Meanwhile, Dedy Darnaedi analyzed the origin of *Dryopteris yakusilvicola* indigenous to Yakushima, South Japan. Since first recorded, it had been suggested that this species was a natural hybrid between *D. sabaei*, whose southern limit lies in the area between the mountainside and the summit of Yakushima, and *D. sparsa*, which is widely distributed in Southeast Asia and commonly seen in the lowlands of Yakushima. Certainly, the mountainside of Yakushima is the only area where both species contact each other. *D. sabaei* is an agamosporous diploid and *D. sparsa* is an agamosporous triploid. While agamosporous diploids and triploids and sexual tetraploids have been known in *D. sparsa*, no diploid types have been discovered in Yakushima.

The origin of *D. sparsa* has not been ascertained. However, it has been verified that this species exhibits low variation rates; is genetically uniform, as determined by isozyme; and exhibits, rare for the genus *Dryopteris*, vegetative propagation by gemmas. Additionally, there are strains that, although similar in phenotype to this form, are not agamosporous. Circumstantial evidence enables botanists to infer that hybridization has occurred several times between diploid and tetraploid sexual types, and, in some of their strains, an agamosporous type differentiated into a type with a low intraspecific variation rates, identified as *D. sparsa* (Darnaedi & Iwatsuki, 1990).

D. sparsa is a species in the genus *Dryopteris*. Agamosporous types such as *D. erythrosora* and its allied species and *D. varia* and its allied species have been introduced into this genus. As a result, types that are extremely difficult to identify as species are abundant in the middle and western parts of Honshu. As a matter of course, some of these types have been analyzed. Recent research has contributed to gradual clarification of the complex intraspecific structure.

S.-J. Lin has carefully tracked the spore formation process of *D. pacifica*. She observed that the spore formation process taking place, even on one leaf, involves variation between different spore cases. Tracking a large volume of prothallia growing from spores collected therein reveals the occurrence of even diploid sexual strains from agamosporous triploid bodies (Lin & al., 1992).

In-depth studies have been conducted on the diversity of reproductive forms of the genus *Dryopteris*, particularly for the Japanese species. In 2015, K. Hori published excellent analysis results for the intricate reticulate evolution of the *D. varia* complex (Hori & al., 2015). These species differentiation analyses have contributed to accurately tracking the species diversity of the *D. erythrosora* complex, including the origin and evolutionary process, whereas previously, said diversity was inferred based on external forms. Elucidation of the evolutionary processes of diverse species based on hypothesis-testing analysis provides a foundation for species diversity studies and makes a valuable contribution to research in this domain. [After the publication of this book in Japanese version, N. Murakami and his group, including K. Hori, are actually developing the research on this topic, yielding interesting papers.]

(6) Phytogeography and plant ecology of agamosporous forms

The above research results have ascertained concurrent differentiation in many species groups of the agamosporous triploid species of *Hymenasplenium*. Although the species differentiation mechanism has not been scientifically ascertained, it is possible to infer that species differentiation occurs by a simple mechanism, as far as can be inferred from the observed phenomena.

Aside from the *Hymenasplenium* (family Aspleniaceae), many agamosporous ferns are known. Species of this form of reproduction are observed in *Dryopteris erythrosora* and its allied species, *Dryopteris varia* and its allied species, the genus *Cyrtomium* (these within the family Dryopteridaceae), and *Pteris* (family Pteridaceae), in addition to *Crepidomanes minutum* (family Hymenophyllaceae). The introduction of agamospory to these types is known to have undergone parallel evolution across phyletic groups. Of course, analyses have been conducted on the genus *Notholaena* and other genera abundant in arid zones in America, as well as on Japanese species.

The analysis of differentiation in agamosporous species developed as a botanical study. It presented a research challenge in species evolution that was expected to conform to scientific rigor. Consequently, it is necessary that the speciation process must be scientifically tracked through rigorous analysis. At the same time, as a backdrop, such study is driven by scientific curiosity about how such evolution occurs. Of course, no empirical answer to this question can be immediately provided. However, inference based on available information can be made in a reasonable manner.

In the second half of the 1980s and the 1990s, botanists associated with our research group examined the issue of endangered species, and they were active in the research of this issue in the most advanced and leading positions in Japan. In the study of *Asplenium hondoense*, their research included ascertaining that *Asplenium cardiophyllum*, a typical endangered species on Hahajima in the Bonin Islands, is a closely related species of *A. hondoense*.

Taking a broad view of agamosporous ferns in Japan reveals that they are found among terrestrial ferns. *Crepidomanes minutum* in the family Hymenophyllaceae is in the category of epiphytic plants. However, it also grows on moist rocks and moss-grown roadsides, as well as on tree trunks. Although no study has been conducted to verify that agamosporous *C. minutum* is terrestrial, the mode of life for species should be borne in mind when noting that *C. minutum* is rather a specific example. As stated earlier, agamosporous species account for 13% of Japanese ferns, while the global rate is approximately 10%. With many fern species, tropical zones are regarded as suitable fern habitat. In tropical zones, epiphytic ferns account for a high percentage, with high species number in the families Polypodiaceae, Davalliaceae, and Hymenophyllaceae. No agamosporous type has been recorded in Polypodiaceae. Given the finding that epiphytic species, such as those in the family Polypodiaceae, account for a high rate of ferns in tropical zones (although this is not the result of a numerically accurate analysis, but a rough and intuitive inference), it is reasonable that in Japan, with its high fern diversity, the rate of agamosporous types is higher than average because Dryopteridaceae and other families dominant in terrestrial species, mainly constitute fern diversity.

By taking a step beyond simple examination of distributions and looking at places where agamosporous species grow, a new fact emerges. They are found abundantly in rural areas and in *satoyama*, or tamed forest areas, rather than in the deep mountains. Ferns in Dryopteridaceae and other families primarily composed of agamosporous species are also distributed mostly in rural areas and do not grow in deep mountains. *Dryopteris erythrosora* and its allied species are mostly agamosporous. Among them, the few diploid sexual species are *Dryopteris koidzumiana* and *Dryopteris caudipinna*. These two species differ from diverse types of *D. erythrosora* that are principally distributed in rural to tamed forest areas in the middle and western parts of Honshu. They are distributed with *D. koidzumiana* anywhere in and south of Kyushu and with *D. caudipinna* on Hachijo-jima and the Izu Peninsula, although this species has been known to grow in other areas as well. Diversified types of *Dryopteris varia* and its allied species are almost all agamosporous. The only known diploid sexual *Dryopteris saxifraga* typically inhabits gorges in deep mountains.

Although it may sound arbitrary, and I am aware that it is not scientific, agamosporous ferns live in places under the strong influence of humans. The reader is advised to remember that any verdant place occasionally referred to as "nature in tamed forest areas" is characterized by an artificially transformed natural setting.

(7) Advantages and disadvantages of agamospory from the perspective of evolution

Living things successfully and rapidly sped up the rate of evolution after sexual reproduction evolved. Science has not answered the question of how sexual reproduction evolved; however, it was very beneficial for living organisms. Along with adaptation to environmental change, new species are formed through repeated gene mutations, gene exchange by each generation within a sexually reproducing population, and gene combinations delicately made within a population. The formation of a new species has been calculated to require time on the scale of one million years, on average. This is dramatically reduced in comparison with the time expected for a new species to evolve through differentiation by simple of gene mutation repetition. Indeed, compared to the first period of more than one billion years of biological evolution, the second period of more than one billion years after sexual reproduction began has seen a clear acceleration of the rate of diversification of living organisms on the scale of geological time.

Despite the benefits of sexual reproduction in accelerating evolution, as many as 13% of fern species have abandoned sexual reproduction after gaining it, producing an agamospory that counters sexual reproduction. In addition, the agamosporous type is prospering in places under the influence of humans. No keys unlocking these mysteries have been discovered. To meet this challenge, I made a somewhat bold inference based on as many verified facts as were available and contemplated what needed to be analyzed to clarify the inferred hypothesis.

Sexual reproduction is a highly efficient method of reproduction for speeding up evolution. In sexual reproduction, the fusion between two cells, namely an egg cell and a sperm, produces a fertilized egg, which is the starting point of the next generation. Compared with asexual reproduction, in which one cell directly becomes the starting point of the next generation, sexual reproduction mathematically requires twice as many resources for producing offspring.

The process of evolution develops to steadily remove waste. In the case of sexual reproduction, the prototype is likely to have begun with the simple union of two cells, likened to homozygosis. The two united cells were identical in shape and size, hence requiring exactly twice as many resources. However, what is required to form the next generation by sexual reproduction is a set of two nuclei and one cell. Indeed, the evolution of sexual reproduction produced heterozygosis and neatly developed into the fertilization of an egg cell by a sperm. The reproductive cell known as sperm has radically reduced cytoplasm, with only a nucleus and a motor organ (flagellum). The egg cell is large in size. Although it stores nutrient content to sustain the initial development of the next generation, its cytoplasm is not enlarged. Nonetheless, compared with producing offspring by one cell, the consumption of two cells for the starting point of the next generation is disadvantageous in terms of resource usage. This is not a preferable condition, particularly at the instant of producing offspring vital for the maintenance of life.

Nevertheless, assuming that the evolution of living things takes place in an undisturbed natural world, it is better for living things to incorporate sexual reproduction despite some disadvantages, which can be minimized by transforming reproductive cells into egg and sperm. Accordingly, metazoa and terrestrial plants have diversified dramatically through evolution by sexual reproduction.

Homo sapiens emerged with highly evolved intelligence, built highly advanced civilizations, and caused radical changes in the natural environment on earth. This caused an artificial impact on natural evolution. On the Japanese archipelago, where nature is characterized by complex terrain and a warm humid climate, diverse genetic resources were supplied and verdant ecological systems developed. These developed fertile soils, leading to diverse living organisms. However, since the dense forests were partially cut down and open farmlands were developed, plants favoring strong sunshine began to grow rampantly, due to the moisture supplied by inherently abundant rain. This implied that areas for heliophytes, in addition to forest floor plants, broadened.

Of course, herbaceous plants previously in locally limited sunny exposures spread to new habitats. In addition, new signs indicate that living organisms, when adapted to new ecological systems, underwent quick speciation processes.

In plants, speciation by way of cytogenetic mutation has been genetically verified. Various examples indicate contributions to speciation made by combinations of polyploid series, lines resulting from natural hybridization, and polyploidization. In ferns, difficult-to-classify species groups in the genus *Asplenium* in the Appalachian Mountains developed from reticulate evolution, as pointed out in a doctoral thesis by Warren H. Wagner Jr. (1920–2000) at the University of California (Wagner, 1953). Cytogenetic evolution triggered by chromosomal mutation is understood as a type of speciation that takes place in parallel with steady species evolution through gene mutation.

The mechanism of evolution has not yet been elucidated in detail. Nonetheless, it is highly likely that the evolution of agamosporous types, which takes place concurrently in various taxonomic groups, resulted from a one-off origin induced by mutation or a relatively simple mechanism, rather than speciation through repeated gene mutation. If this is true, it is reasonable to infer that types that would form a next generation from one cell developed an emergent response to adapt to a newly emerging artificial environment, while minimizing the resource use. Diversification of living things is the most important characteristic of life. This characteristic comes with various exceptions as well as universal phenomena, which exemplifies the versatility of life. It is an extremely important fact for knowing what it is to be alive. It is also a challenge that can be elucidated exclusively by the methods of natural history.

(8) Universal challenges to the biosphere?

The question of whether the evolution of agamospory is a phenomenon specific to ferns must also be studied in detail.

Types that have abandoned sexual reproduction during evolution are not rare in seed plants. Asexual reproduction is not limited to particular taxonomic groups. Various types of asexual reproduction have been reported in various families. Ferns present many examples of the formation of gemmas. Increases in individual numbers by means of extension and division of rhizomes are a common phenomenon observed in plants. Plant cells do not completely lose totipotency. Under favorable conditions, plants can develop an individual body from one somatic cell. Indeed, for several species, humans have successfully raised individual bodies in this manner. For this, the term *cloning* has come into use.

When looking at asexual reproduction of plants in general, although off-topic in the context of this section, it appears that, in asexually reproducing seed plants, there are a considerable number of species that grow thickly in rural areas and other artificially disturbed places. Such plants were, by chance, analyzed by the research group (in the broad sense) at the Botanical Gardens of the University of Tokyo at the time. The genus *Eupatorium* and *Ixeris dentata* (Asteraceae) and the genus *Bohemeria* (Urticaceae) are typical examples of this.

Plants differ from metazoa in the phenotypic characters involved in evolution, such as that cells of vegetative bodies have totipotency and ploidy of chromosomes readily occurs. In contrast, behaviors and other lifestyle specialization often lead to the isolation of individual groups in the species differentiation of animals, which exhibits a noticeable difference in the mode of species differentiation. In relation to this topic, animals show no change in the form of reproduction comparable to the agamospory observed in plants. That said, it is not reasonable to generalize that this type of evolution does not occur in animals.

(9) Expansion of the issue and the subject of curiosity

Asexual reproduction plays diverse roles in the process of evolution. Studies on this issue have been pursued in a diverse manner. However, this book is not intended to describe those studies. My former research group elucidated some facts about the process of evolution. Since then, other researchers have conducted studies along the same line. These studies developed into many directions in an apparently disorderly manner. This was because it became more important to know the connections between phenomena. This comes from the perspective that being alive is not simply a sum of individual phenomena; the fact of being alive is expressed in diverse phenomena. However, investigative analysis making full use of available techniques should be conducted on individual phenomena after a series of observation.

At this stage, necessary research to answer this challenge rests in verifying whether agamosporous types have occurred concurrently and frequently by truly simple mechanisms. *Hymenasplenium* was inferred to have concurrently undergone several events of agamospory evolution. For *Dryopteris sparsa*, botanists infer from auxiliary evidence that the currently living agamosporous species has increased in number asexually through apogamy or vegetative propagation, despite the evidence for inference that natural hybrids were formed several times.

Natural hybrids between different forms, or more specifically, the fusion of gametes between unrelated lines and the development of the zygote, are not rare. However, this may only occur when exceptionally favorable conditions are met to allow individuals of outbred natural hybrids to increase to form a new species. On one hand, there are individual sporophytes from outbred zygotes, and, on the other hand, there are individual sporophytes growing from zygotes produced by ordinary fusion between intraspecific gametes. The mechanism of agamosporous type development, as to whether or not there is any difference in the rate of agamospory between two types of individual bodies and whether or not necessary conditions for agamospory differ between them, still remains unknown. Moreover, whether any laws underlie the diversification of phenomena and whether diversity is on an individual basis also remains unclear.

For this sort of analysis, the universal principles underlying phenomena must be explored. Moreover, studies, in terms only of the analysis of phenomena, imply dealing with each living body as a mass of matter. In this sense, the analysis of individual phenomena itself would be the natural science of living bodies. However, it should be remembered that in this case, the need for research is driven by intellectual curiosity and promoted without an expectation for immediately useful material- and energy-oriented results. Researchers may be confident that their findings will lead to useful applications for society at some time in the future, although who knows when that will be?

Emergent short-term evolution adapted to changes in the nature and accompanied by an artificial transformation should be termed *jumping evolution* (Iwatsuki, 1997), as in the case of the formation of an agamosporous type. Such evolution should be distinguished from natural evolution taking place in line with the eternal development of the natural world. Indeed, plant abandonment of sexual reproduction implies dropping out from the league of successfully accelerating speciation. Nonetheless, this example describes evolution taking place over millions of years, and is not an issue worsening within several decades. For years to come, humanity should devise measures to adapt to evolution in nature as transformed by artificial pressure, while accurately acknowledging the current state of change. Human activities, if continued for the sake of one-sided convenience on the part of humanity, will cause trouble and disharmony to the surrounding environment.

It is superfluous to say that science self-propagates in the domain of science, and that for intellectual curiosity, the right path is to follow a course of self-development. Scientific findings that have righteously evolved for the sake of science are applied to technology or used to manage growth. When utilized for society, science is evaluated as science for society. It should be remembered, however, that history is full of examples in which the thoughtless application of technology resulted in evil.

As a desired course of thought, analysis of scientific challenges should be conducted and then how much these should be utilized for society should be considered. The points in question are whether the study of agamosporous type evolution addresses the question of what it is to be alive, and whether the obtained findings can be used in relation to civil life. For further information on the detailed research process, the reader is encouraged to refer to the works mentioned in this book. I anticipate that this kind of biodiversity analysis will be used in understanding the human environment and responding to its degradation.

If one asks whether the possibility of formation of agamosporous species as jumping evolution under an artificial environment is of any significance given the lack of scientific evidence, I have two answers ready. First, this suggestion answers scientific curiosity and exemplifies how dynamically species differentiation can progress. Its proof rests on providing more reliable scientific evidence to test the hypothesis. Of course, it is also necessary to consider instances in which polyploidization of the nuclear phase or other variations might trigger jumping speciation.

The second response is that this suggestion supports a more thoughtful approach to environmental issues. When I discuss how awful it is to abandon tamed forest areas to desolation, some people ask why allowing *satoyamas*, or artificially created scenery, to return to nature is harmful. For these people, I point out that according to the accepted notion in vegetation science, it will take some 400 years before the scenery, even if allowed to return to nature, becomes stable, while the desolated scenery will continue to persist throughout the period of recovery. Furthermore, given that in an artificial environment, many species have formed through jumping evolution, the stability in 400 years may not return to its prior inherent course of natural evolution, instead incorporating elements not exhibited by natural evolution. Current science is unable to determine with absolute certainty whether scenery created under these conditions is truly stable or not. This discussion is, of course, based on the assumption that jumping evolution is a reality. As inferred in this book, the possibility of speciation through jumping evolution is not low. Speculating on future evolution without regarding its possibility is meaningless.

This research is in its earliest stage, as is the case with all other research projects driven by intellectual curiosity. It has many interesting problems requiring solutions in the future. Although this book refrains from naming these problems, it is appropriate to briefly clarify why this book describes the pursuit of this challenge, despite its premature stage.

In terms of research, concrete examples of what should be termed the diversity of species differentiation need to be provided separately from mainstream species differentiation analysis. Moreover, tracking specific examples will probably not end in simply exemplifying a variety of species. Doing so may lead to a clue that suggests the effects of the artificial development of the human environment on species differentiation in nature. Thus, it is significant to speculate on the possibility of topical extension. In this course of contemplation, evolution, which is a purely biological topic, will lead to very human and social subjects. The flexible nature of the lifestyles of living organisms blends together all the phenomena in the natural world and all the artificial phenomena. This is quite normal. While elucidating this normal fact by the methodology of natural sciences, the expansion of the topic of evolution comes to mind. In this context, analysis made by the techniques of natural sciences naturally paves the way for bridging humanities and science. There lies the true form of natural history research.

TEA TIME 5 What Does It Mean to Know the Names of Plant Species?

As a topic associated with the discussion of agamosporous species, Tea Time 5 looks at species as the basic unit of taxonomy, while referring to what I experienced when I was a graduate student. The story is related to the *Dryopteris erythrosora* complex example described above.

In the late 1950s, when many elementary and junior high school students collected plants as a summer homework project, I and other graduate students attended plant identification workshops held as corporate philanthropy in the last days of school summer vacation. It was a rewarding moonlighting job. I joined the workshop whenever I heard of an offer.

Once, I picked up a specimen of an allied species of *D. erythrosora*. Inadvertently, with the mind of a researcher, I looked at the specimen and mumbled that it was not possible to name these types of ferns right away due to their complexity. An old man, probably helping his grandson do homework, caught my mumble. He abruptly and reproachfully exclaimed that I had no ability even to name the specimen of a common plain plant. The *D. erythrosora* complex is difficult to identify as species simply by form. Knowledge at the time was not sufficient to accurately classify the plants collectively known as *D. erythrosora*, although it was known that several types existed. My mumble was from my awareness of this scientific difficulty. However, the man who reproached me thought that the guy in the instructor's seat was so poorly trained that he was incapable of even identifying a common plant.

Ordinary people were then unaware about how science was ignorant of nature. This is true even today. If I said with certainty that it was a *D. erythrosora* (admitting that it means *D. erythrosora* complex), I would give an impression that I was a capable instructor. Even at the time, I realized that if I were to follow the career of a commentariat, I should accordingly be more practical. However, I had inadvertently uttered my real opinion, despite my inability to accurately communicate the true state of science.

The development of scientific studies analyzing the diversity of the *D. erythrosora* complex represents the state of Japanese study of ferns, which has grown in each era along with researchers' awareness of issues and corresponding analysis techniques. Findings in each period have been compiled into books, which range from the 'Colored Illustrations of Japanese Pteridophyta' (Tagawa, 1959) to an illustrated manual published in 1992, in which I was involved (Iwatsuki ed., 1992), and to an illustrated manual showcasing the latest knowledge of the Nippon Fernist Club (Ebihara & al., 2016–2017). [Pteridophytes of Flora of Japan is in K. Iwatsuki & al. (ed.), 'Flora of Japan' vol. 1 (1995).] Methods of reproduction are considered deeply in the analysis of species differentiation of this complex, as described above by even very recent research examples.

Of course, using current findings, it is possible, in front of specimens collected in a junior high school summer homework project, to find a conclusion based on more detailed study, rather than simply identifying them as an allied species of *D. erythrosora*. However, for junior high school students (and perhaps for the man who looked like a grandfather accompanying a junior high school student), advanced findings would be of little significance. They would not understand the significance of the difference between referring to a specimen by the term *D. erythrosora* versus using the more specific name of *Dryopteris xx*.

This raises questions about the meaning of the summer homework assigned to collect plants, make specimens, and write names on workshop labels. If students are collecting specimens as an introductory step to learning natural history, they should explore these differences. Even if not, they can at least acknowledge that there are diverse types in *D. erythrosora*. On the other hand, perhaps carefully collecting plants, making specimens, and even simply writing indicated names on labels may serve as an introduction to natural history. Of this, I am not sure.

Analytical studies that make full use of advanced analysis techniques have recently enabled researchers to move forward in elucidating the *D. erythrosora* complex. Studies pursued by Hori et al. (2015), as described earlier, also elucidate the structure of the allied species, utilizing the most advanced methods. As the species structure has become clear, what I said when I was a graduate student, namely, that it is not possible to name these types of ferns right away due to their complexity, shines even more at the forefront of science.

Regarding the identification workshop, in retrospect, it would have been better to discuss the difficulty of naming even common plain plants, including why this was so, rather than simply giving names to specimens to help students with their homework. Presently, I regret the limitations of such opportunities to impress children with the true fun of biology.

In recent years, schools have quit assigning or recommending summer homework, no longer asking students to collect biological specimens. Simple identification workshops in which participants hear the names of specimens are not expected to be held. In contrast, even though to a very limited extent, museum activities have come to include substantial learning projects where museum staff ascertain the reality of biodiversity together with junior high and high school students. When in contact with nature, every child shows an energetic smile, a sign of the development of scientific curiosity. This should be nurtured healthily. Instead, the present-day school education system tends to squeeze it to death.

Indeed, very recently, I have heard from learned people not well versed in biology about speeches given by award winners at academic awards-related meetings. One commented, similar to several others, that when he was a high school student, he could not like biology because rote learning was imperative and that it was unfortunate that his teachers did not relate stories as interesting as those presented in the awards meeting. What they said may not be rectified simply by criticizing the entrance exam system or teaching guidelines.

Additionally, when I attended a party after a meeting related to the assessment of a large research project, I had a casual talk with a famous life scientist. I told him that current biology was unable to define species. In a surprised manner, he then shared with the people surrounding him that taxonomy was incapable of defining species. I was surprised that the basics of biodiversity were unfamiliar even to biologists in the broad sense.

Dryopteris erythrosora is important as a supporting feature of Japanese gardens. While on a garden tour, I remember that *D. erythrosora* is an agamosporous triploid created as a result of harmonious living relationships between the Japanese people and nature, contemplate the meaning of the design of the Japanese garden, and am emotionally moved. This is my personal emotional response. Nonetheless, as an undeniable fact, this intellectual emotion evokes a true sense of pleasure, irreplaceable by economic value.

Chapter
3

Passing Natural History Knowledge to the Next Generation

Chapter 2 traced the development and the current state of the field of natural history. Chapter 3 focuses on issues of how knowledge in the field is passed down and nurtured, as well as what essential points natural historians need to impart to ensure the achievements of future generations.

The study of history is significant only when one goes beyond simply assessing past events and considers how history can provide clues for the future. By studying changes that occurred in the past, one can infer how the present will be recorded 100 or 1000 years from now, which motivates improving society for historical posterity. This is the significance of history.

3.1 Teaching Natural History: The History of Inheriting Knowledge in Japan

The people who settled on the Japanese archipelago built knowledge through their way of life. They lived in awe of nature on the archipelago, irrespective of the origin of the knowledge they inherited, and created the Japanese culture over a long period of time. If the Japanese people are truly a mix of groups who arrived on the archipelago at different times from various areas, they likely built a unique culture through interactions between different groups of people and the environment, rather than developing a monoculture as an indigenous ethnic group.

Without having fixed ideas formed by a specific lineage, Japanese people on the whole have lived peacefully and sincerely with nature on the archipelago, humbly observing and learning from nature. Indeed, both documents, such as the 'Man'yoshu' completed in the eighth century, and the natural environment itself, noticeable in the ecology of forest areas tamed in earlier times throughout the archipelago, are records of Japanese people's view of nature.

Nevertheless, the literature does not indicate that education in natural history was systematically provided in ancient Japan preceding the completion of the 'Man'yoshu.' It is unlikely that the characteristics of living organisms and the ecological significance of tamed forest areas on the Japanese archipelago were understood by the average person living on the archipelago. Notwithstanding, not only scientists or artists, but also the average person, composed the poems compiled in the 'Man'yoshu' and created tamed forest areas for use in their lives, without guidance from a political or economic leader. It is highly likely that ordinary people simply lived in accordance with the principle of natural history.

However, the above-described fact does not imply that Japanese people at the time had a high level of scientific understanding. If the average resident on the archipelago, before or at the earliest recorded time, had knowledge of natural history as a common understanding, one wonders what that common understanding was. To make an accurate assessment, it is necessary to examine the relationships between the development of natural history as a science and the process of living in accordance with the principles of natural history as forged in life.

(1) Matters taught at school and passed down in society

Research and Education before the Meiji Restoration The notion that education is equivalent to school has become prevalent only very recently. Humans have fostered intellectual activity and created culture on the basis of the sharing of knowledge accumulated by individuals and the passing down of this knowledge in society as a whole. The continuity of knowledge was ensured by family and social education. Systematic school education to ensure the passing down of knowledge began only very recently in society. It is not correct to say that well-organized school education systems have developed throughout history.

Records from the Sumerian civilization, which were produced around 2000 BC in Mesopotamia, are said to contain a book entitled 'Edubba (= house of tablets = school) period,' which is among the earliest records of school. Edubba was apparently a place of training in reading and writing provided for trainees to become government officials. The earliest higher education institution is said to be a monastery at Taxila in ancient India, founded in the seventh century BC. The monastery is regarded as having granted graduation that was equivalent to a present-day academic degree.

In ancient Greece, the Platonic Academy (founded in 347 BC) was the first school. Encouraging physical exercises, the gymnasium played the role of a college of physical education. It is said that philosophy and other classes were also conducted at the gymnasium.

Meanwhile, in China, *Taixue*, founded in 124 BC, is regarded as the first institution to train bureaucrats. Designations such as *college* and *elementary school* appear in the 'Book of Rites,' one of the Five Classics of China. Accordingly, China appears to have already had an education system in place before the Common Era.

In Japan, some scholars consider that the School of Arts and Sciences, founded in 828 by Kūkai as a place of education for the public, is the forerunner institution of today's Japanese colleges. In Japan, the designation *college* was first used in the College Office, which was institutionalized as a bureaucrat training organization by the Taihō Code (701). (It is also said that, in fact, Emperor Tenji planned to build a school of this kind by the end of the seventh century.) The College Office in nature was a program designed for trainees to qualify for a bureaucratic position, although it provided intellectual education. The institution might have assessed trainees primarily in the context of qualifications, rather than learning itself. Before long, noble families individually developed programs for education. Consequently, the public school system as a place for passing down knowledge did not develop. Individuals from each family bestowed knowledge on the next generation.

Exploring the origin and development of educational institutions will typically uncover the historical records described above. The educational institution in this context is an organization involved in higher education. Such institutions are intended to develop leaders in the governmental, economic, and academic sectors, which is an important function that colleges still carry out today. However, this context does not describe an organization that serves as a core of research activities to create advancement and innovations in culture. Indeed, it was only more recently that the intellectually creative part of colleges was understood. Moreover, colleges began to be engaged in truly academic activities in the field of natural sciences after the arrival of the modern age, when scientific findings and analysis methods began to be passed down. Additionally, a system of education as a means of disseminating a system of knowledge to the general public, whether publicly or privately, was organized long afterward.

It is said that Japan was slow in developing modern sciences because it was closed during the Edo period. However, this is not necessarily true when one looks at history in Japan and Europe side by side. In the domain of natural history, when Linnaeus published the first edition of 'Systema Naturae' in 1735, Japan was in the reign of the eighth shogun Yoshimune. By that time, Japanese people had become active in absorbing Western knowledge, partly under the influence of Yoshimune, who was enthusiastic about Dutch studies. It appears that once the regulation was relaxed, incoming books became rather welcome, unless they were meant to convert people to Christianity. Even books brought by visiting scientists and presented to the shogun did not end in the personal possession of the shogun. They were apparently used by leading researchers. At least the researcher community (or the like) was blessed with conditions that allowed them to absorb the most advanced European knowledge. Moreover, Japanese people clearly possessed the intellectual foundation to digest such knowledge.

As notable examples, Edo was far cleaner than London and Paris, despite being similar megacities, and the literacy rate of Japanese people at the end of the Edo period was higher than that of Europe and America. (Some scholars may be skeptical about this high literacy rate because no statistical data exists. However, even without any scientific evidence, various pieces of auxiliary evidence make this assertion reasonable. The education provided at *terakoya* (literally, temple schools) in the Edo period has recently been elucidated in depth.)

The late Edo period saw the founding of the Institute for Western Culture and a vaccination institute, in which the University of Tokyo has early roots. The Institute for Western Culture was established in 1857 by the Tokugawa shogunate as an organization for studying foreign literature. The Institute was renamed Kaiseijo in 1863, and then Kaisei School, comprising three departments of English, German, and French.

The vaccination institute was first founded in 1858 jointly by 82 Western medicine doctors as a private institute. In 1860, the institute became a government-run institute, aided and appropriated by the Tokugawa shogunate. Thereafter, the institute was renamed the School of Western Medical Science, then the School of Medical Science, and finally reorganized as the Tokyo Medical School. (Built in 1875 and 1876, the Tokyo Medical School was later designated as an important cultural asset and moved onto the premises of the Botanical Gardens of the University of Tokyo where it is used as the Koishikawa Annex to the University Museum, the University of Tokyo.) These institutions used designations such as *professor* and *assistant professor* as staff job titles. In the course of time, the University of Tokyo was founded based on these two institutions. Although there was no institution referred to as a college, it may be safe to say that organizations equivalent to colleges were already present in Japan in the late Edo period.

Nonetheless, at these schools, importance was not placed on academic activities intended for the creation of culture driven by intellectual curiosity. Here, students learned about the advanced cultures and civilizations of Europe and America through literature, while the growth of medicine and medical practice was promoted beyond the boundaries of that achieved by the Dutch and other Western cultures. The social understanding of science was still at the above-described stage even in advanced Western countries.

At this time, research and study of nature's creations was conducted in connection with medicinal plants for medical practice, a matter which should be recorded in the context of natural history. Moreover, in line with the tradition of herbalism, research and methods in fauna and flora study at an advanced level in Europe were introduced in Edo period Japan.

Meanwhile, many youths were apprenticed to outstanding scholars. Many private schools were founded in provincial areas, as well as in Edo. Of course, private schools had roots in training sites founded for trainees to learn martial arts and Chinese characters. However, in the second half of the Edo period, these training sites commonly provided intellectual training. These organizations were not readily open to everyone, as various stories are told of narrow gates to apprenticeship. However, it appears that there were certainly opportunities to grow with the joy of learning under a respected teacher. Information about outstanding scholars was disseminated throughout the archipelago. Moreover, there was a class of people, in addition to the particularly wealthy class, who, although not extremely rich, had the financial base to attend private school.

Gradually, on one hand, some private school students grew to have a progressive awareness, and on the other hand, an increasing number of youths became purely science-oriented. Some private schools provided training in Dutch and other Western studies, while other medicine-focused schools, including Tekijuku, were broad-minded, allowing students to grow into learned non-doctors. In this respect, until the end of the Edo period in Japan, both the provisioning of education at private schools and learning by children at *terakoya* alike was on a voluntary basis.

In the field of natural history, although during the period of national seclusion, many youths actively contacted and received guidance by outstanding foreign scientists, such as Kämpfer, Thunberg, and Siebold. A research basis developed at a level that could adsorb Western culture, based on the prior accumulation of findings in herbalism and other fields. In the second half of the Edo period in Japan, there was a sufficient scientific base to demonstrate world-leading achievements in the production of diverse forms of domesticated animals and plants. There are many outstanding results in the records of natural history from the Edo period (e.g. Ueno, 1973; Kimura, 1974; Nishimura, 1999).

Growth of Natural History at Universities After the Meiji Restoration, a Western-style school education system was established in Japan as part of the nation's efforts to construct an educational structure. The University of Tokyo, founded in 1877, conducted research and provided higher education in a broad range of humanities and scientific fields. Works by the botanist OGURA Yuzuru, professor at Tokyo Imperial University (present University of Tokyo), and others (1940); the Botanical Society of Japan ed. (1982); and the limnologist UENO Masuzō, professor emeritus at Kyoto University (2003), provide a detailed history of research and education in zoology and botany.

In Japan, Western-style research and educational institutions were founded after the Meiji Restoration. The first national university, the University of Tokyo, was formed from three research and education institutions established by the Tokugawa shogunate in the period from the end of the 18th century to the mid-19th century, namely, the Shoheizaka Study Office, the Institute for Western Culture (later Tokyo Kaisei School), and the vaccination institution (later Tokyo Medical School). At its founding, the University of Tokyo was a sort of federation consisting of the Kaisei School and the Tokyo Medical School. The Koishikawa Medicinal Plants Garden became a facility annexed to the University of Tokyo immediately after the university's founding. Similar to how the *terakoya* structure, voluntarily developed in the Edo period, served to establish a compulsory education system in the Meiji period, organizations in which the University of Tokyo had earlier roots were also already in existence in the Edo period.

At its founding, the Faculty of Science of the University of Tokyo comprised eight departments: Mathematics, Physics, Chemistry (Pure and Applied), Biology (Zoology and Botany), Astronomy, Engineering (Mechanical and Civil), Geology, and Mining & Metallurgy. The Koishikawa Botanical Garden became a facility annexed to the University. In the first year, students of all departments attended common classes. In the second year, they took their respective courses. In the Biology course, fourth-year students majored in Zoology or Botany. All lectures were provided in English.

Professor YATABE Ryōkichi (1851–1899) oversaw botany. For zoology, Professor E.S. Morse (1838–1925) was hired with a tenure of two years (Morse, 1879, 1917, 1970, 1983). Biology at the time looked at research subjects largely at the level of individual organisms and was, overall, within the domain of natural history. Domains relating to the universe and the Earth might have been in the Physics and Astronomy departments. Geology and mineralogy might have been researched and taught in the Geology and Mining & Metallurgy courses.

Morse was not trained professionally at a college. However, he had been interested in invertebrate animals since childhood and was a superb collector and observer. To collect specimens, he came to Japan by chance in 1877, the year in which the University of Tokyo was founded. To obtain permission for gathering, Morse visited the Ministry of Education. At this visit, the first Japanese professor at the University of Tokyo, TOYAMA Masakazu (1848–1900, sociologist), requested that Morse take the post of professor in zoology. Morse accepted the offer. Toyama, who would later take the post of the President of the University of Tokyo and the Minister for Education, compiled 'Shintaishishō' (Poems in the new style) (1882) with Yatabe and others. The two probably had many opportunities to exchange information. Possibly influenced by this, Yatabe strived to bring foreigners as hired teachers at the University of Tokyo, and, sometime after the founding of the University, he took an administrative position.

Morse's hiring seems to have benefited the University of Tokyo beyond the domain of zoology. At the time of the University's founding, many hired professors did not deserve the post, such as missionaries in Japan by chance. Morse, in cooperation with Yatabe and others, dismissed some of these hired professors and contributed to newly inviting outstanding expert professors. Among those whom Morse recommended for invitation to Japan included the aesthetician and philosopher E.F. Fenollosa (1853–1908).

Morse is famous for his discovery of the Ōmori Shell Mounds. It is said that when he headed for the Ministry of Education by train to obtain permission for gathering, he viewed scenery through the train's window that caught his attention. To study it more closely, he conducted Japan's first excavation and research project. MATSUMURA Jinzō (1856–1928), who would succeed Yatabe and take the post of the second-generation professor in botany, was an assistant at the Department of Zoology at the time and joined the excavation. The designation of *Jōmon pottery* originates in a treatise by Morse.

Building a marine laboratory for research and education was apparently an early promise made between Morse and Minister Toyama. Indeed, at a somewhat later point in 1886, a marine laboratory was established in Misaki (Miura peninsula, Kanagawa).

The Biological Society of Tokyo, founded in 1878 in response to a proposal by Yatabe and Morse, was one of Japan's first academic societies. In 1885, this Society broke into the Zoological Society of Tokyo and the Botanical Society of Tokyo. Professor Yatabe was recommended for the Chairman of the Biological Society of Tokyo. The Botanical Society of Japan designates 1885, the year it became independently named, as its founding year. In comparison, the Zoological Society of Japan (renamed in 1923) claims that its history began in the year of the founding of the

Biological Society of Tokyo. Thereafter, for a long period of time until very recently, the Zoological Society of Japan and the Botanical Society of Japan have lacked close cooperation, despite both being associations of researchers working in the same basic biology. While friendly rivalry is desirable, turning away from each other is unsightly.

Morse completed the planned tenure of two years. Subsequently, C.O. Whitman (1842–1910) took the post and served for two years. The position of the third-generation professor was occupied by MITSUKURI Kakichi (1857–1909), a Japanese who had returned from his studies in America and the United Kingdom. When the University of Tokyo was founded and began to operate, biology was pursued primarily by observation and documentation. At this time Mendel's treatise on genetics remained ignored by the mainstream academic community in Europe, and at a later time three researchers independently carried out experiments to rediscover Mendel's achievements.

When the University was founded, Yatabe Ryōkichi, at the tender age of 25 and back from studying in America, began to serve as a Japanese professor. He was one of only three Japanese professors among 25 professors in the Faculty of Science and was the first-generation professor in the field of botany. Yatabe took the posts of vice-principal at the College of Science (Faculty of Science during the Imperial University period) and of Councilor for the University of Tokyo. He also served as the Principal of Tokyo Women's Normal School (present Ochanomizu University).

ITO Keisuke was an authority in herbalism whose publications include 'Flora Japonica.' At the age of 58, he served at the Institute for Western Culture, run by the Tokugawa shogunate (the institute was one of the predecessors to the University of Tokyo). When the University was founded, Ito secured the position known as *extraordinary professor*, despite his advanced age of 74, and he commuted to the Koishikawa Botanical Garden to advise garden management and to continue his study. (The job title *extraordinary professor* has apparently not been used since that time.) At the startup phase of the University of Tokyo, the study of botany was inaugurated in two directions: 1. a research and education at the botanical laboratory started by young Professor Yatabe, who had training in America and research and education experience in the field of natural history and by 2. Extraordinary Professor Ito, who had a traditional foundation in herbalism and was also familiar with Western plant taxonomy.

With the promotion of research at the University, academic research results were published in university journals. For botany, the Botanical Society of Japan published the first issue of the 'Botanical Magazine, Tokyo' in 1887. The journal contained original papers, primarily reporting on new species and, in the early period, described international research trends in the Japanese language. While the institution was organized into a university, its forerunner, the Institute for Western Culture, was an organization originally intended to learn advanced Western cultures. The University at first concentrated on importing advanced cultures, which was a commendable course of action.

However, since some individuals had strong curiosity about the leading-edge research pursued in Europe and America as early as in the Edo period, research conducted at the University soon competed at an international level. As Japanese scientists became able to document new species, Yatabe published a proclamation "A few words of explanation to European botanists" in English (1890) in 'Botanical Magazine, Tokyo' (now, 'Journal of Plant Research'), which announced his enthusiasm about Japanese botanists' playing a leading role in elucidating Japanese flora. The discovery of the sperm of gingko and cycad by HIRASE Sakugorō (1894) and IKENO Seiichirō (1895) was an internationally acclaimed achievement of botany at the time in Japan. They were later awarded a 1912 Imperial Academy Award.

In the field of botany, outstanding natural historians such as Hirase (1856–1925), MAKINO Tomitarō (1862–1957), and MINAKATA Kumagusu (1867–1941) emerged, although they had no formal college education.

The level of biology in Japan moved to the global forefront due to the observation of gingko sperm in 1894 by Hirase, which he reported in an 1896 paper. Driven by strong scientific curiosity, Hirase began to observe fertilization in gingko, and through continual observation, he successfully observed gingko sperm. Ikeno helped this research and many say that Hirase was awarded an Imperial Academy Award owing to Ikeno's strong support. At the time, Ikeno was in the position of assistant professor (later to become professor) at the College of Agriculture after studying at Kaisei School, then Preparatory School, and graduating from the College of Science of Tokyo Imperial University. He exemplified the essence of academia. Ikeno respected the priority of Hirase's observation, providing support for Hirase's treatise, rather than to writing a competing treatise. In relation to the Imperial Academy Award, he adhered to the belief that without Hirase's discovery, he would not have been able to make his discovery. It is believed that Hirase was the first person to observe gingko sperm swimming on a microscope slide, while Ikeno identified it as sperm.

Hirase resigned from the position of assistant at the University of Tokyo one year after his discovery of gingko sperm, moved to Shiga Prefecture to serve as a teacher at the Hikone Middle School, and subsequently took the post of teacher at Hanazono Middle School in Kyoto. As a middle school teacher, he achieved no notable research results. According to Minakata, who is widely known as a natural historian, Hirase continued his research, in cooperation with Minakata, by observing the life cycle of whisk ferns (Minakata, 1925).

The provision of a Western-style college research and education framework was a part of the Westernization of Japan. Areas specifically relating to natural history were also in line with the trend. Immediately after returning from America, Yatabe Ryōkichi served as the first-generation professor in botany and gave lectures at the college in English. In addition to studying and teaching botany, Yatabe frequented Rokumeikan, promoted the romanization of Japanese, and published 'Shintaishishō' as a coauthor with Toyama and INOUE Tetsujirō (the first Japanese professor in philosophy at the University of Tokyo). Yatabe was a trendsetter of the new era. Although he was in an important post, he left the University of Tokyo in the middle of his career. He later served as the principal of the Tokyo Higher Normal School (later Tokyo University of Education, present University of Tsukuba). He died at the age of 47 due to a water accident.

When he was a student at Cornell University, Yatabe apparently studied general botany, including physiology. At the University of Tokyo, his area of research was taxonomy. His successor, Matsumura was also a researcher in plant taxonomy (although he also studied zoology). Another assistant professor, ŌKUBO Saburō, who was a graduate of the University of Michigan and son of the then Governor of Tokyo Prefecture ŌKUBO Ichiō, researched the flora of Izu. He resigned when MIYOSHI Manabu, after studying abroad, was hired as an assistant professor (1895). Miyoshi followed the tradition of documenting new species. At the same time, he coined a Japanese term (生態学) for ecology and contributed to the preservation of protected species. He also aimed to study a broader domain of botany, in addition to taxonomy.

Botanical research became diverse at the University of Tokyo in response to international advances in biology. In addition to taxonomy, research was actively conducted in the fields of physiology, cytology, genetics, embryology, morphology, ecology, and more. Meanwhile, instead of placing importance on the traditional perspective of natural history, scientists in the field of zoology assigned priority to analytical research involving experiments, as the first Japanese professor Mitsukuri, who at a young age learned zoology in America, contributed to the growth of experimental embryology. This tradition continued with the subsequent development of the Department of Zoology at the University of Tokyo.

In Japan, botanical research began, in a narrow sense, with taxonomy that was developed in line with the mode of research in Europe and America at the time. Before long, towards the 20th century, biological research began to be conducted in a hypothesis-testing manner through experimental analysis, beyond observation and documentation. In laboratories, biological research gradually specialized due to the development of analytical equipment to deeply analyze individual phenomena.

The tradition of natural history from the Edo period was retained in the Botanical Course at the University of Tokyo, as demonstrated by Professor Ito who participated in research, while adhering to the tradition of herbalism and adapting the Western taxonomic hierarchy to a Japanese style. Japanese researchers also took pride in their research, as indicated by Yatabe, who declared to botanists from inside and outside Japan that Japanese researchers would lead the way in the research of Japanese plants.

Teachers in the early period of the University of Tokyo had studied and experienced advanced research in Europe or America. There were exceptions, however. For instance, Makino made research contributions through self-education. He was hired as an assistant in 1893, promoted to a lecturer in 1912, and continued research until age 77 (1939). During this time, he used the facilities of the University, except for a very short period during which he was barred from campus due to friction with Yatabe. Records say that he gave lectures in practical training matters, popular among students, and contributed to training successors in the field.

In several works, Makino states that he was not recognized by academia. In fact, he enjoyed his status at the University of Tokyo for most of his life. He was a member of the Japan Academy and was intimate with University of Tokyo students. Nonetheless, his interaction with non-professional naturalists throughout Japan was important. The network he built, with contributions from TASHIRO Zentarō at Kyoto University described later in this book, has fostered the special strength of natural history in Japan, at least in the field of botany.

Imperial universities were also founded in Kyoto, Sendai, and other cities, which also conducted natural history-related research. Kyoto Imperial University was founded in 1897. Around 1921, when zoological and botanical departments were established at this university, scientific research was in an era of specialization distinct from when the University of Tokyo was founded. At its foundation in 1897, Kyoto University had the College of Science and Engineering, from which in 1914, mathematics, physics, and chemistry courses spun off as the College of Science (renamed the Faculty of Science in 1919). In 1921, the departments of zoology and botany were established, along with the departments of space physics and geophysics. After the first decades of the 20th century, biology became compartmentalized into specialized research areas, as represented in the structure of the Biological Course of the University of Tokyo.

The Department of Zoology of Kyoto University comprised three courses of Animal Taxonomy & Morphology, Animal Physiology & Ecology, and Embryology, which conducted active research and education in ecology and other fields. They probably looked at experiment-oriented analytical research then promoted at the University of Tokyo.

The Department of Botany had two courses at the time of establishment, Plant Physiology & Ecology and Cytology. The formal establishment of the Plant Taxonomy Course occurred somewhat later. In 1929, KOIDZUMI Gen-ichi (1883–1953), who was an assistant professor under the first-generation professor KOORIBA Kwan (promoted to professor in 1936), shared the responsibility for opening the course. Koizumi was a taxonomist with interest in geographic botany. He contributed to continuing research and education by training superb successors. He also invited TASHIRO Zentarō (1872–1947) as temporary staff to promote collaboration with non-professional naturalists. Tashiro at the time was a middle school teacher in Kyushu, contributing to the research and study of local plants. Until the Second World War, many non-professional naturalists were in an academic exchange with professional collegiate researchers under the auspices of Makino in Kanto and Tashiro in Kansai. These non-professional naturalists were an inexhaustible group of superb talent and played a substantial role in producing critical information for elucidating the flora on the Japanese archipelago. These collaborations exemplify how the Japanese people's inherent awe towards nature can produce superb academic results.

There are other excellent examples of natural history study pursued outside formal universities in the Meiji period as well. In botany, whether at the University of Tokyo or the second oldest institution, Kyoto University, the focus of research was vascular plants. However, in Japan surrounded by seas, seaweed was traditionally used as a resource, and information on seaweed natural history had long been stewarded by society. OKAMURA Kintarō (1867–1935) used this information to pursue the study of marine algae in Japan. Already in Europe and America, study in this area had produced results to some extent. After graduation from the Department of Botany, Tokyo Imperial University, Okamura was engaged in research at the graduate school for some time. Subsequently, he found employment with the Fisheries Education Center of the Japan Fisheries Association (later Fisheries Training Center, Tokyo University of Fisheries, and Tokyo University of Marine Science and Technology), a post suitable for his research of seaweed. Okamura, almost solely through his own efforts, elevated the study of seaweed to a level comparable to the forefront of international research. In 1936, he completed his great lifework, 'Marine Algae of Japan.'

This tradition of seaweed study was inherited by Hokkaido University and Tokyo University of Education (later Tsukuba University). Excellent graduates from these universities have continually conducted research, constantly producing the world's leading findings through the study of marine algae in Japan.

In the study of macrofungi (commonly known as mushrooms), KAWAMURA Seiichi (1881–1948), a graduate from the Department of Botany, Tokyo Imperial University, commenced a systematic fungus taxonomy in Japan upon his employment with Chiba College of Horticulture (later Faculty of Horticulture, Chiba University) and after working at the Imperial Household Forestry Bureau. He compiled his lifetime research results into the 'Color-Illustrated Manual of Japanese Fungi' (1954–55), an illustrated manual of large mushrooms. Further research of whole fungi later evolved from this.

In the field of fungi, pathogenic fungi were studied in the field of medicobiology. This field of plant pathology, initiated by Department of Botany, Tokyo Imperial University graduate SHIRAI Mitsutarō (1863–1932) and others, was regarded as a domain of practical science and pursued exclusively at the Faculty of Agriculture of the University.

Westernization, or modernization of Japan, after the Meiji Restoration was a social phenomenon. Similar development also occurred in the domain of science. Influenced by advanced Western civilization on one hand, and the collapse of the neighboring nation Qing on the other, Japan followed the path of wealth and military strength through a nationwide effort. As a result, Japan leaned towards an extreme worship of the West and as a reaction, superficial nationalism. Japanese people did occasionally demonstrate their innate abilities that had been fostered through their long history. However, dazzled by the single-minded pursuit of Western-style material- and energy-oriented wealth, they made the way of living in harmony with nature a thing of the past.

The trend of elucidating the biota on the Japanese archipelago by Japanese scientists grew into research and study of the biota within the entire territory of the Empire of Japan, colonized as a result of the First Sino-Japanese War, the Russo-Japanese War, and Japan's annexation of Korea. Japanese scientists energetically researched the biota on Sakhalin, Kuril, the Korean peninsula, and Taiwan. Their research made substantial contributions to the research of species diversity on a global scale. Furthermore, there was an emerging drive to view living organisms on South-Sea Islands as part of the resources to be developed in line with circumstances that would evolve into the Second World War. At this time, however, Japanese scientists were not yet equipped with the abilities required for full-fledged research in tropical Asia.

Japanese researchers were at the forefront of surveys exploring the origin of domesticated plants through karyotype analysis. They had been achieving steady results in breeding improved crop varieties using advanced genetics. In the early years of the Showa period (1926–1989), Japanese researchers contributed to the research and study of living things as a global vanguard in research quality.

Schools for *Samurai* Children and *Terakoya* In the Edo period, feudal lords founded domain schools as public education institutions for the *samurai* class. The curriculum of each domain school may have differed from domain to domain, according to the enthusiasm of each feudal lord. Meanwhile, private *terakoya* were run for commoners and peasants. These institutions provided elementary to secondary education throughout Japan. Domain schools evolved into various forms according to the economic strength and educational policy in each domain and depending on the presence of outstanding masters. Education provided at domain schools for children in the *samurai* class consisted primarily of martial arts training and reading the Nine Chinese Classics. Some of domain schools transformed into middle schools and high schools (under the old education system) in the Meiji period. In the second half of the Edo period, some capable *samurai* children, especially non-heir sons, moved to Edo or other cities looking for outstanding teachers. They became resident scholars who devoted themselves to learning, as they had few chances of taking over as head of the family. Though few, they did exist and were allowed to dedicate themselves to study.

The *terakoya* system, run voluntarily by local people, developed spontaneously to allow students to master reading, writing, and arithmetic. These learning facilities served to improve the literacy rate and raise the intellectual curiosity of Japanese people. Nonetheless, *terakoya* placed importance on the practical skills of reading, writing, and arithmetic to meet daily needs and were not primarily intended for intellectual enlightenment. Various textbooks had been published and were in widespread use. Even though it was practically oriented, intellectual training fostered human intellectual sensibilities. This style of cultural enrichment they exhibited related to how Japanese culture developed into a proper form in the second half of the Edo period.

Neither domain schools nor *terakoya* involved a scientific perspective in elementary and secondary education. In the Edo period, Japanese people polished their emotional sense towards nature through their everyday lives, while those with particularly strong scientific curiosity learned Dutch or other Western studies. However, it is noteworthy that the prevalence of *terakoya*, even in provincial areas, helped the general public improve its literacy rate and raise its intellectual curiosity, thereby substantially contributing to a growing cultural and academic awareness among Japanese people. The peace throughout the Edo period, which spanned 260 years, allowed citizens to create sophisticated culture in their lives.

Some scholars surmise that the scientific curiosity of Japanese people did not rise in the Edo period, or the era of feudalism, because intentional science education did not take place. This may be superficial thinking. Specifically, the school enrollment rate of *terakoya* was exceptionally high compared with other countries in the world. Even if *terakoya* provided only practical training, there must have been substantial benefits in terms of cultural enrichment of Japanese people on average. It is only natural that the intellectual curiosity of people, even in a feudal society, would grow as a consequence of more than 200 years of peace and thriving intellectual activities. Since the times of the 'Man'yoshu,' the nature-oriented lifestyle of residents on the Japanese archipelago was manifest. The advanced levels of domesticated animals and plants in the Edo period can be explained as a consequence of the nature-oriented lifestyle that was underpinned by the high level of Japanese culture at the time.

Thus, in the Edo period, natural history in Japan would likely already have been at a reasonable level of development. The foundation that enabled this development was a deep understanding of nature rooted in herbalism. The prevalence of learning infrastructure also supported people's growing scientific curiosity.

When universities were founded in the Meiji period, a handful of individuals who gained the opportunity for higher education could then learn science. It is rightfully understood that this opportunity for a small number to achieve higher learning raised cultural awareness across all of society, rather than for simply a tiny percentage of people. Thus, when universities were founded in an institutionalized context after the Meiji Restoration, the social infrastructure was already in place to cultivate individuals with interest in science.

Natural History in Universities How to smoothly incorporate natural history challenges into college research is a difficult question to answer.

Science education at colleges places an importance on training students in scientific thinking. Accordingly, students learn the existing system of knowledge from its early beginnings to today's forefront of knowledge. It is, in a sense, an appropriate system for intellectual learning. However, in recent years, there has been a strong trend towards viewing colleges as institutions for training humans who are instantly useful in society. There is a notion that humanities faculties that are not immediately useful for society should be restructured. If this notion is applied to scientific faculties, expectations will be high for basic research that can be applied to useful technology, rather than for research driven by intellectual curiosity. Colleges as higher education institutions place importance on the transfer of ready-made knowledge and assign a low priority to fostering intellectual curiosity and creative problem-solving abilities, although they are expected to develop society's intellectual leaders.

The knowledge accumulated in the field of natural history is not suitable as a subject of systematic discourse. As a subject of secondary education, it inevitably tends to name facts one by one. Natural history knowledge is difficult to transfer in the form of a system of knowledge. Rather, it is a subject that stimulates the intellectual curiosity of students by means of direct contact with nature and self-supported learning. Consequently, classroom exam questions to assess academic abilities would encourage rote learning, which may be criticized as being not scientific. In college curricula, natural history knowledge is considered as essential basic knowledge in biology. However, despite the knowledge transfer that takes place, it is difficult to stimulate intellectual curiosity and provide for scientific thinking if education is delivered primarily through lectures and other forms of classroom study. Natural history can yield fruitful results in science education when it provides students with practical training and activities using authenticated specimens, stimulates scientific curiosity through physical work, and offers experiences through the excitement of analysis.

It was once pointed out that Japanese colleges were not teaching the theory of evolution. Nowadays, the theory of evolution is termed *evolutionary biology*. Several colleges have begun to offer lectures in evolutionary biology. The College of Arts and Sciences at the University of Tokyo has established a sub-program, the Evolutionary Biology course. I myself have led an omnibus lecture on the theme of evolution at Kyoto University. Additionally, as a faculty member of Rikkyo University, I also organized an omnibus lecture, "Evolution," as a liberal arts program and compiled the contents into a book (Iwatsuki & al., 2000).

The reality of biological evolution has been argued in various fields of study. It is difficult to summarize and organize it into a lecture of 15 lessons of evolutionary biology. Nowadays, a single teacher leading such a lecture would be out of the question. In present-day curricula, organized based on positivism, it proves difficult to determine how to teach evolutionary biology at college. To what extent should it be theory-oriented or should it teach practical examples using real objects? Meanwhile, the question arises: what if the subject is viewed from the perspective of natural history?

More than half a century ago, when I was a college student, TOKUDA Mitoshi lectured on the theory of evolution at the Faculty of Science at Kyoto University. At the time, he already had compiled his lectures at Kyoto University into 'Theory of Evolution' (1951), published as a book of the Iwanami Zensho series. When I attended his lectures around 1956, he often finished his supposedly two-hour lecture in about 30 minutes. Perhaps, he was busy thinking in preparation for 'Two Branches of Genetics II' and 'Revised Theory of Evolution.' That said, his enthusiastic lecture on the theory of evolution should have impressed students to a reasonable degree, aside from the criticism about the influence of the breeding technology by Michurin. What the students gain from a lecture may differ from person to person. Nonetheless, one essential aspect of lecture presented by a researcher is the influence of the researcher's personality on the audience and the facts they convey.

Once Kyoto University was strongly oriented towards natural history, as known by the trend it exhibited soon after the Second World War. A group of researchers led by IMANISHI Kinji, active during the War and having founded the Seihoku Research Institute at Zhangjiakou in Inner Mongolia, returned to Kyoto after the War to start the Society for Natural History and publish the journal 'Nature and Culture.' Leading active members of this society were FUJIEDA Akira, NAKAO Sasuke, and UMESAO Tadao. Their activities provided a prominent guidepost for the development of a subsequent expeditionary culture, achievements whose inclusion is never negligible when presenting an overview of natural history.

Their activities marked the beginning of the tradition of exploratory expeditions deployed during the post-Second World War years. To prepare reports on their expeditions to the Himalayan region, they set up the Fauna and Flora Research Group, an organization whose name is indicative of the group's connection with natural history. Physical anthropology research, regarded as a model of natural history study, commenced under the leadership of Imanishi.

Kyoto University picked up the theory of evolution early on as a lecture topic in its curriculum, whose significance at a time when lecture topics were not transient can only be understood in the context of the above development. (ITANI Junichiro, I, and others brought the theory of evolution back in the curriculum at Kyoto University in the early 1970s when the student political movement was at a peak. It was then unavoidable to take the form of an omnibus lecture involving several lecturers.)

(2) Natural history in museums and related facilities

In Europe and America, museums and other similar facilities play a significant role in natural history study still today. This is because society shares an understanding that museums and other similar facilities basically pursue studies and have a suitable framework for promoting research.

However, according to some survey results, many botanical gardens run as museum-related facilities in Europe are literally botanic research facilities still today, while in North America, only approximately 10% of facilities referred to as botanical gardens set research as their primary purpose.

In Europe, many botanical gardens are national institutions or facilities attached to a university. They were originally founded to serve research purposes. In contrast, in North America, many of them are privately-run facilities. This clearly shows differences in the way of thinking about the responsibility assumed in the maintenance of research institutions and other organizations. In Europe, national and public museums faced staff cutbacks made in the second half of the 20th century. Interested parties in other countries were asked to make governments know the substantial international contributions they made. How much the requests made by researchers in other countries influenced the intentions of governments and public entities remains unknown.

Many private botanical gardens are run on admission revenues, and quite a number of them strive to attract visitors. Naturally, such facilities tend to entice visitors by novel attractions. How people respond to such attractions differs from country to country.

Founded in 1859, the Missouri Botanical Garden is the oldest in North America. The Garden declined with time, and in the early 1970s, it was a poor, small botanical garden with only two researchers who had academic degrees. In 1971, Dr. P.H. Raven (1936–) took the post of director. Since then, the Garden grew rapidly as a research facility. In the mid-1980s, the number of researchers with academic degrees exceeded 50. Before long, it found itself having grown to become the most active botanical garden in North America, both in research and in knowledge dissemination, and it advanced to be known as being on par with Kew Gardens in the United Kingdom. Presently, the governments of Missouri and Saint Louis share the Garden's expenses, helping the Garden ensure management stability, despite its status as a private institution.

To operate an institution stably in the United States, it is necessary to use outside funds. There are strong social trends in America for giving donations to academic or cultural projects. Such donations to private colleges are the mainstream. If the significance of a project is understood, it is not difficult in the U.S. to use outside funds. Of course, for this purpose, the operator is required to provide sufficient explanation of the benefit of the project. If the details and significance of the project fail to be understood, the institution will face financial difficulties.

In Japan, in the settings of the Meiji Restoration, academic and cultural activities became entities maintained and developed at public institutions. It was not a system in which people made donations to support academic and cultural activities. In the adopted framework, culture was funded by tax revenues, although *sumo* was supported by patrons. Nonetheless, it is somewhat strange that culture would be promoted not by the initiative of the people, but by the government.

In 1994, when Plantopia, Fukui Botanical Garden was constructed in Asahi (present Echizen), Fukui Prefecture, WAKASUGI Takao, who had a passion for botanic research in the region, assumed the post of founding director and prepared for the Garden's establishment. Because it was a public facility, there was no need to solicit donations every year. However, founded by a provincial town government, the Garden's budget was small. Even if the public was aware that the Garden had a small budget, generous donations to a public institution were not expected. Wakasugi relied on the connections that he had built through his previous activities. He requested his many friends in Japan (many of them being non-professional naturalists) to supply plants from their respective regions. He garnered help from an exceptionally large group of friends and could successfully implement the planting design. The opening of the botanical garden had good media coverage in the Chubu district. Because of the lack of regional botanical gardens, Plantopia had some 600,000 visitors enjoying it in its first year, well in excess of expectations.

In the second year, the number of visitors decreased sharply, perhaps because of a failure to entice visitors to revisit the Garden. As a solution to bolster the project, the town government appointed a new director from the town office and changed the garden policy to attract visitors. Attractions included a barbecue site set up in the Garden. The number of visitors did not increase, however. Several years later, Wakasugi returned to the post of director. He implemented proper botanical garden activities in a stabilized manner. The Garden is not expected to attract a huge number of visitors;

however, it does not seem to face the crisis of closing due to mounting cost pressures either. The initial boom invited a fatal misunderstanding and an empty dream. The reality is now accepted. Later, Director Wakasugi retired and is now Director Emeritus of the Garden. The small, yet substantial, botanical garden now operates as a facility run by the consolidated larger town government.

Not all public facilities necessarily enjoy management stability. For a botanical garden, it is necessary to have a clear direction for how the local people recognize its exhibits. Such a clear perspective enables the garden to support the local community in lifelong learning. National and public facilities, not only limited to academic and cultural institutions, have had a strong bureaucratic mindset, which has been substantially corrected in recent years. This is also true with museums and the similar facilities.

Until very recently, many people seem to have had an impression that museums are only places where unfamiliar items are displayed in a dimly lit building, with each exhibit accompanied with an unintelligible explanation. These ways of displaying exhibits were common at facilities built imprudently during the high-growth period. Such facilities, with few visitors, were criticized as wasteful construction.

Likely after the bubble economy collapsed, and the world cooled down, museum staff became motivated to pursue activities that were suitable for museums. Not only did they attempt to provide commentary attached to exhibits with an air of authority, but they began to rack their brains on how to devise tools to stimulate visitors' interest in exhibits, organizing their activities for this purpose.

Active research by outstanding researchers and collections of valuable authenticated specimens are characteristic of museums and other similar facilities. These facilities use their collections effectively to support lifelong learning and to serve as a think tank. Museum staff are supposed to serve these roles properly.

The fundamental activity of museums is knowledge dissemination, which is possible if visitors are brought voluntarily into the learning process, as an alternative to compulsory education. The results of this depend on how much the interest of visitors is awakened to voluntary learning. In this sense, examples of activities employed at *terakoya* are good practices to follow.

In the Meiji period, school systems were established, including a compulsory education system. Compared to *terakoya*, education for the purpose of contributing to wealth and military strength to catch up with and overtake the West might have had greater effectiveness to some extent. However, Japan was slow in encouraging motivation for voluntary learning among people in general and in improving public science literacy, as demonstrated by the slow enrichment of museums and other similar facilities. In fact, that slowness might have actually been desirable for some people, although they were probably not so concretely aware of the desire.

Museums and other similar facilities are facing institutional reforms now underway for economic purposes. Some of them are being transformed from national or public facilities into incorporated institutions, while others are placed under a designated administrator system. These institutional reforms may be necessary. However, the first requirement for this process is that the parties concerned recognize the role of museums and other similar facilities by taking part in the activities pursued there to ensure that all systems operate efficiently.

The motivation of museum staff may encounter two scenarios in terms of investment. In one situation, investment would not be made to institutions with highly motivated staff because they are thought to perform well if left to their own devices. In this scenario, resources would be allocated to low-performing institutions. In the other situation, resources would be allocated to institutions with highly motivated staff to further encourage their motivation. The question is which is more beneficial for society. Solutions to improve science literacy, thus furthering natural history and the support for lifelong learning, are of much interest.

3.2 Natural History Specimens

Natural history-related institutions include colleges, museums, and their affiliated facilities. When it comes to museums, ordinary people may imagine a collection on display. Indeed, museums certainly collect artifacts with the aim of exhibiting them. Research activities pursued by curators are the essential foundation of institutions such as university museums and national museums, whose principal aim is research, and even in public and private museums, whose principal aim is information dissemination and support for lifelong learning. These institutions characteristically differ from colleges and other ordinary research and education institutions in that museums manage and maintain a large volume of natural history specimens relating to natural history research. Some institutions use natural history specimens for exhibition or preserve and manage them as subjects of research. Other institutions, such as affiliated zoos, aquariums, and botanical gardens, keep or cultivate living things to use for exhibition, as stock preservation materials, or as research materials.

All objects present in the natural world can serve as natural history specimens. Specifically, research specimens are supposedly collected randomly. In reality, specimens in the collections of museums and similar institutions are not necessarily those purely collected randomly as materials for research purposes. True random specimens selected from the natural world are rare. When using specimens as research materials, it is necessary to bear in mind how they were collected.

Objects that can serve as specimens are all things present in the natural world. They may be living things. Earth science materials and other non-living things are also included. Not only terrestrial items, but also lunar rock and other things collected from space, recently obtained with the increasing opportunities, are regarded as valuable specimens for related research.

'Kōjien' gives three definitions for the word *specimen* (標本), one of which is, "an individual body or part of it suitably processed and preserved for research or education purposes in biology, medicine, mineralogy, and other disciplines." Including specimens for exhibition at a museum, in practice, is not a cause for concern. Such exhibition is intended for social education, which is a kind of education. That said, as defined in the dictionary, the primary uses of specimens are for research and education.

(1) Specimens that represent nature

Herbarium specimens are for research and education. With all objects present in the natural world as potential specimens, natural history specimens normally exclude artifacts. Factory-made products are not collected by natural history museums because they are not natural objects. Artifacts are collected within the science and technology category of museums, which are known as science museums. However, it is not uncommon that the distinction between natural objects and artifacts is obscure. Particularly, in the present age in which human influences are observed everywhere on earth, very few things are free from human influence.

Living organisms growing wild in a subject area are normally considered to be natural. Domesticated animals and cultivated plants are considered as forms artificially created by breeding and are not regarded as constituents of the biota in a subject area. Even though they were once natural objects and are currently living species, they are no longer considered natural objects for the purpose of natural history due to the involvement of human practices.

The use of wild species for eating, medicinal purposes, and other human practices, which interferes with the survival of the wild organisms is severely criticized as destruction of the natural environment. In contrast, humans also artificially grow species for eating or other purposes (domesticated animals and cultivated plants). Systematic slaughtering of these species is viewed as part of legitimate production and economic activities. Religious customs of not eating specific animals and the avoidance of eating meat by vegetarians are exceptional cases. In living in harmony with nature, one may make use of wild species, by scientifically controlling the population and, as a consequence, maintaining the survival of wild species. However, such a way of living is criticized by some people for killing wild living things, which are not artificially bred, for the sake of humans. This type of criticism comes from people who are actively involved in killing noxious animals on a daily basis. Moreover, the general public's ideas about killing wild species differ completely when considering forest animals and marine organisms.

Undomesticated wild organisms that have moved from their original habitat due to human practices are known as introduced species. They are not constituents of the local biota, even if they have survived for generations and their current location has become their native habitat.

Meanwhile, there are species known as archaeophytes (introduced plants in pre-historic time). These plants are considered to have been introduced artificially, whether intentionally or unintentionally, by our ancestors in the primitive age before literal records. They are recorded as a constituent of the local biota, rather than as an artifact. Although not left on record, they were actually introduced along with human activities. This inference may originate in considering human activities before civilization as activities of the wild animal species, *Homo sapiens.*

Whether the above interpretation is correct or not, it is vital to correctly understand these archaeophytes according to their histories to accurately understand the form of living organisms under natural conditions in a subject area. In this regard, rather than excluding artificially influenced objects from natural history materials, it is necessary for the healthy advancement of natural history to broaden the scope of natural history materials and include artificial materials to a reasonable extent.

Indeed, some museums have authenticated specimens of cultivated plants in collection, and many botanical gardens exhibit specimens of cultivated plants. It is not rare to see, in such gardens, horticultural plants beautifully set out as principal constituents. Rather, these types of botanical garden exhibits have increased to the majority in recent years.

In either case, natural history specimens represent all natural objects present in the natural world. Natural objects in this context should be understood in the broadest possible sense. It is desirable to collect and preserve specimens in the category of natural history even if, to some reasonable extent, they take on artificial elements or, in some cases, they no longer exhibit natural qualities but are usable for information purposes.

(2) Research based on specimens

Specimens are absolutely necessary for natural history research. Anything can be material for research. However, it is also necessary to sort out specimens from the perspective of natural history materials.

Museums and other similar facilities collect, manage, and maintain specimens with the aim of preserving research materials. However, natural history specimens in the voluminous collections of museums and similar facilities encompass more than those collected solely for natural history research. A huge volume of materials collected for specific research topics may include those unsuitable for museum collections. Many of these are discarded, except for those particularly needed for recordkeeping purposes, as sample materials acquired for specific research may be biased in their distribution in the natural world and lack universality as natural history materials.

Collected specimens should convey information to the greatest extent possible. As a specimen, an individual or a part of it should convey the form of the individual as multilaterally as possible. A plant specimen, even if sampled from part of an individual, should represent as many traits as practicable, containing the maximum diverse parts of a plant body, such as flowers, blossoms, and fruits, and keeping in mind the time of collection. Of course, preparing a tree specimen complete with roots, stems, leaves, flowers, and fruits results in a big amount of material that makes storage difficult. To achieve a largest possible collection, individual specimens should be limited in volume.

For living things, natural history specimens are required as biodiversity research materials. It is desirable, if possible, to collect living things from the entire area of distribution to ensure the materials encompass diverse individual variations, as shown by the examples presented later. In the case of a tree, it is a good practice to use a living specimen as material knowledge of an individual body. Acquiring a whole body as a natural history specimen is not often needed.

To meet the aforementioned purpose, the preferred size of a plant specimen is one that can be mounted on a mat board. International specifications have been established for specimens. Mat boards for specimens may somewhat vary in size. However, specimens prepared in this format can be most efficiently stored, used, and presented, especially for large amount. Moreover, controlling specimen sizes within a reasonable range is advantageous in that several specimens sampled under identical conditions may be exchanged with other institutions for a more efficient use of materials.

Of course, specimens collected as described above are not necessarily valid as materials for every research project. They are expected to be useful for research and education in general. Needless to say, even less sophisticated research is important for natural history. For specific research, specific suitable materials required must be collected. This is an essential challenge associated with every kind of research. It is selfish to say that materials not directly useful for a current research project are useless.

As large an amount of specimens as possible should be collected as described above from places under every possible set of conditions in every possible region. Obviously, the range in which one full-time researcher can collect is limited. Once many people were engaged in the collection of specimens on a full-time basis by way of surveys in remote areas, known as exploratory expeditions. There exist some individuals who collect materials almost on a full-time basis still today. Those who collected plant materials from unspoiled fields in a kind of exploratory expedition used to be called "plant hunters." Not a few specimen-collecting experts are non-professional naturalists. Collections of excellent specimens owe much to the efforts made by these laypeople.

Local collectors occasionally send specimens to specialists at a college or museum, requesting identification. For accurate identification in such cases, they send specimens of individuals with traits as uniform as possible. It is not rare that specimens sent for identification serve as type specimens for documenting a new species. These kinds of specimens are often invaluable for research purposes.

At museums and other similar facilities with a large amount of specimens, the collections are often not systematically organized, because of the large stacks of specimens acquired for a collection purposes, accepted as identification requests, and donated from collectors (or the bereaved families of collectors). Scarce types and individual bodies with unique variations inevitably attract the interest of collectors and are picked up for identification. Commonly, the natural state of these species is far from the distribution represented by these specimens. Needless to say, with this in mind, researchers attempt to elucidate the truth to the greatest extent possible from studying specimens in collection. What specimens tell us and what they do not will be clarified later in this book.

The forms of natural history specimens are diverse across objects. The basic form of plant specimens is that of the aforementioned herbarium specimens. Some of these are large herbarium specimens or supplementary immersed specimens of voluminous seeds or succulent plant bodies on a case-by-case basis. Insects are often made into specimens with wings unfolded. A large number of vertebrate animal specimens are conserved as stuffed or immersed specimens. For birds, eggs and nests are also collected as specimens. Specimens are conserved in diverse forms according to the research topics for each group of living organisms. A common requirement is that basic information on the state in which specimen was gathered is noted as passport data (also known as label data among natural history researchers). In addition to the collection location (most recently, GIS data is often attached), the date and time of collection, and the name of the collector, or the so-called passport data, should also contain, to the greatest extent possible, the habitat and traits (colors and odor) not observable in specimens.

(3) Use of specimens

One important role of specimens, in relation to research at colleges and other institutions, is their use in education for training successors. This book discusses specimens in the natural history research organization (2.3), yet they are essential for the succession of research as well. Training successors has two primary aspects. One is to attract the interest of talented young people to natural history, and the other is to provide motivated youth with opportunities for training in basic research.

Inciting the curiosity of youth for the various challenges associated with natural history can be accomplished most effectively by allowing them to become aware of and moved by the mystery of nature. To this end, natural history specimens exert a more powerful effect than millions of words, as described in the section on museums exhibitions.

For someone with an intention to professionally study natural history, training means to explore hands-on research, develop an original awareness of issues through learning, and devise analysis methods. For these purposes, specimens are essential as research materials, as described in the section on research.

For the popularization of natural history, school education, museums, and other similar facilities, which support lifelong learning, play their respective roles. **Education programs provided at schools**, from kindergartens to colleges, that facilitate the understanding of nature by school children and students are all associated with natural history. To pass down this intellectual heritage, schools spend many hours on classroom study, because of the emphasis on intellectual training and the overwhelming volume of material to teach. Even more problematic, elementary and secondary schools are bound to act as cram schools for entrance examinations to the extent that gathering knowledge has the highest priority, and fundamental education to foster true humanity tends to be neglected. School education appears to place an importance on preparing students for success in society rather than on nurturing them. If this is true, schools are no longer places for opportunities to motivate students to more fully understand nature. This is an extremely dangerous situation for humanity living as an element of nature.

School curricula are not in a hopeless situation. Failure in implementing their ideal form is the point of concern. Furthermore, many teachers working in elementary and secondary education have an insufficient basic understanding of nature. Additionally, the present situation is problematic in that teacher-training institutions do not provide adequate education for helping students to correctly understand nature, and they lack a fully-fledged framework required for providing such education.

Ultimately, this problem traces back to the fact that the population of aspiring natural history researchers is small, and therefore, superb teachers working for these education programs are not supplied to teacher-training colleges. However, this type of circular reasoning and complaining should be avoided. Therefore, it is necessary to reawaken to the role that natural history plays in training successors. For this reason, the collection of specimens should be enriched, and at the same time in elementary and secondary education, as much effort as possible should be directed towards training the next generation to have a simple curiosity about natural phenomena.

Concrete measures can provide children with more opportunities to have contact with nature in practice. Basically, if it is difficult to have direct contact with nature, children should have more opportunity for contact with specimens. This is clearly indicated by the strength of the impressions that children experience when they, at a museum, see a replica of a dinosaur they only knew by book illustrations or on a television screen. In another bitter reality, this precious impression may not last long, resulting in a failure to raise their curiosity, but this is another matter. Recently, I have recognized that children participating in museum activities more eagerly broaden and deepen their interest in nature than their parents might have expected. For this reason alone, I hope that there will be increasing opportunities for the younger generation to develop this type of scientific curiosity before they mentally mature.

Outside of colleges, preparing natural history specimens for school education is a difficult issue. Enriching local museums and facilitating collaborations in nature education between museums and schools is part of the solution. One minimum requirement for a learning environment that fosters curiosity about nature is contact with genuine materials.

In light of the role of natural history in society, exhibitions of natural history specimens are highly significant. Support for lifelong learning by museums begins with exhibiting natural history specimens. For museums and other similar facilities that are willing to popularize natural history and effectively support lifelong learning and think-tank functions, specimens in collection are of great significance. For the general public, a museum is a place where impressive specimens are on display.

One important challenge for a zoo or botanical garden (as well as for a museum) that exhibits live materials, is how to produce a striking exhibition to attract visitors. The most traditional and frequently used technique is to exhibit rare creatures not viewable anywhere else. However, sole reliance on this technique can devolve into an exhibition that simply attracts visitors. How to organize an exhibition to fascinatingly show the forms of living things and their true nature should be explored.

Another technique of exhibition is required to show the forms of living things by way of specimens, instead of showing them alive. In recent years, museums and other similar facilities have contributed to disseminating information about the social challenge of the sustainable use of biodiversity. Exhibitions aligned with this objective have incorporated various innovative ideas.

Once museum exhibitions simply displayed rare items with somewhat difficult-to-follow explanations. The impression that museums made on the general people was something like a place where unfamiliar items were displayed in a dimly lit hall for visitors to study by reading explanations. Regarding the attitude on the part of the exhibitor, it cannot be flatly denied that they tended to consider information dissemination an opportunity to inform visitors of the value of the materials in collection and to tell visitors to come and study. (Every public institution used to have an intention to lead people.)

Many Japanese museums have come to correctly understand their role in recent years. They are making efforts to arrange exhibitions of their valuable specimens so as to introduce visitors to unfamiliar aspects of nature and encourage them to have curiosity about the mystery of nature by having contact with actual items. Beyond a simple display of rare items, they intend to go further by helping visitors become aware of and have interest in the mystery of nature through displayed materials. I and others explain these activities by the phrase "from exhibition to **performance of exhibition**." This is an endeavor to evolve exhibition into demonstration that involves performance.

Traditionally, the term *demonstration* refers to the use of a figure or table as an aid or the use of a supplementary simulation or experiment to teach a subject that cannot be fully explained by classroom study or a lecture. In contrast, the term *performance* in music or theater refers to playing or acting, with a focus on its dynamic aspects.

Museums implement performance of exhibition to realize their intention of actuating materials and letting them tell stories with the help of humans. A performance of exhibition encompasses a broad category of actions. For example, the instructor and the instructed share curiosity while actually picking up natural objects in the open so that the individuals with more knowledge and less knowledge alike can be moved by the natural objects. A similar activity can be organized using museum collection. Furthermore, museum materials may be taken out to a local venue for an exhibition delivery service (demonstration). It should not end in a simple transfer of knowledge from the haves to the have-nots. It is important that such activities encourage self-motivated learning rather than simply teach the participants. This may be done by more strongly awakening the senses to the excitement of natural objects and their specimens, deepening the induced scientific curiosity, and utilizing necessary knowledge to decipher the mystery of nature. Intellectual curiosity is fostered only where active learning takes place.

Specimens are absolutely necessary for performance of exhibition services and, as before, for making the most of the significance of static exhibitions. Commonly, materials collected for research purposes are also used to hold an exhibition. In an exhibition, especially a performance of exhibition, deterioration of or damage to specimens must be taken into consideration. This implies that there is a need to collect specimens specifically for exhibition, as well as the need to collect those for research purposes. Valuable specimens for research purposes, if noticeably deteriorating, provide only a limited amount of information, and, if damaged heavily, can no longer serve as specimens. Of course, some specimens may well serve for both research and exhibition. However, it seems that museums will increasingly need to collect specimens for exhibition purposes as consumables that allow for damage, in addition to those for research purposes that must be treated carefully, with their essential features intact for preservation. When necessary to define the clear distinction between these uses, the difference between them must be clearly borne in mind.

(4) Natural history specimens and cultural properties

In this chapter, the term *specimens* refers solely to natural history specimens. A specimen is defined as an object from the natural world in the collection of a museum or similar institution that is artificially managed and maintained. However, regarding materials in the collection of museums, the public commonly calls cultural properties to mind. Indeed, while cultural properties are protected appropriately under the Act on Protection of Cultural Properties, natural history specimens are often roughly understood as items collected and stored according to the preferences of people involved in natural history.

In Japan, the Act on Protection of Cultural Properties is a law discussed and established after the valuable mural paintings in the sanctuary hall of Hōryū-ji were burnt down when the hall was set on fire in 1949. The Act classifies cultural properties into six categories of tangible cultural properties, intangible cultural properties, folk cultural properties, monuments, cultural landscapes, and groups of traditional buildings. It is intended to protect cultural properties designated as any of these categories. Before the establishment of the Act on Protection of Cultural Properties, the Act on Conservation of Historic Sites, Scenic Beauties and Natural Monuments, the Act on Conservation of National Treasures, and the Act on Conservation of Important Works of Art were established. However, these laws were abolished upon the enforcement (in 1950) of the Act on Protection of Cultural Properties, which covers items subject to the former three laws. Things included in historic sites, scenic beauty, natural monuments, and those relating to natural history fall under the category of monuments among the categories of cultural properties.

MIYOSHI Manabu (1861–1939), who studied in Germany, introduced the concept of *Naturdenkmal* (natural monuments) in German to Japan (Miyoshi, 1915). Natural monuments are selected from extant living organisms. The concept of cultural properties has been increasingly understood by the general public because superb works of art and artifacts have been recognized as cultural properties and designated as national treasures or important cultural properties. Natural monuments have also become familiar to a considerably large portion of people. However, legislation to protect natural history specimens has not moved forward.

When the 2011 Tohoku Earthquake and Tsunami occurred, a number of museums and other similar facilities were severely damaged. To repair damaged specimens, museums throughout Japan took part in the repair efforts. Since then, a trend towards definitive positioning of natural history specimens has manifested by the manner in which the Science Council of Japan has set up a subcommittee to discuss the matter. Expectations are high for the national government to recognize the significance of natural history specimens and to implement measures for responsible conservation of them in a satisfactory manner. Natural history specimens that were damaged during the 2011 disaster were saved from the worst outcomes because officials in charge of cultural properties interpreted cultural properties in the broad sense to repair damaged specimens. Some budget was allocated to repair natural history specimens. Moreover, museum staff cooperated in the repair work across individual institutions (e.g. Fuse & al., 2012).

(5) What researchers can or cannot expect from specimens

Natural history specimens are essential, if only for the study of living organisms alone.

To know the reality of life, researchers must understand the diversity of things. For this purpose, one shortcut is to recognize, by using real materials, the current diversity of extant living organisms. It would be best to recognize the diversity of living organisms by observing them in their real habitats. However, to compare all life on earth by observing real materials, it is necessary to collect and keep specimens of living things on hand from many parts of the globe, as it is difficult to visit every habitat of living things on earth. Of course, natural history specimens are mostly non-living. How much information a researcher can read from specimens depends on the ability of the researcher.

Biological study uses particular living organisms as models while adhering to the basic principle that they are diverse living organisms. To conduct analytical research of a model organism, a pure unambiguous lineage is selected as a material. For this purpose, the research material is raised or cultivated in a laboratory and maintained under relevant controlled conditions. After all, an analysis that uses a model organism results in knowing the model. To universalize the findings of the analysis as characteristics of a living organism, the researcher must know where the analyzed phenomena fit from the perspective of biodiversity. For this purpose, it is essential to be familiar with the reality of biodiversity. Moreover, for comparative research of diverse organisms – diverse in terms of lineage and habitat – it is necessary not only to analyze specific individuals as models, but also to contrast and compare diverse materials. This kind of research cannot occur without the use of preserved specimens.

(6) Basic research on biodiversity

Biodiversity research uses observable models. Likewise, it is essential to position the model in actual biodiversity. In this process, the researcher must pinpoint the appropriate point of view to avoid the risk of circular reasoning.

In research that analyzes a particular phenomenon, a pure-line model organism is used. In contrast, research on diversity should use materials that have diverse variations. This research should be conducted alongside research on model organisms. Diverse variations in this context include all the aspects of diversity, such as genetic diversity, species diversity, ecosystem diversity, and diversity of the traits that constitute living individuals. For species research, both inter-species affinity and intraspecific variation are analyzed. For this purpose, all information on genetic diversity is required. Lineage analysis deals with overall species diversity. For this analysis, it is essential to look at similarities and differences at various ranks, from inter-species to inter-phylons affinity. Needless to say, the presence of species is positioned in ecosystem diversity. Unless ecosystem diversity is analyzed, the accurate reality of species cannot be known.

While analyses that use living models may provide valid information, it is essential to grasp the state of each extant individual body on earth in an integrated whole for comparative analyses of taxonomic groups. For this purpose, present-day science can analyze a large number of specimens to know the various lifestyles of actual living organisms.

Indeed, specimens of diverse living organisms on earth have been observed and compared to accumulate information on global presence and location of living organisms. Scientific findings obtained this way have been organized into all currently known biological species, numbering between 1,500,000 and 1,800,000. These findings, in large part, are composed of major taxonomic groups of terrestrial plants and vertebrate animals. Insects, marine invertebrates, and fungi are still in an early state of very little knowledge about their species diversity even today. Moreover, until just very recently, only a fraction of research has begun on the diversity of deep-sea organisms. Determination of the degree of deep-sea diversity will depend on future research.

(7) Research using specimens versus research on living organisms

Around the 1930s, biosystematics was strongly advocated, claiming that research based on dried specimens prepared for research purposes was limited and that unless analyzed through comparisons of living organisms, the reality of diversity would not be elucidated. This claim pointed out the limitations of research by specimens. Since the rediscovery of the laws of inheritance, analyses incorporating genetic methods rapidly advanced, species affinity was traced by karyotypic analysis, and the species differentiation process was clarified. Thus, the claim was, in a sense, based on the belief that cytotaxonomic analysis was the most powerful means of species research. Certainly, to obtain chromosome information, living materials were essential, and to know the reality of ecotypes, as a new challenge, comparative cultivation provided valid results. Those conducting research based on dried specimens did not negate the essentiality of analyses of living materials.

Nevertheless, they did not think that research solely based on living materials would be enough to eliminate the need for specimens-based research.

Indeed, there are many examples of research based on specimens providing information that cannot be produced from living materials. The following is a somewhat dated example in which I was involved, also described in the 'Natural History of the Ferns' (Iwatsuki, 1996) in the Natural History series. This section will focus on topics relating to the research utility of specimens.

Crepidomanes minutum in the family Hymenophyllaceae is a widely distributed fern species. On the Japanese archipelago, this species is distributed as far as to Hokkaido to the north. As the plants in this family are distributed primarily in the tropical zones and the southern hemisphere, *C. minutum* is one of its more northernmost examples. According to records, its area of distribution is wide, from East Africa to the west to Polynesia to the east. It had been widely known that variation in the phenotypic characters of this species was too complex to classify, as indicated by the many names assigned to diverse forms as early as in the 19th century. However, certain distinguished types are largely unidentifiable. This was surmised through a comparison of phenotypic characters by Copeland (1933, 1938; Tea Time 9), a great contributor to research on this family.

British botanists Bell (1960) and Braithwaite (1968 and 1974) studied the Polynesian *C. minutum* using living materials and analyzed its reproductive mode and chromosomes. They recognized sexual and apogamic types of *C. minutum*. Their conclusion was that *C. minutum*, in the broad sense, could be discriminated into two species, with the two types differing in phenotypic characteristics in addition to in reproductive form, according to the characteristic of bearing gemmas.

With interest in research on the family Hymenophyllaceae, I extended the scope of my research to *C. minutum* in a joint project with YOROI Reiko who had achieved research results on the prothallia in this family. In our research we paid special attention to whether *C. minutum* could be discriminated into two types by combining reproductive form and phenotypic character. As Copeland stated, it is desperately difficult to distinguish the plant into two types, at least by observation of external morphology of specimens in stock.

It takes a tremendous amount of energy to collect living materials that exhibit intraspecific variations from every part of the globe and to observe their reproductive forms and chromosomes. Immediately embarking on such research is an intimidating task in light of research validity. However, using a large number of specimens in stock from every part of a wide area of distribution, it is possible to carefully compare variations in phenotypic characteristics. Consequently, the use of specimens to observe reproductive forms provides preliminary results that enable the researcher to obtain the necessary data.

Based on this preliminary observation, we predicted that it would be impossible to identify the two types that Bell and Braithwaite reasoned existed (Yoroi and Iwatsuki, 1977) [My report on the results attracted attention at the 1978 International Tropical Botany Symposium held at Aarhus University in Denmark (Iwatsuki, 1979)]. Our advantage was that we could closely observe materials growing in Japan. Taking this advantage and analyzing materials obtained from different areas that represented the variations, we proved that the species structure of *C. minutum* cannot be divided into two types. In this process, information collected from specimens played an important role. The point is that the attitude of the researcher determines what information can be observed on specimens. Through this research, we demonstrate the utility of herbarium specimens in elucidating reproductive forms and other traits that once were considered knowable only from living materials.

Additionally, research on the species structure of *C. minutum* has evolved to further incorporate ever-advancing analysis methods. Taxonomic groups with intraspecific variations that have crossbred frequently were elucidated, resulting in understanding the complexity of this species structure due to repeated differentiation and convergence (Nitta & al., 2011). Bell and Braithwaite believed that *C. minutum* could be discriminated into two types because they only dealt with a limited amount of living materials observed in the field. This problem can be corrected by using many specimens collected from various areas. Moreover, in diversity analyses, results obtained from particular materials of one species should not be universalized to the entirety of the species in question. This is a common mistake concerning genetic diversity.

Let us look at research on *Hymenasplenium* (Aspleniaceae) conducted by N. Murakami and others who used the same method as described above. He began to be deeply involved in research on *Hymenasplenium* when tackling his college thesis research by studying the correlation between the reproductive forms and phenotypic characteristics of *Asplenium hondoense*. At the time, while different colleges operated their respective programs differently, students in the departments of biology at the University of Tokyo and Kyoto University conducted research for graduation from college in the second semester of the fourth year. Under the current education system, the second semester starts in the autumn. During the second semester season, students who receive practical training in the biodiversity of wild plants have difficulty in collecting living materials. Of course, some appropriate living materials are available; however, for some research and in the case of some unavoidable conditions, it is necessary to conduct analyses in accordance with seasonal conditions. Murakami was supposed to work on the family Aspleniaceae due to his research theme at the graduate school he was to enroll. Consequently, the theme he selected was an analysis of the species diversity of *Hymenasplenium*, which he could pursue mostly using dried specimens.

Around 1980, *Asplenium hondoense* was recognized as one of the widespread species in the tropics of the old world. Of course, names had been given to identify several types; however, there was no organized set of information adopted by most researchers. As in the case of *Crepidomanes minutum*, an apogamic type of *A. hondoense* growing in Japan was identified through prothallia observation (Momose, 1967). Murakami first distinguished the reproductive forms of *A. hondoense* and examined their distributions within the species using a large stock of specimens. Statistically significant examination of results in phenotypic characteristics was obtained, which definitively distinguished sexual and apogamic types. Consequently, it was ascertained that the differentiation of reproductive forms had led to speciation.

Research on *A. hondoense* subsequently evolved. However, its development deviates from the context of this section (see 2.3). The point is that a pile of specimens can be used to predict organism-specific traits, such as differences in reproductive form. This book refrains from declaring that it is impossible to know living conditions from dead specimens. Instead, it stresses that while observation to determine the reproductive forms of *A. hondoense* throughout the world is tremendously laborious, organized specimens enable a fourth-year student to preliminarily determine different reproductive forms of variants from all over the globe, within a single winter semester.

That said, obtaining diverse information from specimens is not so easy. Determining reproductive forms using specimens is not technically difficult. However, the work requires manual dexterity, plus patience. The results of this work were impressive to a considerable number of researchers. However, few researchers outside of Japan determine apogamic types by this method.

Determination of specimens is intended only for predictive purposes and is not meant to be conclusive. I thought this method was applicable to leptosporangiate ferns, except in the case of several primitive families. However, contrary to my expectations, these materials persistently led to incoherent results. A reanalysis showed that the family Lindsaeaceae differed from other fern families in the spore formation process [sporogenesis]. Subsequently, research conducted to verify that in this family determination results should be read differently was valid (Lin, Kato and Iwatsuki, 1990). Thus, I encountered an example with results that could be added to the diversity formation process. Facts should be observed to devise a valid analysis method. For final conclusions, predictions made by a simple method must be proven by a more reliable method.

Most of natural history specimens are dead materials. However, they can provide various information about the characteristics of living organism, as shown in the above example. Moreover, for species that are widely distributed and produce diverse variants, materials can provide trait and morphological information for which living materials are not readily available. Researchers using such methods are expected to be sufficiently qualified to extract relevant information from specimens. Fossils are remains of dead living bodies. Fossil specimens allow researchers to elucidate various facts about living things once present on earth. Furthermore, advances in the DNA determination technique for materials extracted from fossil specimens have been achieved. Of course, what researchers can do is not limitless. Nevertheless, it should not be ignored that what can be extracted from specimens substantially depends on the ideas and the skills of the person who extracts the information.

Meanwhile, collecting living materials is not always easy. It is often not efficient to travel all over the world solely for determining a particular trait. When deciding where to collect materials in their living state, researchers rely on information available on specimens. Of course, databased information services enable other types of information searches, a topic in the following section discussing database compilation. Meanwhile, specimens validly function as a foundation to verify databased information.

In the above example, Murakami ultimately led an attempt to conduct comparative research of *Hymenasplenium* materials throughout the world by actually cultivating all identified species. Of known species, the last required one was growing in Brazil. In the autumn of 1992, I presented a keynote speech at an international botanical garden symposium held in Rio de Janeiro. On this occasion, I visited the specimen storage room at the Jardim Botânico do Rio de Janeiro to look into specimen data and found that the species in question was growing in a place tens of kilometers to the west of Rio. The distance was a one-day trip from Rio. As I had decided to stay one more day, I could obtain detailed information on the place of collection, gain cooperation from the staff of Jardim Botânico, and collect living plants of the species with ease. The place was not suitable for a foreigner to travel alone due to security reasons. The Botanical Garden provided me with safety arrangements, which was most appreciated. This is another example of the utility of information from specimens for carrying out necessary research tasks. Many researchers in this field use information from specimens to obtain details about the locations of living plants.

To build a general picture of the current state of biodiversity on earth through basic research, it is essential to conduct field research in every part of the world. However, not every task can be finished on site in the habitat. It is necessary to bring collected specimens to research institutes with organized materials for careful comparisons and determinations to elucidate their current state. Of course, information on traits represented by specimens is limited. Common-sense ways of scientific analysis are to know a general picture based on specimens and to analyze, in detail, issues arising from the general picture, addressing each issue appropriately, and at the same time, obtaining living materials on an as-needed basis.

For this purpose, a regional or global revision for each particular taxonomic group (an overview of a particular taxonomic group, listing all relevant species growing in a target area and describing their characteristics and species differences based on all existing information) is compiled, with a record of fundamental specimens. To review the compilation results, previously studied specimens are reexamined on an as-needed basis, and the reexamination results are checked against new information to repeatedly revise and update findings.

Compiled, these results have built current scientific knowledge about the types and locations of living organisms present on earth. Needless to say, a final conclusion is yet to come. Research is constantly underway. As an important part of this ongoing process, all specimens used for research need to be managed and maintained as fundamental information. Valuable specimens must be managed with responsibility. Therefore, use of materials is subject to specific rules to avoid damage caused by people unfamiliar with the handling of specimens.

Research based on field surveys and preserved specimens is strongly oriented towards building a general picture of biodiversity. It is an integrative type of research that combines findings from individual researchers. Various issues arising from this process are not solved simply by integration. Individual issues need to be analyzed deeply in an investigative manner. In this way, findings are compiled from a fresh and integrative perspective and can be utilized for the next revision of global biodiversity knowledge to continue to contribute to improved human scientific knowledge.

(8) Nomenclatorial type specimens

When understanding specimens in natural history, specifically in biology, it is important to recognize that specimens in collection include nomenclatorial type specimens.

Scientific names of organisms are used commonly as conventions throughout the world. Since the early 20th century, international nomenclatures have been progressively established for animals and plants, independently and concurrently. These are written codes. However, as a scientific area, nomenclatures are also moral rules that lack penalties. Although internationally recognized, they are not bound by treaty or international law. Nonetheless, nowadays, scientists throughout the world follow the code of nomenclature. Names in violation are ignored by most scientists and will disappear.

According to the nomenclature, the specific name of a taxonomic group, such as a species or a subspecies of living organisms, is given to a particular specimen. All individual bodies in the same taxonomic group as the particular specimen are referred to by the specific name (type method).

The species that I first described as new is *Abrodictyum boninense* (Iwatsuki, 1957) in an abstract of my master's thesis. This species is a peculiar fern in the family Hymenophyllaceae living on the surface of the trunk of a *Cyathea spinulosa*. Before I published, it was considered as a form of *Abrodictyum cumingii* widely distributed in the tropical zone of Southeast Asia and was commonly referred to by this scientific name. With *A. cumingii*, leaf cells are orderly arranged into a long ellipse, composed of, at the most, two lines with short, vein-looking cell walls (most species in the family Hymenophyllaceae have leaf cells arranged into one layer). However, close observation reveals that only the form growing on the Bonin Islands, unlike examples from other habitats, has leaf cells not orderly arranged, although three to four lines of leaf cells are formed roughly into a long ellipse. Despite morphological similarity between these plants, the Bonin Islands form definitively differed from *A. cumingii* in leaf cell shape and arrangement, which are the features of this species. There was no doubt that the Bonin Islands form had specialized. Therefore, I recognized the Bonin Islands form as a new species, using trait specialization as an indicator, the basis of *Abrodictyum* genus recognition.

Abrodictyum cumingii is a species that Presl described based on Cuming's collection from the Philippines. Its holotype represents the types widely distributed in Southeast Asia. Therefore, it was necessary to give a new name to the Bonin Islands form. (Although I did not have a chance to see the holotype of *A. cumingii* when I described the new species *A. boninense*, I examined and determined the real thing later in life.) I designated the most appropriate specimen among those in the collection of the herbarium at Kyoto University as a type specimen. Consequently, plants known as *A. boninense* are those identified as identical to this type specimen. As of now, the forms widely distributed in Southeast Asia are *A. cumingii*, while only the Bonin Islands form is identified with *A. boninense*. Conversely, all Bonin Islands specimens are identified as *A. boninense*.

As an aside, the sheet used to designate the type specimen has nine bodies affixed to it. The nomenclature sets out that all plants on the sheet used to designate the type specimen are recognized as holotypes. Consequently, in this case, the nine plants serve as holotypes. In cases where several individuals are affixed on one sheet, it cannot be said that the specimens are completely free from the error mixing different species. If these specimens are designated as holotypes, the individual body that was given the name must be determined. In the case of *A. boninense*, the largest body plant at the center bottom was photographed and appeared in the original paper. In many such cases, a specimen is recognized as the holotype in the narrowest sense, assuming that the intent of the original author was unambiguous. That said, in some instances, trait descriptions disagree with the photographed specimen. Use of the type method does not always ensure that scientific names are given mechanically and uniformly. Because of this, naming remains subject to difficult problems.

The above-mentioned practice enables the researcher to automatically determine to which form the original species name was given, even when an early broad scope of species is later divided into several species according to subsequent research results. For example, in Japan, when the species referred to as *Asplenium unilaterale* was distinguished into two species using reproductive form and other characters as indicators, it was necessary to determine which form should be given the original scientific name without confusion. The original scientific name *Asplenium unilaterale* Lamarck was publicized in the 'Encyclopédie' (extended edition) by Lamarck (the scientist famous for use/disuse in his evolutionary theory) based on specimens collected from the Seychelles in the Indian Ocean. Accordingly, the form growing on the Seychelles should be called by this scientific name. The apogamic type distinguished from it was given the new scientific name, *Asplenium hondoense* Murakami & Hatanaka.

It is a general rule that only one holotype specimen exists. Two or more holotypes can result in confusion. Conversely, however, the one holotype might be accidentally lost. In many instances, holotypes have been lost due to the chaos of war. As a partial solution for such cases, substitute type specimens may be provided. The nomenclature sets out useful suggestions on how to prepare substitutions (neotypes) in the event of loss. Understanding isotypes differs between animals and plants (with different codes of nomenclature formulated for animals and plants). However, this book refrains from providing a detailed explanation about isotypes.

Type specimens are valuable materials intended to maintain the correct names of living organisms, valid throughout the world. For the growth of a living organism database and the bioinformatics that uses it, the use of correct names given to diverse living organisms is the minimum fundamental requirement, as described later in this book. Type specimens essential for this purpose are, in many cases, together in collections with other specimens for research convenience. Indeed, these specimens are meaningless unless they are used for research. Accordingly, they are managed to ensure both convenience in usage and conservation as precious materials. The fact that natural history specimens include these precious materials is understood only by a fraction of researchers, which is a matter of some concern.

To underscore the point, a nomenclatural type specimen does not imply that it is ideal representative specimen of the species. It is the specimen used to name the species for the first time. Consequently, it may be an extreme variant of the species. Some researchers appear to misunderstand that a type specimen shows the nature of the species. The term *type* in this context is limited for nomenclatural purposes. It is necessary to have a correct understanding of this reality.

(9) Natural history specimens as environmental indicators

Since the exacerbation of human impact on the natural environment, and specifically the beginning of the use of endangered species as models to track the dynamic state of biodiversity, appreciation of the value of specimens as an environmental indicator has greatly increased.

While records of fauna and flora have survived as literal information, locally produced records may contain incorrectly identified species. It is not uncommon that old records in their original condition may not be useful due to changes in species recognition. However, preserved specimens in the collection of a museum or similar institution enable researchers to reexamine the identification of species and to confirm when and where a particular species occurred. Of course, as with fossil records, it is not possible to reexamine all information about the past. Nonetheless, contrary to what one might think, rare species facing extinction are often on record.

To record the dynamic state of future biodiversity, specimens play a significant role. In particular, specimens of endangered species, as well as species in their living state (ex situ conservation), are desirable as research materials that may be required in the future. Present-day science is incapable of conserving all endangered species in their living state in a facility.

(10) Storing, managing, maintaining and researching natural history specimens

Specimens are expected to represent individual bodies and species. Therefore, when collecting specimens, researchers should select representative individuals and species. In practice, specimens are not intentionally selected materials that represent individuals or species. Conversely, they are non-selectively collected materials to some degree, a fact which has reasonable significance. As such, when studying them, it is essential to compile data with the character of the specimens in mind.

As an organizational characteristic, museums and other similar facilities in the domain of natural history store specimens and study them. The collection, organization, conservation, and management of specimens are all vitally significant in developing a foundation of research in this field. The preservation of materials as specimens used for primary research guarantees that results will be reviewed, deepened, and improved.

Historically, precious specimens were often preserved with a detailed chronicle of their collection. For example, specimens from renowned collectors bear information, such as the locality of collection. Commonly, museums or similar facilities were founded primarily for storing these types of prominent collections. In Japan, the Yamashina Institute for Ornithology is a research institute whose founding was based on the collection of the late Dr. YAMASHINA Yoshimaro (1900–1989). Needless to say, the institute currently has a broader range of collections and runs diverse research projects. The Makino Herbarium at Tokyo Metropolitan University, an institute founded based on the collection of MAKINO Tomitarō (1862–1957), is currently active in generally collecting plant specimens.

Specimens cited in a treatise, or authenticated specimens, in addition to the above type specimens, serve as naming records and data from prior studies that can be used as materials for reexamination. For relevant facilities these are important specimens that they are obligated to preserve. Of course, unstudied specimens might retain unknown features and are therefore also precious research materials.

Museums and other similar facilities that preserve a large number of specimens must implement careful management. That said, specimens are not always used on a daily basis. More than a few specimens remain in unused stacks for years. During 1969 and 1970, I worked on research at the

Natural History Museum, London and the Royal Botanical Gardens, Kew. As materials for this research, I asked these facilities to take from storage specimens collected in Thailand more than 50 years ago. I identified a large number of specimens, which served as precious materials for our research project on the flora of Thailand. When I later visited these herbariums, the specimens I identified were organized in storage with covers for their respective species. Until identification, specimens in storage wait for a researcher, bearing only a label with their collection information. It took half a century before the aforementioned specimens became meaningful as research materials. In challenging competition to become a pioneer in the world of science, some may think that the storage of unused materials for half a century is very wasteful. Nevertheless, researchers in natural history often need to be prepared for such time lags.

Nature in present-day Thailand has been far more affected by development than at the time of the collection of the Thai specimens that I studied in the United Kingdom. Take Thailand's highest mountain, Doi Inthanon, as an example. When I and other researchers conducted our first collecting trip on this mountain during 1965 and 1967, it took one week to walk from the Mae Klang waterfalls to the summit. Still today, I clearly recall that we were fascinated by the wonderful beauty of tropical evergreen forests at the time. In contrast, nowadays, a road leads up to the summit of Doi Inthanon, which allows people to travel from the downtown area of Chiang Mai to the summit within about two hours by car. The vegetation has dramatically changed, as well. I am not sure whether it is possible today to collect the same species that we collected in our first trip.

In the 1990s, our research team had the opportunity to conduct five surveys in areas covering the Sa Pa mountains in Northern Vietnam to the Mekong Delta in the south (I participated in four of them). In these surveys, we were unable to collect all the fern species described in the 'Flore général de l'Indo-Chine' 7-2, which was published between 1939 and 1951. Interestingly, our observations also led to some new findings that had not been elucidated until then.

These facts indicate that the natural environment is being rapidly transformed by human action on a global scale. The need to track this transformation in comparison to historical collections is an issue to be raised by present-day natural history. Database compilation of specimens, with tracking and inference based on the optimal use of the data, may effectively serve to contribute directly to solving social challenges like this. This will be discussed in the following section.

(11) Studying living things – universal principles and diverse appearances of life

Museums and other similar facilities have specimens in possession primarily for research purposes. Specimens are essential materials for natural history research. It cannot be said that the presence of specimens ensures sufficiency (sufficient condition). However, without specimens, biodiversity research cannot be conducted (necessary condition).

It has become common to analyze model organisms to accurately know the particular characters and phenomena they exhibit on the presumption that living organisms have diverse features. To do so, materials must be free of variation, also in terms of genetic configuration. A specific line is used, noting genetic diversity, on top of ensuring the use of materials of the same species. To deepen the analyses that use these materials, how the obtained information is universalized and generalized for the species in question, and for living organisms in general, depends on knowledge of biodiversity.

The above procedure is essential for the analysis of life phenomena. The research method required for this process differs from that required to obtain findings on items of uniform quality, such as mechanical products.

Like the theme of a fugue, this book repeatedly states that only collaboration between investigative and integrative research is the high road to explore what it is to be alive. For species diversity research along this exploration, natural history specimens should be recognized as imperative materials. Expectations are high for natural history to make substantial contributions in this regard.

3.3 Natural History and Bioinformatics

(1) Biodiversity informatics

While natural history is in the title, this section first focuses on living things as a concrete topic, rather than on the integration between general natural history and information science.

The term *bioinformatics* is a composite word compiled from *bio* and *informatics*. The term *bio* means living thing, as used in the term *biology*. The term *informatics* means information science.

The term *bioinformatics* refers to all the domains where life phenomena are tracked by information science methods. Nonetheless, it tends to be viewed particularly as research in the domain of molecular biology. This is because of the backdrop that, in practice, bioinformatics research has progressed in combination with genetic information as molecular biology evolved. In light of scientific inevitability, there are high expectations, possibly essential expectations, for the domain of biodiversity to advance information from science-based analyses.

Numerical taxonomy is a method that was used in a certain period of time to objectivize the taxonomic characteristics evaluation by processing large amounts of numerical data. Meanwhile, for more fundamental phyletic evolution research, analyses incorporating bioinformatics methods may play a more basic role.

Theory of Evolution and Evolutionary Biology Research on biological evolution was initially recognized as a theory, as in the theory of evolution. Subsequently, through verification by tangible facts, the evolutionary theory became acknowledged as evolutionary biology. Incidentally, nowadays, more than a few colleges have evolutionary biology as a subject in their curricula.

At the root of the above transition was an increasing level of validation of past events as scientific facts, on top of the theoretical demonstration of evolution. This occurred due to increasing the number of examples of the empirically-proven reality of evolution, resulting from mounting fossil evidence and advances in fossil-based analysis methods. However, the evidence was not limitless, as obtainable fossils survive by chance. Therefore, it is not possible to reproduce the entire history of past through the study of fossils. Another limitation is that the discipline's growth depends on the quantity and quality of fossils that happen to be discovered.

Naturally, the next step is to look for a scientific method to infer past events based on current facts. First, it is preferable to explore the possibility of looking at historical facts on a material basis. In this regard, a technique that has come into use in molecular phylogeny points to this sort of desired method.

Living things transfer life from parents to offspring under the controls of genetic information. When genetic information manifests itself during the embryology process, phenotypic characters develop, with genetic information later modified by the environment. This is the exquisite formula of nature by which biological evolution has been crafted. In molecular phylogeny, researchers compare basic sequences of nucleic acids that convey genetic information, estimate genetic distances from differences known by the comparison, and infer lineage links between extant living organisms. The next step is to formulate a difference inference method that incorporates a feature designed to compile every piece of information known to date and infer lineage according to the formulated method. Subsequently, when inter-line distances and relationships can be traced, it should be possible to infer emergent trait variations by comparing taxonomic characteristics that led to the differences between phyletic groups.

In practice, methods of scientifically predicting phenomena believed to occur in the future based on currently available information have advanced in the field of weather forecasting. Prediction of a future phenomenon is subject to unknown, perhaps major, changes under unforeseeable conditions. Nevertheless, attempts are made to make long-term forecasts months ahead of time, in addition to predictions of tomorrow or next week's weather. Moreover, forecasts are becoming increasingly accurate. For weather forecasting, detailed meteorological information on current weather is input into a supercomputer, based on which future developmental changes with time are predicted. If all meteorological conditions are input, it is possible to forecast weather at the next instant with a high degree of certainty. Indeed, to meet this prerequisite, meteorological conditions for weather forecasting purposes are imperatively recorded using extremely detailed measurements.

Forecasting for long-term climate change is also under way as an extension of weather forecasting. This topic in the context of global warming has become a primary focus for global environmental issues. However, when it comes to forecasting global climate change, some researchers still dissent from numerical forecasts made by international organizations, such as the Intergovernmental Panel on Climate Change (IPCC). Even so, scientists are obliged to make accurate forecasts as it is their responsibility to explain how people's action today will affect the future earth and to verify the degree of forecast accuracy. Because of this, it is necessary to predict not only natural phenomena, but also future human activities as essential data. Consequently, forecasting is difficult due to the inclusion of information on human's degree of self-restraint.

The same is true of global environmental change brought about by biodiversity. For this reason, scientists should explore to what extent it is possible to know the past and predict the future by collecting biodiversity information.

Prediction of the influence of biodiversity on people's lives is important as a practical challenge closely related to our lifestyle. For ease of understanding, look at this challenge from the perspective of science for society. Compilation of biodiversity information stimulates scientific curiosity, as a purely scientific challenge. But it is also inherently relevant as a scientific challenge that will bring results to help meet social demands.

Some scientists hold fast to their opinion that lineage will remain unknowable by natural science analysis methods, reasoning that it is impossible to fully trace past events. Indeed, more than a few of these skeptics are even among researchers studying phylogenetics. Certainly, as they say, it is impossible to fully and empirically reproduce past events, unless a time machine exists.

Nevertheless, it is possible to accurately reproduce events of the immediate past to a reasonable extent by exploiting all kinds of records. This implies a need for an as accurate as possible understanding of currently available information. Use of this as a basis for inference can enable a researcher to reproduce past events, or at least to infer possible profiles, to some degree of accuracy. If the future can be predicted, it should be possible with even greater certainty to trace back the past.

Tracing Phylons Scientifically What methods exist to trace lineage scientifically? Until the 19th century, lineage could only be inferred by comparing taxonomic characteristics, which were primarily comprised of morphology. In this comparison, theoretical tracing to investigate the nature of form was pursued, as in the evaluation method used to contrast taxonomic characteristics with the ontogenic process of morphologic characteristics. However, this sort of theoretical tracing could not deliver scientific evidence. Naturally, this approach largely depended on the intuition of the researcher, based on their wealth of experience. Nevertheless, researchers using this method lacked the means to verify whether their intuition-based inferences were correct.

In the first half of the 20th century, studying the relationships between chromosomes and genes was suggested as an approach to provide evidence useful for tracing lineages, and chromosomes began to be used as an indicator to trace phylogeny. After the mid-20th century, DNA became the focus of attention as the carrier of genetic information, and methods of biological analysis of this macromolecule advanced dramatically. Along with advances in DNA research, the neutral theory of molecular evolution was advocated (Kimura, 1968 and 1983). This led to a remarkable research progress in terms both of theory and mass data storage to establish molecular phylogeny.

In parallel with the research progress in this domain, there were rising calls for a global database of biodiversity information and the establishment of bioinformatics based in such a database. Meanwhile, there were hopes for using this information to meet social demands in the context of the environmental issues mentioned earlier. This progress moreover stimulated the scientific curiosity of some researchers as to what extent the origin and development of organism biodiversity could be verified.

Genetic information imparted by parents manifests itself during embryogenetic stages and growth processes through interactions with the environment, resulting in the features of adult bodies. The relationship between genetic information and features of adult bodies becomes clear if a researcher accurately understands genetic information, completely elucidates how it reacts with the environment, and details how it links with embryogenesis and growth during the process of growing an adult. The capabilities of present-day science are far from elucidating a general picture of the potential of genetic information. Researchers have only elucidated a small part of the reaction of how the genetic information manifests adult features according to the environment.

Underpinned by the rapid growth of analysis technology, research is in progress to trace how the manifestation of characteristics induced by genes has changed during the evolutionary process. This research has become referred to as evo-devo (a term coined by combining the initial parts of the words *evolution* and *development*). Advances in this field should be described as typical in the investigative domain of natural history. A later section [3.3 (3)] will refer to evo-devo.

The difficulty with demonstrating lineage is not that there is no solution, but that when the necessary information is collected, there will be infinite probabilities of inter-line relationships. Rather than saying that asserting the unknowability of lineage will simply lead to an argument without substance, I would say that driven by scientific curiosity, people naturally explore possibilities if they are present. Indeed, scientists are currently taking steps towards the dream of empirically tracing lineages.

International Organization for Plant Information (IOPI) and 'Species Plantarum'

I have taken part in several international projects, believing in the need to promote bioinformatics research. Among others, I actively participated in the International Organization for Plant Information (IOPI).

To organize and computerize information on a large number of stored plant specimens and to offer the information as a convenient database for international use, major herbariums in Europe and America discussed the founding of a liaison organization. This was the beginning of IOPI. They began to discuss the formation of an international organization and an informational publication, such as one on global flora, based on botanical data collected by the organization.

In November 1990, Ghillean T. Prance, then the director of the Royal Botanical Gardens, Kew, and Peter H. Raven, then the director of the Missouri Botanical Garden, promoted a meeting at Kew to discuss the founding of IOPI. These two institutions took the hegemony in the move towards IOPI.

At the time in Japan, the Makino Herbarium at Tokyo Metropolitan University was working on an advanced project of compiling a plant specimen database. However, perhaps the promoters did not contact the Makino Herbarium due to a lack of information, or perhaps the Herbarium was unwilling to take part. In any case, I was the only Japanese participant at the meeting. (Around the time, there were continuously very few Japanese researchers participating in natural history-related international meetings.)

Invited by the two promoters, I decided to attend the meeting. Fortunately, immediately before learning of the meeting I was staying in Paris and could adjust my return trip schedule to make time for it. At the time, as the director-general of the XIV International Botanical Congress (IBC) to be held in 1993, I endeavored to attend as many meetings of botanists as practicable with the purpose of public relations for the Congress. At the meeting, I took the chance to inform the participants of the progress in preparation for IBC, having made, in advance, a request to do so to the two promoters.

The international voluntary association IOPI started successfully as an academic association. I was designated and joined as one of the founding board members. News of the meeting was aired in NHK's morning TV interview program immediately after my return to Japan. I strove to disseminate the information in the academic community. However, I failed to start a groundswell among Japanese researchers to actively participate in the IOPI. Moreover, their motivation was also hindered by a failure to raise funds for that purpose. Nevertheless, I was sympathetic towards the project and attended board meetings and other events thereafter to the greatest extent possible.

The plan to organize records of flora in electronic format was discussed by the organization. However, in practice, the project progressed very slowly at that time. A large number of the participating members were enthusiastically willing to compile records of global flora in print. The subcommittee set up for the compilation of flora records was ahead of other committees in their activity. Australia, willing to cooperate in this kind of international project, undertook editing 'Species Plantarum: Flora of the World.' In 1999, an introduction to the editing policy was published. Since that time, compilations of 11 families were published, and the global project seemed to have taken off favorably. However, it soon ran out of funds. Moreover, it became impossible to continue the project in Australia, the location of the editing office. As a result, the project has been at an impasse since several years after the inauguration.

There is another reason that IOPI focused its efforts on 'Species Plantarum' and failed to achieve continuous growth even in the publication. In 2001, the Global Biodiversity Information Facility (GBIF) was inaugurated, which affected IOPI projects. GBIF is described below.

(2) Global Biodiversity Information Facility (GBIF) – Biodiversity Research and Its Social Contribution

In recent years, areas of natural history that study living things have tended to strongly be linked to the category of biodiversity. Sustainable use of biodiversity has become a social challenge, and the Convention on Biological Diversity and other international treaties have been executed in this regard. The Conference of the Parties (COP) is held every two years. Through these activities, the dynamic state of biodiversity and its future condition are discussed in relation to humanity, and projects aligned with the discussion are implemented. Consequently, scientific research on biodiversity is needed as basic information for these discussions. Science is always an activity driven by human scientific curiosity. On the assumption that the achievements of science can greatly contribute to human life, our hopes rest on promoting science as a foundation for this social contribution.

Around the end of the 20th century after the Convention on Biological Diversity came into effect, the Megascience Forum, which was set up as part of the Organization for Economic Cooperation and Development (OECD), discussed the organization of biodiversity information. In 1998, the Forum wrote up a recommendation encouraging the compilation of a global biodiversity-related database. Japan expressed its willingness to support this proposition and committed itself to the same funding contribution as the United States.

In 2001, the founding meeting of GBIF was held in Ottawa, Canada. It was decided that an organization would be formed, and its projects would be promoted for a trial period of three years, followed by assessment by other international organizations as to whether or not to deploy its activities further.

The Structure of GBIF By the end of the 20th century, biodiversity information was recognized as important, and various arguments were made as rationale for this assertion. Not only personal research projects were pursued, but academic associations were formed and activities under them were concurrently organized. Several voluntary international organizations also began biodiversity projects. The aforementioned IOPI was one of such organizations beginning to produce concrete results.

At this time, activities relating to the processing of biodiversity information were all pursued by voluntary organizations, or NGOs. Academic associations generally take the form of an NGO, except in particular nations. In contrast, GBIF is an international organization operating on funds paid by governments who signed an agreement. GBIF started with funding from 13 countries, principally including OECD members, which became voting participants of the Governing Board. Initially it was decided that Japan and the United States would pay 700,000 dollars every year, and Germany, 500,000 dollars.

The organization of GBIF is somewhat anomalous. In addition to the voting participants, associate country participants that do not pay funds and associate members, such as international organizations and academic associations that have expertise, take part in the organization's activities. [Incidentally, the International Union for Conservation of Nature (IUCN) had already been active in a similar fashion.] The Governing Board, in principle, continues discussions until it reaches a unanimous conclusion. At Governing Board meetings, associate country participants and associate member organizations have a voice on a par with the voting participants. In practice, nothing is decided by majority vote, except for personnel affairs. NGOs also participate in the discussion on an equal footing with governments. This is inevitable at meetings like this that are held based on scientific facts. In discussions about concrete activities, opinions of delegates from academic associations led by proponents of the work in this field are often stronger than government delegates, who are subject to periodic personnel relocation. Due to this characteristic, GBIF's activities are reasonably well organized.

Activities of GBIF From the perspective of natural history, this book briefly describes what GBIF is doing in practice. For official information, refer to the GBIF website (<http://www.gbif.org/>, the secretariat located in Denmark provides it in English).

The effort to establish an international organization like GBIF arose due to the reality that while there were many voluntary organizations individually exploring ways to build and use a biodiversity database, no organization competent enough to integrate the overall data into one database on a global scale existed. At the same time, the provisioning of biodiversity information was understood to be an urgent social challenge as it would serve a foundation for international projects in the field. Consequently, the challenges that GBIF needed to address were to integrate existing databases in accordance with a uniform format on a global scale, organize information worldwide, and explore database utilization methods accordingly.

For database integration, it was faster to discuss how to facilitate the provisioning of existing databases than to decide on the ultimate format in advance, as converting provided data into a specific format was relatively easy. After the startup, because project achievements would be evaluated in the third year, it was necessary to establish a substantial track record two and a half years before a committee would begin evaluating the projects. However, the development of an overall GBIF management structure, with subordinate units under the Governing Board, and the placement of a secretariat would take a considerable amount of time prior to the implementation of concrete action in a workable framework. The members of the Governing Board were unable to gather frequently because they were delegates from the governments of funding countries.

Thus, GBIF somehow took off even with this temporal restriction. Not every associate member NGO was willing to provide data, although they actively made statements to act on their own initiative. For GBIF, the immediate task was to organize its own data. GBIF first asked research institutes in participant countries to file their existing data. In this way, data digitization progressed, and GBIF somehow collected data from more than 200 million entries by the project evaluation in the third year.

Although it was intended to collect all kinds of data relating to biodiversity, the data quickly available by the above-described method were mainly those on specimens in storage at museums, in addition to observation data. In particular, major museums in Europe and America had computerized specimens to a considerable extent. In addition, there was a growing trend to strive for database enrichment, in tandem with the establishment of GBIF.

Notwithstanding the achievement of these reasonable results, the data digitization was not free of problems, which persist today. First, the quality of the digitized data is a matter of concern, as data quality is the responsibility of those who digitize the data. In many cases, the work is the simple mechanical computerization of information from specimen labels. Information deficiency, if any, is not addressed, and it is unlikely that the data undergo re-evaluation. Consequently, while some data are rich, the quality of other data does remain questionable. They form an indiscriminate mixture. However, it is difficult to provide a framework for re-evaluating the provided data. The reality is that data quality depends unilaterally on the morals of the person or organization who digitizes and provides the data. This is true not only with specimen data, but also with the accuracy of observation data.

While data digitization is an urgent task, another important challenge is to ask society how the obtained data can be used. Nowadays, making the beneficial utility of the data visible, as a return on funds invested, is an urgent challenge. For natural history, a reasonable amount of energy has been poured into constructing databases. Nonetheless, examples of readily visible social contributions made through the use of these organized databases are few. Of course, in the extended scope of natural history in general, obvious examples of social contributions include improved accuracy in weather forecasting and advanced exploration in earthquake prediction. However, when it comes to biodiversity-related information, there is no example of a brilliant contribution through use of the data.

GBIF has actively tried to explore what facts can be inferred and what benefits can be derived from the obtained data. However, in terms of database usage, the demonstration of tangible results to society has not been successful. Biodiversity information covers myriad types of data. The specificity of diverse components (ranging from species to individual bodies and from particular cells to diverse ecosystems) concerns the true nature of biodiversity. Therefore, use of limited amounts of data enables researchers to make scientific inferences only to a limited extent. This has led to poor results achieved by the use of databases, despite the active efforts to construct them. There is no denying that insiders have made excuses that limited amounts of data, if used, cannot produce results. However, because it was organized as a large international project, GBIF is not allowed to make such excuses.

In very recent years, research results based on the GBIF data have been relatively frequently published. While these research projects have been implemented mainly on a trial basis, a prospect exists that excellent research results will be achieved using the GBIF data.

My Participation in GBIF Japanese government nominated me to serve the GBIF as a founding officer, and I was elected deputy chair of the governing board, for which I served in the first two terms.

At the founding meeting, GBIF Ebbe Nielsen Prize was conceived to honor young researchers contributing to biodiversity informatics. Dr. YTOW Nozomi at the University of Tsukuba was elected as the first winner of the prize.

As an aside, as the deputy chair, I was first assigned to work as the chair of the field research committee, tasked with selecting a place, from four candidate countries, to locate the secretariat. There were three members on this committee, Dr. Simon Tillier from France, Dr. Peter Arzberger from the United States, and myself. In a forced march arranged to suit my schedule at that time, we hastily traveled to the four candidate cities of Amsterdam, Copenhagen, Madrid, and Canberra during Japan's holiday-studded Golden Week. At a social gathering held in one country, we were asked why a botanist was the chair of the committee, while a zoologist would normally take charge of this sort of mission.

As a result, Copenhagen was chosen as the secretariat location. I felt this was a fair conclusion, and I have appreciated the strong support thereafter provided by Denmark. Nevertheless, I wondered how the results of our preliminary research influenced the voting of government delegates, a question I often have in mind at international conferences.

Although I helped form the organization and start its activities, after the GBIF began, I frequently experienced difficulties in following the real programs of the promoted projects.

I had the idea that a project for facilitating the compilation and utilization of a biodiversity database was a current, urgent and important research challenge, in terms both of natural history study and its social contribution as the foundation of biodiversity informatics. I was worried about the reality that there were only a limited number of researchers, specifically in Japan, who were willing to focus their efforts on the above-described project. Therefore, to attempt to contribute to this project within my capability, I assisted IOPI and took part in starting up GBIF.

However, I was neither familiar with informatics, nor had any experience in devoting myself to research in that field. Therefore, it was natural that I encountered topics that I could not understand during discussions on technical issues. However, once involved in the management of these projects, I could not withdraw early. In anticipation of the emergence of individuals familiar with informatics and aware of issues concerning this challenge in Japan, I was involved in the Governing Board for two terms.

As a result, against my initial expectations, I completed my two-term tenure in the Governing Board under some influence from the attitude of the Japanese government. In 2005, having passed the board tasks to the next generation, I retired from my role in the GBIF Governing Board. However, this does not mean I have come to neglect biodiversity informatics. How I take part in it as a researcher is another story.

GBIF and Natural History in Japan At first, Japan intended to participate actively in GBIF. However, subsequent policy changes prevented its contributions to GBIF from increasing. As a result, Japan failed to meet its assigned share of contributions due to budget limitations and ceded its position as a voting participant. However, Japan continues to participate in the activity of GBIF as an associate country participant, paying funds to the greatest extent possible and still making second largest financial contributions after the United States.

Needless to say, Japan participates in GBIF beyond simply funding. Since the inauguration of GBIF, gaining understanding from related agencies and policy board members, subsidies have been granted to this field, although the amount is very small. Accordingly, natural history researchers have become willing to take part in a database construction.

Early on, some researchers were interested in this kind of activity. However, at small museums, administrators tended not to welcome it because the results seemed to make no direct contributions to their institutions. In tandem with the initiation of GBIF, researchers affiliated with small and medium-sized museums increasingly provided databases compiled through individual personal efforts, reasoning that GBIF was a government-supported international cooperation project. Subsequently, data provisioning from Japan gradually increased from initially very small contributions to considerable amounts. Consequently, data from Japan has made valuable contributions to the GBIF database, as most of them are of high quality.

On the other hand, larger institutions tended to be slower in compiling databases of their materials and more unwilling to provide their data. If anything, they should be sorry that their levels of contributions remain low.

Since its establishment years ago, GBIF has accumulated a considerable amount of data. Research papers based on them have been published at a favorable pace. While the amount of data filed from Japan has steadily increased, Japan has still been slow in terms of its use of the data. Excellent bioinformatics research reports drawing on the accumulated data are anticipated in the future, as such reports will help the significance of research in this field become widely known.

The study of evolution elucidates not only biodiversity, but also the very evolution of earth, which natural history is truly meant to study. Analyses conducted for this purpose trace the truth by making the most of the obtained information and their interpretations in an integrative manner. To finalize analyses from an objective point of view, it is essential to apply informatics methods.

Weather forecasting at its dawn tended to rely on manual labor, with almost poorer results than intuitive forecasting. (At the time, forecasts by elderly people were relatively highly accurate.) Nowadays, data processing devices have rapidly evolved, and science for making estimates has noticeably advanced. Making full use of large amounts of data, forecasts have rapidly improved in accuracy. For biodiversity informatics, it is essential to promote research activities using the stored data, while, at the same time, having high expectations for further data organization. Claiming that no satisfactory treatise can be produced because of the less-than-necessary amount of accumulated data is only an excuse. Using available data, leading-edge research should and can be conducted right now. Given the considerable amount of data in storage and rapid advances in peripheral devices and statistics, promoting bioinformatics research using these data and based on tangible specimens and observations is one of the most fascinating contemporary challenges in natural history.

(3) Evolutionary developmental biology (evo-devo)

Evolutionary developmental biology is also known by the punning term *evo-devo*, coined from *evolution* and *development*.

Evolution was introduced to the field of biology originally as the theory of evolution. As natural scientists, researchers in biology expected evidence for the phenomenon of evolution. However, phenomena that involve a time series, requiring millions of years in the case of species differentiation, were difficult to prove by experiments. Nonetheless, since gene analysis techniques became available, researchers have attempted to prove evolution in a hypothesis-testing manner.

Meanwhile, initially, embryology was a research field that studied the initial stages of the life history of individuals, from fertilized egg to adult. Modern biology has successfully achieved results in empirical research, focused on specific events observed in life histories. However, the scientific curiosity of researchers in this field seems to have been stimulated by the implication that changes occur over time in individual bodies. When it became technically possible, they began to analyze the entirety of an individual body, by using the body organization.

It was once hopefully thought that ontogeny could recapitulate phylogeny, as Haeckel advocated in his recapitulation theory, to ascertain the truth of evolution. However, the clarification of the role of genes has revealed that for genes, and the proteins they produce, to manifest as characteristics despite similarities within themselves, homeobox genes must emerge with the capability to control the works of other genes during the zygote stage of development. In this way, the diversity of living bodies is substantially affected by how the homeobox genes work to operate genetic switches.

Molecular biology that deals with genes has experienced amazing technical advances. It has even become possible to color the proteins of genes with this switching capability to tangibly photograph how they actually function within a living body. This technique makes embryogenesis visible.

Using the advances in analysis technology as described above, evo-devo researchers elucidate genetic similarities and differences with the goal of tracing lineages by analyzing how living things have diversified according to the manner of control acting in the expression of phenotypic characters. Some closely related lines exhibit significantly different forms, while some other remote lines are similar in phenotypic character. In many of these cases, the mechanism to control the development of characteristics plays a role, rather than direct genetic differences.

Molecular phylogeny methods have enabled researchers to improve the demonstrability of research that traces distances between lines based on genetic similarities. Large amounts of research data have been not only accumulated, but also used to make advances in lineage inference methods. The application of evo-devo methods has improved the demonstrability of research to trace evolution.

Research in the domain of evo-devo has been vigorously conducted on animals. Examples are presented in the Natural History series by Kuratani (2004 and 2017). This section will describe research examples for plants. It must be borne in mind that there are developmental pattern differences between animals [triproblastica] and plants [vascular plants]. In the differentiation and growth processes of metazoa, a fertilized egg undergoes a sequence of cell division, blastulation, and gastrulation. In contrast, [vascular] plants retain embryonic properties at the shoot and root apexes. Even within an individual that is thousands of years old, like Jōmon Sugi (*Cryptomeria* tree [on the Island of Yakushima]), shoot and root apexes exhibit the same phenomena as in the initial developmental period. In the domain of evo-devo, the term *body plan* is used instead of the traditional term *organization* to refer to how bodies of metazoa are arranged. In recent years, this term has also come into use to explain plant structures.

The ABC model of flower development is an example of research on plant morphogenesis, which received the 1997 International Prize for Biology. Groups of Elliot M. Meyerowitz (1938–), an American biologist, and others and Enrico S. Coen (1957–), a British botanist, and others jointly published their research on *Arabidopsis thaliana* and *Antirrhinum majus* (Coen & Meyerowitz, 1991).

A flower has outer perianth segments (= sepals), inner perianth segments (= petals), stamens, and a pistil arranged in this order from the bottom to the top. Some flowers have spiral arrangements of these parts, while others exhibit radial trimerous or pentamerous symmetry. It was found that for these features to manifest, specific genes named "MADS-box genes" were functioning. Three groups of the MADS-box genes A, B, C were identified. It was verified that outer perianth segments formed solely by the function of A gene groups, inner perianth segments by A and B, stamens by B and C, and the pistil solely by C.

The functions of these genes are referred to as the ABC model. These genes are not specific to flowering angiosperms. They are also found in flowerless ferns and mosses, as well as in gymnosperms. The functions of these genes in ferns have not yet been elucidated. However, the genes that control the morphogenesis of flowers are thought to have served functions needed by plants before the evolution of flowers and subsequently evolved to function in the flowering process.

Another example of this work comes from the research group of M. Hasebe and others at the National Institute for Basic Biology. In recent years, researchers have elucidated various facts about the functions of the genes that regulate the morphogenesis of plants. Among them, Hasebe and others have consistently analyzed these functions contrasted to how the diversification of plant organization is related with phyletic differentiation. Their idea can be read in Hasebe's work (2015), although this book also presents many zoological facts as examples.

TEA TIME 7 Humboldt's Kosmos

Humboldt and Natural History Among the first people who incorporated modern scientific methods into the field of natural history was F.W.H. Alexander von Humboldt (1769–1859).

In the first half of his 30s, Humboldt, in league with Aime Bonpland (1773–1858), extensively surveyed fauna and flora in South America. They compiled their survey results into a tome of 35 volumes. As part of this project, they conducted basic research on plants and published many scientific names with the author names of Humb. & Bonpl. Those who are involved in the research of plant species diversity, including me, first become familiar with Humboldt's name upon encountering one of these scientific names. In this connection, I only understood him as one of many who contributed to species diversity research.

After having studied botany and geology at university, Humboldt was first appointed as a mine inspector. Before long, he undertook research on nature as an explorer. He stayed in South America for five years to conduct surveys. His activity extended beyond simply compiling survey results within the scope of plant species diversity. Probably driven by his genuine curiosity, or perhaps led by the scientific curiosity he developed through interactions with others, including Goethe, Humboldt, as a cultural figure, contributed to the growth of natural sciences in Prussia (Germany). Specifically, ecology and geography record him as the father of these fields.

Results obtained through one's field research are significant. While the results of research on specimens in collection are also meaningful, one gains results from a creative point of view through direct contact with nature. Humboldt looked at how plants live in nature. He became aware of the commonality of the lifestyles of diverse plants from the perspective of phytogeography. The basic perspective of present-day geography and ecology emerged robustly from these observations, even though, naturally, no one paid attention to it at the time. Obviously, he tried to view living things from the perspective of natural history.

As a refined cultural figure in the later part of his life, Humboldt released a tome of five volumes from 1845 to 1862, entitled 'Cosmos' (in German 'Kosmos: Entwurf einer physischen Weltbeschreibung'). He wrote this book after reaching old age, with the last volume published posthumously. The subtitle is translated as "A Sketch of the Physical Description of the Universe." Although basically from the perspective of natural sciences, he tried to understand nature on earth based on a theological worldview, and he contemplated relationships between the natural environment and humanity. He viewed nature from the typical perspective of natural history, which was his cosmic view, or thought on *kosmos.*

Incidentally, the renowned Prussian politician and linguist Wilhelm von Humboldt (1767–1835) was his blood brother.

Cosmos Flower Cosmos plants are distributed in tropical high plains in the New World including Mexico. Antonio José Cavanilles (1745–1804) named the plants *Cosmos* under the International Code of Botanical Nomenclature. Cavanilles was a botanist who, following a career as a researcher in Paris, returned to Spain and worked as director of the Royal Botanical Garden of Madrid. In this, he had transactions with Humboldt.

The word *cosmos* means orderly universe, an antonym of *chaos.* The reason that Cavanilles likened the impression of cosmos flowers to the orderly universe is occasionally explained by the fact that he noted the orderly flower structure of the plants. The plants were introduced into Japan during the Meiji period. The Japanese name for cosmos consists of a kanji character for autumn and another for cherry.

Cosmos versus Kosmos In English, the term *kosmos* is rarely used.

Meanwhile, the original title of Humbolt's book is 'Kosmos.' He wrote the book in German, in which cosmos is translated as *kosmos.* Apparently, in countries with languages where many German-origin words are in use, such as Danish, the plant name is spelled as *kosmos.* Nonetheless, it seems most countries use cosmos, containing the letter c.

In Japan, when alphabets are used to refer to this term, *cosmos* is widely used. When it is necessary to give a philosophical explanation, *kosmos*, originating in Greek, is used explicitly, probably due to the use of *cosmos* lending a general impression of plant name. Users of the term *kosmos* feel a considerable attachment to this spelling. Incidentally, thinking about this in connection with Humboldt's title may be far-fetched.

The term *kosmos* originated in Greek. The oldest example of this term in literature is found in the fragments of 'On Nature' (no surviving complete copy) by Heraclitus (c. 540 BC–c. 480 BC). Meanwhile, the concept of *kosmos* goes back to the thought of Pythagoras of Samos, who influenced Heraclitus. From these figures to Plato and Aristotle, the Greek view of the universe is summarized as the pattern of cosmos versus chaos before the creation of the universe. Based on the worldview at the time, *kosmos* referred to the orderly universe (celestial world). Of course, scientific knowledge in this context differs from present-day knowledge. It is not productive to discuss the Greek view in the context of science. That said, the term *cosmos* encompasses a view of the celestial world as a stable entity.

The Greek word *kosmos* was Latinized to *cosmos* due to writing differences between the two languages. *Kosmos* in Greek was the antonym of chaos. Then, the heliocentric theory brought about Copernican revolutions, making *kosmos* a term of etymological significance. Nonetheless, *cosmos* is in use still today with a connotation of the universe. Additionally, derived words such as *cosmetic*, *cosmology*, and *cosmopolitan* are not obsolete.

Perhaps, Humboldt used the title 'Kosmos' to indicate his intention to look at the earth as present in the universe from an integrative perspective. It is likely that when the pursuit of investigative research began in specialized branches of science, he attempted to compile what he saw by integrating available, leading-edge knowledge based on his research experience. He probably came to this perspective through contemplation, and the experience of honestly coming face-to-face with nature, observing and analyzing it to make full use of methods available at the time.

The Vision of the International Cosmos Prize: Harmonious Co-existence between Nature and Mankind While writing this book, I received the 2016 International Cosmos Prize. This international award is bestowed by the Commemorative Foundation for the International Garden and Greenery Exposition, Osaka, Japan 1990 (Expo '90 Foundation), established with the aim of inheriting the vision of the International Garden and Greenery Exposition held in 1990 at Tsurumi Ryokuchi in Osaka. Since 1993, the prize has been given annually to an individual or a team.

Although it is an academic award, the International Cosmos Prize is not intended to honor an individual who has achieved noticeable results in a specific, designated research field. The vision of the prize is described in the prize records:

"The prize will be awarded for research and work that has achieved excellence and is recognized as contributing to a significant understanding of the relationships among living organisms, the interdependence of life and the global environment, and the common nature integrating these interrelationships. It should be characterized by a global perspective which tries to illuminate the relationships between diverse phenomena in keeping with the concepts and principle of 'The Harmonious Coexistence between Nature and Humankind.'

"From the above viewpoint, importance is attached to the following:

(1) The body of achievements should show an inclusive and integrated methodology and approach, in contrast to analytic and reductive methodologies.
(2) The achievements must be based on a global perspective. If the focus is on a particular phenomenon or specific area, it must have universal significance and applicability.
(3) The achievements should offer a long-term vision which leads to further development, rather than solution to limited problems."

In this book I endeavor to describe the present-day picture of natural history and remember that Humboldt, who used the word *Kosmos* in the title of his lifework tome, was an epoch-making giant in the field of natural history.

Humboldt strived to compile the results of his research, pursued by comprehensive and integrative methods, by taking notice of relationships between nature and humanity. I frequently recall that his achievements were the essence of natural history.

Chapter
4

Learning Natural History: By Lifelong Learning

The preceding chapters presented an overview of the current state of natural history, along with its historical development. Precedented on the current state of natural history, and in anticipation of its further development, Chapter 4 pursues idealized learning. What should one learn, and by what methods, to achieve results in this field of research?

Natural history research is not only promoted at colleges and research institutes but is also vigorously conducted at museums and similar facilities. This chapter contemplates how to study natural history while examining the current state of natural history research in Japan.

4.1 Natural History in Japan

This book has described the study of natural history primarily in the context of Japan. However, no scientific issue is specific to a certain country or ethnic group. An issue is meaningless unless understood globally.

When it comes to science, locality should not exist. Nonetheless, a local focus is unavoidable within individual research projects about the natural environment. As this book was originally prepared in Japanese, it is worth mentioning in what respect Japan leads and lags behind the rest of the world in natural history research, as well as what it should pursue to continue to progress.

Japan has developed its own culture by taking cues from the cultures of other advanced countries. In the domain of natural history, it has followed a similar history. In the Engi era (10th century) during the Heian period, Doctor FUKANE Sukehito compiled 'Honzō Wamyō' (Japanese names of medicinal plants). Using 'Xinxiu Bencao' (Newly revised Materia Medica) from the Tang dynasty as an example, the book identified Japanese names of medical plants used in Tang, an advanced country at the time. Thereafter, Japanese herbalism grew through observation of living organisms in Japan, while following the style of Chinese herbalism.

Herbalism originating from China was well-suited to develop on the Japanese archipelago. During the Edo period, natural history from the West came to Japan like a new breeze, and existing knowledge was reorganized to align with the Western system. ITO Keisuke translated Thunberg's 'Flora Japonica' and published it in Japanese (1829). IINUMA Yokusai's 'Sōmoku-zusetsu' [consisting of volumes for herbaceous plants (1856–1862) and woody plants (1977), as edited by KITAMURA Sirō] was compiled following the Linnaean system. In the domain of natural history, as with the development of other aspects of culture, the useful tradition of importing advanced concepts from advanced countries and digesting them in a Japanese style thrived.

Regarding interactions between humans and nature, the Japanese people were not accustomed to viewing nature as an exploitable resource. They lived in harmony with nature, as an element of nature. They thought deities resided in all nature's creations, and in interacting with nature's creations they held an attitude of equality, at least until about the Meiji Restoration [cf. the postscript].

In the second half of the Edo period, the Japanese people succeeded in breeding diverse races of domesticated animals and plants. This did not result from geniuses developing new ways to effectively use resources. Rather, individuals living in harmony with animals and plants succeeded in breeding diversity by being obsessed with the diverse forms of living things and fascinated by their diverse lifestyles. Therefore, it is likely that they found diverse forms even in completely unexploitable species of organisms. Take the discrimination and identification of diverse forms of whisk ferns as an example. These efforts were achieved without a thought as to their use from a material- or energy-oriented resource perspective.

Since the Meiji Restoration, when the Japanese government adopted the material- and energy-oriented way of thinking to catch up with and overtake Western civilization and enforce measures for wealth and military strength, the traditional way of viewing things and the lifestyle of the Japanese people veered substantially. In the process of incorporating cultures of advanced Western countries, the Japanese people failed to assimilate Western culture into Japanese culture. Instead, Japan was almost swallowed by Western culture.

Natural sciences have advanced from natural philosophy since Aristotle's time. Nowadays, humanity's wealth and safety are maintained by natural sciences. However, in the field of natural history, the methods of modern natural sciences remained unutilized because the subject deals with an excessively large volume of information. As late as in the second half of the 20th century, biology and earth science began to use scientific methods as their principal techniques. However, the pace was still slow in the field of natural history. Nonetheless, given the rapid advances in natural sciences, natural history research, which discusses concrete findings about real things, holds high expectations for the future advancement.

Perhaps in addressing a macroscopic challenge like combining natural history with modern science, Japanese scientists, who imported modern science after its development, can have a more objective and constructive perspective than Western scientists who grew with modern science. A new perspective should be explored to unify sciences.

For this purpose, it is essential to train creative talent. Education in school should ensure the succession of natural history. Museums and similar facilities should promote people's contact with nature in nurturing creativity. Natural history knowledge is inherited by the acquisition of methods for using knowledge, in addition to learning accumulated knowledge. Learning not only through school education, but also through lifelong learning programs should be implemented. Education should not only nurture unique geniuses, but also promote the accumulation and enhanced dissemination of fundamental knowledge in society overall.

Whatever the case may be, we are living as an element in nature. While the perspective that humans, as the noblest of all animals, can utilize natural materials as resources in a sustainable manner, humans are living in nature as an element of biodiversity. In this situation, hopes rest on the promotion of natural history from an understanding of nature that includes ourselves.

To understand the most recent developments in natural history research in Japan, one can read the Natural History series, which includes this book, published by University of Tokyo Press. The first of this series is 'Natural History Museums in Japan' by ITOIGAWA Junji, Professor Emeritus at Nagoya University, published in 1993. Publication of this series has continued on a nonperiodic basis until this book, the 50th and last volume. The series covers a wide variety of topics. According to the authors' affiliation, their research activities are conducted mostly at colleges (34 authors), and notably at faculty of sciences (10 authors). (Despite their current affiliation with a faculty of science and engineering, education, integrated human studies, or graduate school of environment and information sciences, some authors are from a faculty of science or similar institution. Moreover, as in my case, other authors, after affiliation with a faculty of science, worked at the Open University at the time of writing or use Professor Emeritus as a title, having retired from a full-time post.) In present-day Japan, university faculties of science make the most vigorous contributions to natural history research. To examine whether universities are or have actually played a leading role as an institution for learning, including the training of successors, this section examines the backdrop to university roles since the Meiji Restoration, when modern institutions of higher research were established.

When thinking of learning, school generally comes to mind. Likewise, when talking about education, education at home or social education is rarely the first thing one discusses. However, perhaps these are based on biased interpretations of humans as intellectual animals. Certainly nowadays school education is the main way of passing down of knowledge. However, humans begin learning in the womb, and humans continue to influence the next generation even after death. This post-mortem influence is not only derived from great people who leave their names to posterity, but also from people on the street. Everyone may say that his or her deceased grandmother used to say this or that. Intellectual achievements are acquired through learning. Unlike genetic information inherited inside an individual body in a closed manner, intellectual achievements mature in society through transactions between individuals. The results are accumulated and inherited in society.

That said, to discuss systematic topics in natural history research and studies, it is naturally necessary to begin with discussing universities and research institutions. This chapter addresses how a topic of research conducted at a university relates to individuals' lifelong learning or the passing on of knowledge. Incidentally, lifelong learning or the passing on of knowledge takes place throughout the course "from the state of a fertilized egg to a ghost," as described in Tea Time 10.

In Japan, an overwhelmingly larger proportion of research is conducted at colleges than at institutions dedicated to research. A large factor concerning this may be the number of personnel involved in research. Colleges and graduate schools are education institutions. Teachers at a national university corporation are educational personnel. Nevertheless, most teachers at the Graduate Schools of Science of the universities consider their position more as researcher than as teacher. I might have felt this atmosphere relatively strongly because for a long period of time I was affiliated with universities with doctoral courses, such as Kyoto University and the University of Tokyo.

College teachers are engaged in higher education. Their role is not only to pass down existing accumulated knowledge, but also to train the talent whose skills develop to acquire and build additional new knowledge based on existing findings. In other words, students are not expected to be widely informed individuals with encyclopedic brains. Education institutions are required to train individuals to contribute to intellectual creation on the frontline of society. For this reason, teachers should be creative researchers. Such researchers are outstanding educators. Therefore, in the college teacher screening process, the quality of the candidate as a researcher is strictly assessed as the minimum qualification. In fact, some people taking up a post at a college tend to aim at achieving excellent performance as a researcher, rather than to be active as an educator. In extreme cases, they even use the term *odd jobs* to refer to duties carried out for an education-related committee.

This tendency is also observed in the field of natural history research. The National Museum of Nature and Science should exemplify a natural history research institution by Western standards. However, not long ago, the performance of the Museum in the domain of botany, for example, was not rated as highly by the academic community as research at colleges. Many faculty members considered themselves as researchers rather than staff in charge of higher education. Parallel with this, museum staff were poorly motivated to lead Japan's research. Indeed, when national research institutes jointly formed the Graduate University for Advanced Studies, the National Museum of Nature and Science did not join the project. In this case, there were certainly organizational issues. That said, the attitude of researchers affiliated with the Museum at the time was also a contributing factor.

To understand the very recent situation of natural history research and education in Japan, one must go back and examine the backdrop to the situation.

A considerable amount of information entered Japan even during the Edo period, which was known to be in a state of isolation. Those who were interested in natural history research used and digested incoming information to apply it to their research on nature in Japan. In the Edo period, despite the impression that only Dutch studies were imported, other major researchers visiting Japan (supposedly Dutch) included non-Dutch researchers, such as Kämpfer and Siebold from Germany and Thunberg from Sweden. While known to have opened a window to the Netherlands, Japan exchanged plenty of scientific information with the rest of Europe, the only advanced region at the time. Accordingly, the entire body of natural research in Europe was almost completely introduced and transferred to Japanese researchers, driven by the similar scientific curiosity of individuals visiting Japan to have accurate knowledge about the Japanese archipelago. Additionally, Japan had a rich history of research in the field of natural history, as represented by the basic botanical research in an undeveloped Far

Eastern country conducted by Thunberg, who was one of the most profound scholars at the time.

It is also remarkable that the second half of the Edo period saw the development of diverse forms of domesticated animals and plants, in parallel with the building of scientific knowledge in Japan. Somewhat later in Japan's Meiji period (1868–1912), Luther Burbank (1849–1926) in America and Ivan V. Michurin (1855–1935) in Russia developed diverse domesticate races using their personal gifts in horticulture. In contrast, in Japan in the second half of the Edo period, many ordinary (non-professional research) people, including feudal lords and merchants with interest in domesticated animals and plants, were successful in developing new races of domesticates. The range of animals and plants selected to develop diverse forms was not limited to so-called useful animals and plants important for livelihood but extended to those appreciated simply for their diversity.

Take whisk ferns, rare plants on the Japanese archipelago, as an illustrative example. Whisk ferns have been cultivated since olden times and diverse forms seem to have been developed due to their interesting peculiar forms. Various forms of this species were organized and published in 1836 as 'Matsubaran-fu' (Illustrated manual of whisk ferns), which described and illustrated 120 forms (in Japan, the only known wild species of whisk fern is *Psilotum nudum* (L.) Beauv.). Whisk ferns are non-flowering, not fancy, peculiar, and far from useful in terms of food or medicine. However, even in these plants, dilettantes driven by intellectual curiosity had strong enough interest to drive publication of an illustrated manual. Likely, only the Japanese people found enthusiasm in the cultivation of such peculiar plants as whisk ferns as early as 200 years ago.

Specifically, the Bunryuzan variant illustrated in 'Matsubaran-fu' has sporangia at branch ends, differing from general whisk ferns, which have axillary sporangia. This feature is morphologically interesting. Indeed, a renowned recently published textbook in plant morphology describes this variant with a photograph.

While natural history research in Japan favorably followed a path to elucidation, the cultural enlightenment following the Meiji Restoration boosted scientific research at higher research institutions, including colleges. Chapter 3 describes how natural history research conducted at Japanese colleges flourished during this time, focusing on the research on living things.

(1) Natural history at universities

Since universities were formally established in Japan in the Meiji period, the nation has rapidly absorbed modern science on the foundation formed during the Edo period, developing internationally competitive sciences. Indeed, Japanese researchers diligently worked on their observations of nature in Japan in collaboration with naturalists throughout the country, as reflected by Yatabe's declaration (Yatabe, 1890), saying that Japanese researchers would lead research on Japanese plants. The phrase *throughout the country* implies the inclusion of Taiwan, the Korean peninsula, the Kuril Islands, and Micronesia, according to the circumstances of the times. The researchers were not particularly collaborative with the nation's colonial policy. Rather, they were purely driven by scientific curiosity in expanding research subject regions and achieving results.

Universities were never staffed with a sufficient number of researchers. However, along with the expansion of research subject regions, accurate and detailed surveys were conducted in the first half of the 20th century. In those days, the proportion of researchers involved in natural history was considerably large. Universities implemented basic research projects without placing priority on any specific areas. In various disciplines, they produced internationally acclaimed results.

Nonetheless, despite its apparently satisfactory development, natural history research in Japan was never free of problems. One unmistakable problem was that research was often conducted based on the preference of the individual researcher. Honor for scientific research rests solely on priority in discovery. Naturally, researchers compete with other researchers to ascertain and publicize truth at the earliest possible point of time. One typical example is the notation of new species. It is often said that in the past, researchers did not disclose materials even to their colleagues or make specimens open to the public until the publication of the discovery.

This biased way of conducting research places the research environment under strain. Having recognized HIRASE Sakugorō's achievements, IKENO Seiichirō made efforts to honor his work. Although this was spoken as a touching tale, collaborative work between natural history researchers had become subject to bias.

In Japan, every domain of basic research was devastated during and immediately after the Second World War. This was also true in the field of natural history. Japan expanded its territories of occupation in Southeast Asia and adopted a national policy to direct efforts towards resource surveys. However, the policy lacked capable human resources and research-supporting frameworks. During wartime, progress was seen at most only in organizing literature. Immediately after the Second World War, no overseas literature was imported into Japan, people's lives were excessively poor, and colleges could not provide conditions suitable for conducting research.

Students studying away from home were even allowed to transfer freely to a college close to their home, to which they could easily commute. This was partly due to the arrested distribution of staple foods. (If a student's home was in Tokyo and the student was studying at a college in a provincial area, he or she could transfer to the University of Tokyo. Some cast an envious eye on this. However, no one would choose a dormitory's supper of crude sugar instead of a bowl of rice.)

Nevertheless, revival of academic research occurred when possible in the aftermath of the defeat, as economic recovery became robust. Both professional researchers at colleges or other institutions and non-professional naturalists alike resumed making steady progress in natural history observation. In 1953 Kyoto University sent a large survey team to the Himalayas and the Hindu Kush. In 1952, when I was a third-year high school student, I joined the Tanba Koganegadake plant collecting meeting organized by the Kansai-based Phytogeographical Society.

Soon after the mid-1950s, a group of several people, including myself, visited Mt. Mimuro in Hyogo Prefecture, where I had an unforgettable experience. The place was not very inconvenient to visit. At lunch break, I opened the food prepared by the inn where we had stayed. What I found was only a rice ball as big as the head of a baby, with no side dish. As might be expected, it was very tough to eat. Quite some time later, I first visited Beijing in 1980, when the Gang of Four in the Cultural Revolution was subjected to a trial that was broadcast on television. In China still in chaos after the Cultural Revolution, I often saw researchers returning from the dining hall to their offices with large amounts of rice in a basin-like bowl. In times when they were living poorly, people everywhere were accustomed to relying on rice as a sole nutritional source. Even without gourmet food, researchers were strongly motivated by their work.

In the post-Second World War political climate, it was unimaginable for Japanese researchers to someday be allowed to conduct surveys in China. Therefore, Japanese researchers working on biota began to research the flora in the Himalayan region, located opposite of Japan in the Sino-Japanese region and closely related to Japan in terms of plants. Subsequently, they extended the scope of their research to tropical zones in Southeast Asia. In the 1960s, they built a foundation and capacity sufficient for deploying their research programs on a global scale. They could contribute to elucidating local biota of overseas countries largely depending on the Grant-in-Aid for Overseas Scientific Research aided by the government was available for them (this expenditure item was created in 1963).

The Center for Southeast Asian Studies, Kyoto University was established in 1965. At the time, the subjects of area studies were humanities and social sciences. However, under the mission of area studies, the aims of the Center originally included natural scientific research. We participated in the area studies in the field of biota research. Although the contribution we made was not great, we learned a lot of things, which were effectively incorporated into our later research in various ways.

I became involved in college research and education at the beginning of the second half of the 20th century. I did so as a student and a graduate student between 1953 and 1963. After 1963, I worked as a faculty staff, began my career as a support staff member, and reached the post of laboratory manager (promoted to professor in 1972). I have held various positions in college research and education.

By the time that I first joined the research at the Faculty of Science of a college, the field of biological taxonomy was predominantly viewed as unfashionable in college research. While the field of ecology was flourishing at the time, taxonomy appeared to be a domain of the past. Some researchers voiced this outspokenly. When, after the vigorous research activities pursued along with expanding subject regions during the pre- and mid-Second World War periods, research expenses became limited and the research subject regions were confined to the reduced national territory, many of those who were involved in natural history-related research at college were at an impasse as to how they should develop natural history for the time being. They were bewildered, looking for a way out. Diverse challenges must have unfolded before them. However, for researchers, who at the time were strenuously working on fundamental surveys in untapped regions, the pending analysis challenges were probably unclear.

I also had a similar impression. When I became a college student, I had an intention to pursue other subjects that seemed important to me. However, as I learned more, I became increasingly interested in the biology of diversity. As a graduate student, I decided to pursue a phytogeographical course. A sense of self-responsibility for this choice made me build my career by constantly and necessarily seeking reasons for the significance of my research. My course of life that led to the selection of my major is described elsewhere (Iwatsuki, 2012).

Early in the 20th century, the laws of inheritance were verified as universal in the kingdom of organisms. Until then, biological research was persuasive, based on logical inference by observation and documentation. With these new developments, it became gradually possible for biology to incorporate physicochemical analysis methods. Since then, phenomena inherent to living things have been scientifically elucidated little by little using methods designed to delineate universal principles underlying phenomena through hypothesis-testing exploration.

In the study of organisms, exploration focusing on molecular-level analyses was readily understood by physicists and chemists and helped biology gain a secure position in science. Among those who expected scientific evaluation, this brought about trend of making light of the method of inference based on observation and documentation. This was a drive to construct a new biological science in contrast to traditional biology.

The scientific elucidation of causality in phenomena inherent to organisms spurred the development of technology for the effective use of biological materials, based on the principles it clarified. Since this was a technology related to living things, the term *biotechnology* came into general use.

Currently, the specialization of science has become noticeable even in the world of biology; importance is placed only on proven results; and the application of the results to technology has become an urgent issue. Under these circumstances, in fact, research based on down-to-earth surveys and underpinned by observation and documentation tends to be held in low regard. Now is the time to reexamine the truth about what natural history research is for naturalists in the history of scientific research in Japan. While being aware of this need, I remember that however rapidly biology evolved in the second half of the 20th century, there is very little that natural sciences can answer to the question of what it is to be alive.

Scientific research results are compiled after a specialized problem has been analyzed by a hypothesis-testing analysis through an experiment or some similar means; evidence that substantiates the hypothesis is gained; and a solution to the problem is obtained. This is valid when addressing an extremely specialized challenge. However, it is not immediately useful for reaching an inference that approaches the whole picture of a problem. Nevertheless, it is not rare that the application of an ascertained fact to technology produces great benefits for human life. This way of problem-solving is also very valuable for the social significance of science. In extreme cases, science even makes substantial contributions to warfare.

From the perspective of researchers in science, social science papers occasionally fail to take the form of a treatise. Although not very clear, many social science papers simply infer probable situations using limited information, refraining from describing a logically proven state. That said, natural history-related research is similar in many cases, as experienced on a daily basis. For example, to have a bird's eye view of the entirety of species diversity within the range of currently available knowledge, one would be able to infer the whole picture based on the recognized facts, which are very limited. Notwithstanding this, naturalists describe taxonomic hierarchy using determinate expressions. In the domain of science, it is difficult to correct misunderstandings involved in activities that are regarded as non-scientific.

In science, a major advance in research methodology naturally opens a way to analysis in a new field. However, this does not imply that the research challenges that can be met with long-established methods have all been completed. For example, species diversity research has progressed only to the extent that the number of known species is surmised to be only a very small portion of species on earth. For this research, the need to introduce new analytic points of view and research methods cannot be stressed enough. However, it is also necessary to steadily pursue research based on traditional methods. Unless this is done, as a reality faced by science, there will be no healthy progress in the science of living things.

(2) Research by non-professional naturalists

This heading may sound somewhat strange, but people known as non-professional naturalists have made substantial contributions to the domain of natural history in Japan. This book uses the term *naturalist* in a wide sense, sometimes with an adjective. When describing natural history in Japan, the contributions made by those who have a strong scientific curiosity to explore nature, yet do not make their livelihood by natural history research, can never be overlooked.

In the course of development in natural history research in Japan since the Meiji period, one thing can characterize Japanese naturalists. This is the ideal collaborations, growing along with the modernization of science, between professional naturalists, who devote their lives to research at a college or other institution, and non-professional naturalists, who love nature where they live and continuously conduct basic surveys. It is now appropriate to consider what contributions non-professional naturalists have made to natural history in Japan.

Research on the flora of the Japanese archipelago continued from the Edo to the Meiji period, building on the foundation of herbalism. In the Meiji period, associations of plant lovers were formed in many local areas mostly under the leadership, or at least the influence, of MAKINO Tomitarō and TASHIRO Zentarō. (This tradition seemed to develop after private schools in Edo period, represented by Shuhoujyuku by ONO Ranzan.) The herbalism from the Edo period had likely transformed substantially in the wave of modernization that came from the West. Notwithstanding the national restructuring known as the Meiji Restoration, people with an acute sense of curiosity about nature in their vicinity were active in many parts of the archipelago, irrespective of the epochal division made by the restoration.

In the Kansai area, there were many plant lover associations influenced by Tashiro, who helped Professor KOIDZUMI Gen-ichi at Kyoto University. Regionally representative non-professional naturalists pursued research through close exchanges of information as well as materials with college researchers. As in the case of Makino, research conducted at the University of Tokyo and Kyoto University was impossible without the cooperation of those who worked on detailed natural history research in their respective localities. At the same time, leaders of research activities deployed in various parts of Japan were able to know the state of most advanced research by closely cooperating with college staffs. Research for elucidating natural history on the Japanese archipelago was conducted through an ideal collaboration, drawing on the respective strengths of professional researchers at colleges and people conducting close observations of the dynamic state of nature in their vicinity, as driven purely by scientific curiosity. These favorable relationships improved the research quality of non-professional naturalists.

The countless volumes of works by Makino provide records on these activities. Writings of others associated with him also provide records from other perspectives. His autobiography, written late in his life, (Makino, 1956 and 2004) contains interesting stories about his personal history. Tashiro's activities, along with those of other people at the time, are recorded in detail and objectively in 'The Diary of TASHIRO Zentarō' (K. Tashiro, 1968–1973).

Makino set out to collecting meetings held in many parts of Japan and gave instructions on site. His style of instruction was beneficial in nurturing a growing number of plant lovers. Makino began 'The Journal of Japanese Botany' (launched in 1916), which was highly reputed as an international journal that actively reported on new species and new records. Discussions in Japanese that appeared in the journal were used to popularize natural history and helped train plant lovers in Japan. Makino was a government official serving most of his life as a staff at Tokyo Imperial University. 'The Journal of Japanese Botany' was not a journal published by an academic association. It initially was Makino's self-published academic journal. Since Volume 9 published in 1933, the private company Tsumura Laboratory (present-day Tsumura & Co.) has been the publisher, with Dr. ASAHINA Yasuhiko, an authority in pharmacognosy, as editor in chief (subsequently, Dr. HARA Hiroshi, Dr. SHIBATA Shōji, and ŌHASHI Hiroyoshi took the post in this order).

Following the above journal, in 1932 the Phytogeographical Society, formed mainly by faculty staffs at Kyoto University, launched 'Acta Phytotaxonomica et Geobotanica.' This journal also played the role of allowing college staffs and graduated students to quickly publish new species. Half of the pages in this journal were allocated to print articles in

Japanese. By attracting readers in Japan in this way, it expected to achieve economic independence. It was like a coterie magazine in the literary world. The Society's management policy continued long after my enrollment in college. Until impact factors began to determine an academic journal's importance, and in parallel with the major course of the Society's activities of holding annual meetings for researchers to present their leading-edge research activities and to publish academic journals, the journal was positively used to popularize their research and promote their exchange with non-professional naturalists by organizing collecting meetings on a periodic basis. I have heard that according to the founding principles, the Society endeavored to operate independently as long as possible, as dependence on subsidies would result in the discontinuation of publication when the subsidies become unavailable.

Nonetheless, before long, this policy became out of the step with the times. In 2001, the Phytogeographical Society merged with the Japanese Society for Plant Systematics. 'Acta Phytotaxonomica et Geobotanica' is presently a purely academic journal published exclusively in Western languages as an in-house publication of the Japanese Society for Plant Systematics, which until then had no academic journal. Separately, the Society publishes the Japanese magazine 'Bunrui' solely for its Japanese members. Publications from the Society continue to be financed principally by membership fees without subsidies from any third party, as in the times of the Phytogeographical Society.

The Kanazawa Shokubutsu Doukokai was founded under the initiative of faculty staffs at Kanazawa University. In 1952, the Society launched the 'Hokuriku Journal of Botany,' which was later renamed the 'Journal of Phytogeography and Taxonomy.' Publication of this journal continued in a similar spirit to 'Acta Phytotaxonomica et Geobotanica' in its old days. The Doukokai was renamed as the Society for the Study of Phytogeography and Taxonomy and merged with the Japanese Society for Plant Systematics in 2018. The title (in Japanese) of the journal was inherited by the Japanese journal published by the Japanese Society for Plant Systematics.

Knowledge of the biota on the Japanese archipelago increased primarily by research at colleges. Nonetheless, in that process, basic information was provided by non-professional naturalists who were closely investigating local biota. The Japanese people have a deep interest in the biota of their local area. Their sentiments have been recorded since the times of the 'Man'yoshu.' In later times, their interest in nature and the environment developed into exploration driven by scientific curiosity pursued in tandem with emotional appreciation. On one hand were those who had interest in the diversity of animals and plants in their vicinity, wished to know the names given to them, and had hobby-like pleasure in the regional characters and the wealth of living things; on the other hand were research-oriented individuals with scientific curiosity to produce accurate records of biota, trace its dynamic state, and observe lives exhibited in the biosphere. The Japanese people thus exhibit a wide range of tastes.

Indeed, when research quality improved in the Showa period, records of regional flora were successively launched in many parts of Japan. College researchers took part in the efforts to produce numerous quality reports. These included many outstanding works, such as UI Nuizō's 'Flora Kiiensis' (1929, plants in Wakayama Prefecture) and MAEHARA Kanjirō's 'Flora Austro-Higoensis' (1932, plants in Kuma and other areas in Kumamoto Prefecture).

Of course, some of the people involved in surveys looked to a greater or lesser extent for unrecorded forms, with hope for the joy of discovering a new species. Researchers were not completely free of the tendency to reward the discoverer's ambition by naming even a type of variant after them. It is understandable if the given name was recorded and published in a proper manner. However, occasionally, unpublished names were given by some researchers as lip service to the discoverers, which went into use locally and in some other places as well. Regrettably, these cases gave other general researchers in science the impression that the above researchers were engaging in an irresponsible behavior.

Contributions of Non-Professional Naturalists in Japan

To understand the contributions made by naturalists in Japan, I would like to delve into the topic of non-professional naturalists in more detail. Knowing the lifestyles of this type of naturalists is essential for understanding natural history in Japan.

In his letter known as *Resume*, MINAKATA Kumagusu used the term *literary men* [described by the term *literati* by TSURUMI Kazuko (1979 and 1981)], which was in use in the Victorian era in Great Britain to idealize a life with enough personal fortune to independently conduct research, as in Darwin's case.

Apparently, there are still those who believe that natural history is something in which amateurs find pleasure. Needless to say, as a healthy hobby, anyone can experience joy in life by having interest in nature in his or her vicinity. It is difficult to define the word *amateur*. Hence, I have intentionally used the term *non-professional naturalists* to refer to those who seriously observe nature outside the academic community and are not professional researchers at a college or other research institution. The word *fan* in this context is also inappropriate.

Amateur is contrasted with professional, an abbreviation of which is *pro*. Profession means a vocation. A pro means a person engaged in a vocation. The person follows the occupation for their livelihood. In Japan, the word *amateur* is used to refer to a person engaged in an activity without compensation in the context of sports. [In the sporting world in Japan, the Japanese term *nonpuro* (derived from *non-professional*) is distinguished from *amateur*.] In the sense of without compensation, an amateur has something common with a volunteer. However, the word *volunteer* strongly implies a person who offers service. In either way, with these words, the level of skill in the activity in question is not a matter of concern.

Throughout his lifetime, Darwin was not engaged in any professional post. Consequently, his status could be defined as amateur according to the above discussion. However, no one would call him an amateur evolutionary biologist. If Darwin were an amateur evolutionary biologist, there would be no one who can be called a veteran evolutionary biologist. That said, today, marking more than 150 years since the publication of 'On the Origin of Species,' details of his treatise can be criticized, and it is clear that his theory is not perfect. However, no one would claim that he was amateur for this reason.

Taking a more familiar example to me, in Japan the most prominent non-professional naturalist would be MINAKATA Kumagusu. In the domain of natural history, no one considers him to be amateur or low-skilled.

According to popular belief, MAKINO Tomitarō was a jobless writer. He had permission to learn in the campus of the University of Tokyo during the period of 1884 to 1889, although subsequently he was temporarily shut out of the university due to being at odds with Professors Yatabe and Matsumura. In 1893, at the age of 31, he was hired as a research assistant at the university (then Imperial University). In 1912, he was promoted to lecturer at the then Tokyo Imperial University. Since that time until 1939, when he was 77, he was a faculty staff at the University of Tokyo, for a total of 47 years.

Accordingly, Makino was absolutely a professional researcher affiliated with a state-run college. Moreover, he was at the center of the academic community in Japan because he was a member of the Japan Academy in his later life. When the Person of Cultural Merit program was established, Makino received its first award for his career and achievements. He was also posthumously presented with an Order of Culture. The assessment of research results is separate from position in rank at college, as exemplified also by HIRASE Sakugorō [3.1 (1)], a research assistant at the University of Tokyo who received an Imperial Prize, the Japan Academy Prize.

Stories do circulate that due to his poor educational background there was an anti-Makino sentiment at the University of Tokyo and that the university gave the cold shoulder to him as a full-time instructor. The anti-Makino sentiment refers to the fact that he was shut out of the university for more than three years because he was at odds with Professor YATABE Ryōkichi over, among other reasons, the publication of books on Japanese flora, and MATSUMURA Jinzō also presumably followed the anti-Makino policy. Presently, there is no means to ascertain the real interactions at the time. However, careful reading of Makino's works reveals that he was unique, independent from the current development of science, and as an individual he would not lend support to those who were focused on the creation of new research frameworks. As a community, universities are not the organizations where only groups who agree and proceed in a specific manner work together. When a new university system formed, more efforts were directed towards new research frameworks.

Matsumura's achievements have been compared with Makino's by some scholars. However, in these arguments, Matsumura's efforts in training FUJII Kenjirō, who later founded a cytogenetics course at the University of Tokyo, and in leading the modernization of botany, which resulted in the discovery of sperm of gymnosperms by Ikeno and Hirase, have been ignored. Makino's achievements were appealing to the media. In contrast, immediately after making the outstanding discovery of gingko sperm, Hirase left the school, an event which has not been fully examined historically. The person who collaborated in research with Hirase after his separation from the school was MINAKATA Kumagusu, a person completely outside the academic community.

Yatabe was hired as professor at the young age of 25. However, probably due to his enthusiasm for Europeanism at the time and his activities outside the domain of science in the narrow sense, he was dismissed from the University of Tokyo after an early active period contributing to science policy. He eventually died in an accident. Records say nothing about why he was dismissed or what happened to him at that time.

Highborn Assistant Professor ŌKUBO Saburō was a son of ŌKUBO Ichiō, who served as President at the Institute for Western Culture, eventually as junior councilor in the last days of the Tokugawa shogunate, and as the Governor of Tokyo Prefecture after the Meiji Restoration. Although Saburō had studied in the United States, he was not on par with MIYOSHI Manabu when he came back from studying in Germany. Before long, he left the university. Ōkubo has been almost lost to oblivion after leaving the university, probably because he made no prominent achievements, unlike Makino, who returned to the university after a three-year absence. (His name will eternally mark the plant name *Xiphopteris okuboi* (Yatabe) Copel., a fern in the family Grammitidaceae.)

After three years of being shut out, Makino could return to the Imperial University as a full-time researcher, probably because he was rated highly as a researcher in the Meiji period. In another period around the age of 30, Makino made creative contributions to plant taxonomy during an intensive few years while serving as a lecturer at Tokyo Imperial University at the turn of the century in the Meiji period. Around that time, Makino published articles on many new species with easy-to-understand descriptions. His understanding of species diversity was accurate. His papers, including phyletic positioning of individual species, were in line with the world's leading-edge findings at the time. His achievements were extraordinary for an individual who began studying virtually by self-education. In Europe and America, there have been also people who achieved results through self-education, without affiliation to any research institution. Such achievements might be easy to access for those reading and writing in their respective native languages. However, more than a few of these types of researchers produced reports confusing to later generations. It is laudable that Makino's achievements produced at his zenith are almost free of such defects.

Makino's research shined with its accuracy in its contributions of monographs. Accurate descriptions of diverse species are vitally important as a foundation of species diversity research and to underscore the essential significance of natural history, although accurate descriptions of species are not simply sufficient for the elucidation of species diversity. To search for the universal principles underlying diversity, it is necessary to recognize species diversity alongside a phyletic backdrop. Makino's accurate observations provide important clues for tracing lineage. However, although his research revealed his acute sense of classification into genera and families based on species rank descriptions and an understanding of the magnitudes of species differences, he lacked the perspective of exploring biodiversity in an integrative manner.

In his later years, Makino's activities ranged far from presenting his original research results to the world. In 1926, he launched 'the Journal of Japanese Botany' to show his willingness to lead the era. However, his papers lacked the quality seen at his zenith. Of course, he subsequently continued to be as successful as before in disseminating the botanical knowledge described earlier and published 'Makino's Illustrated Flora of Japan.' Nevertheless, he was noticeably eccentric, as exemplified by his open declaration of antagonism against the International Code of Botanical Nomenclature. It was certainly his personality that attracted people's interest. Makino lived a freewheeling life, if not as free as Minakata, which might have served as an iconic image for the Japanese notion that scientists were better if they were insane. This is perhaps why he has been occasionally described as a non-professional, even though he consistently made a livelihood serving at the University of Tokyo.

While the above developments took place at the Department of Botany, the Department of Zoology saw a comparable relationship between Professor ISHIKAWA Chiyomatsu and SAITŌ Hirokichi, which was concisely described by Professor ENDŌ Hideki at the University Museum of the University of Tokyo (Endō, 2015). Saitō was a researcher on Japanese dogs outside the academic community. He wrote the academic clique at the Imperial University made light of his accomplishments and that Professor Ishikawa at the Imperial University was ignorant about dogs. The academism of university researchers rejected Saitō, which disgusted him. However, before long, people began to seek advice from Saitō anything about dogs (in a similar manner to 'Makino's Illustrated Flora of Japan' being all the rage at the time), while the author remained merely a full-time lecturer at the University of Tokyo.

Besides these examples at the University of Tokyo, those who substantially contributed to the field of natural history in Japan included more than a few unpaid non-professional naturalists, although, unlike Darwin, they had other jobs for their livelihood. This situation is rare in the more analytic areas of biology. Therefore, it is occasionally pointed out that feudal lords were often involved in the domain of natural history (Kagaku-Asahi ed., 1991). Perhaps, to demonstrate their abilities, the domain was preferred by those with strong scientific curiosity, who also had the means of easy living. According to the general understanding, one characteristic of the field of natural history was that one can investigate and observe without expensive laboratory instruments. Even without the need for expensive instruments, huge costs can be incurred in investigations and the management and maintenance of collected materials. Indeed, some people spent an enormous fortune for those reasons. However, for no special reason, these expenses were regarded as a part of wealthy people's luxurious hobby. Moreover, even if the achievements by these non-professional people were substantially contributing to the world of scientific information, it cannot be denied that these kinds of consistently observational and descriptive contributions tended to be considered amateur achievements.

Those who regarded natural history as a domain that could be studied as an amateur hobby might perhaps have considered, consciously or subconsciously, that natural history could be explored even outside colleges and research institutions by collecting materials and conducting observations with simple instruments, without the need for expensive laboratory equipment. In these assumptions, it may be possible to understand the nature of their attitude towards natural history. In Europe and America, art has been refined owing to support from discerning patrons. Unlike those domains that are directly beneficial to society, historically, the development of pure sciences, which are more like an intellectual game, relied on support from wealthy people, as well as from political or economic powers. This might have also been true in the history of natural history. Nowadays, natural history is facing the question of whether to remain in this situation.

Established physicochemical methods are used to analyze events. In contrast, to understand nature as an integrated whole, it is necessary to use not only investigative, but also integrative methods [2.1 (5)]. Laboratory equipment is required for the sake of analysis. To discuss obtained data, it is advantageous to work in a group of acclaimed researchers. Advances in science inevitably require specialization of professional researchers. However, for accurate chronological recognition and recording of nature in a region, not only professional researchers, but also non-professional naturalists make contributions by their enthusiastic efforts and achievements. In Japan, people's understanding of nature owes much to contributions by non-professional naturalists, underpinned by the traditional view of nature of the Japanese people.

Of course, natural history cannot avoid being modernized. The scientific significance of this domain lies in its method used to view nature as an integrated whole and to understand it in an integrated manner. Natural history may be criticized for being an amateur hobby, if not aiming beyond simple documentation of diverse phenomena and stopping short of noting interrelationships between them. Taxonomy brought into Japan in the Meiji period already aimed at structuring a taxonomic hierarchy with lineage in mind. In this regard, it is true that some non-professional Japanese naturalists were not in line with the philosophy of natural history.

Non-professional Japanese naturalists cannot be fully understood simply by comparing the examples of Darwin and feudal lords' biology. To understand them, it is helpful to imagine the times of the 'Man'yoshu' [edited in 8th or at latest the beginning of 9th century]. This collection of poetry contains many guilelessly composed pieces on nature by a wide-ranging class of people. In this respect, the 'Man'yoshu' is rare as an ancient record in the world. The anthology clearly portrays the depth of interactions between the Japanese people and nature, in which they not only sought practical benefits, but also allowed their heartstrings to be pulled by nature. By reading the 'Man'yoshu' from the perspective of natural sciences, today's scientists would understand that the contributions made by non-professional naturalists had been forecasted in these ancient collections of Japanese poetry. Of course, the anthology lacks systematic ecological discussions because ecology did not yet exist in the 8th century. Nevertheless, considering the provision of first-class information on nature at the time, those who contributed to building that knowledge should be regarded as outstanding non-professional naturalists (Hattori & al., 2010).

Since that time, the people of the Japanese archipelago conserved biodiversity by creating *satoyama* [see the postscript] to live in harmony with nature, thus saving all medium-sized and large animal species until the end of the Edo period. By maintaining *satoyama* forests, they concretely contributed to biodiversity as non-professional naturalists. The manner of environmental conservation on the Japanese archipelago, which has maintained *satoyama* forests as buffer zones between deep mountains and rural areas, is not a topic detailed in this book, although I will discuss it elsewhere (cf. the postscript).

In the second half of the 20th century, researchers became aware that excessive development had put biodiversity in danger, and its conservation became an urgent issue for earth. Consequently, to make society aware of this reality faced, they studied the actual biodiversity situation using endangered species as models. In the 1980s, slightly behind Europe and America, Japan began to research endangered species. Research on vascular plants was at the vanguard of this trend. However, the number of researchers capable of working on this topic was far smaller than those working on plant diversity of the Japanese archipelago. Fortunately, in every part of the archipelago, non-professional naturalists were active in closely observing the dynamic state of living things on a daily basis near their homes, in a tradition likely from the times of the 'Man'yoshu.' From this, full-time researchers involved were able to put together the vast amounts of data provided by non-professional naturalists to compile an internationally prominent red list within a short period of time (e.g. Subcommittee for Plant Species, Committee for Editing the Red List of Plant Species and Communities in Japan, 1989). Since then, the Environment Agency (and later the Ministry of the Environment) has led basic surveys and the conservation of endangered species in Japan. In this, full-time researchers (Japanese Society for Plant Systematics) and non-professional naturalists have been collaborating in an ideal manner, at least in activities relating to plants.

In biological species diversity research, while countless research needs require laboratory analysis with expensive research equipment, building basic knowledge about the living things present on earth also presents urgent challenges. This challenge cannot be met quickly by a limited number of researchers. Hopes for this accomplishment rest on scientific contributions made by numerous full-time researchers, as well as those made, although in a different manner, by non-professional naturalists, in the vein of Darwin and feudal lords.

In focusing on the above-mentioned contributions made by non-professional naturalists, one should contemplate the lifestyle implications to maintain *satoyama* forests in everyday life, which helped conserve biodiversity. The Japanese people did not learn about ecology to realize *satoyama* by dividing areas and conserving biodiversity. They realized this ecologically preferable state because of their activities and philosophies in everyday life. However, although it is easy to explain that as a result, *satoyama* forests grew and biodiversity increased, it was, in a sense, simply an expression of the harmonious relationship between nature and the people living on the Japanese archipelago. Following this example, it should then be possible for *Homo sapiens* on earth to live sustainably by enhancing scientific literacy among society and consumers. In this regard, fostering people's interest in natural history, as well as promoting research by professional researchers, is expected to improve scientific literacy among the wider public. The significance of the evolution of *Homo sapiens* as an intellectual animal lies in their ability to encounter nature, thereby developing a lifestyle that evolved with humans as one element of nature.

The focus on non-professional naturalists may align with the taste of the Japanese media, which highly values contributions to science by those who do not make a livelihood from science. Neither Mendel nor Darwin, both of whom provided a foundation for modern biology, made a livelihood by biology. Mendel was a priest (during the Age of Exploration, many botanists were churchmen). Darwin spent his life pursuing research, contemplation, and writing, without taking a job. How much they contributed to science is more significant than what their jobs were.

It is highly valuable that Japan has a huge population curious about natural history, which serves as an energetic source of Sunday naturalists. In this country, unlike in Europe or America, there have been few patrons who offer their property to support scientific research, a trend that still continues. However, the contributions that naturalists have made to building basic scientific information are comparable to this type of donation in that they provide their intellectual property at no charge. It is a bitter reality that the building of intellectual property is fading into obscurity.

---

(3) Natural history research at museums in Japan

When a research framework was established in the Meiji period, TANAKA Yoshio (1838–1916), who learned under ITŌ Keisuke and later served as a bureaucrat, and SHIRAI Mitsutarō, a professor at the College of Agriculture, Tokyo Imperial University, strongly asserted the need to conduct research at museums in tandem with leading-edge research at colleges. This was the time when national industrial exhibitions, modeled after international exhibitions held in London and Paris, were helping disseminate industrial knowledge.

The present-day National Museum of Nature and Science (formerly National Science Museum, Tokyo) is the oldest natural history museum in Japan and asserts that the Museum was founded when its current building was constructed in Ueno Park in 1877. That said, records say that its predecessor was an exhibition hall of the Bureau of Natural History, which the Ministry of Education established in Yushima Seidō in 1871 and that an exhibition was held the following year under the name of the Ministry of Education Museum. In 1875, the Museum, already in service, was renamed the Tokyo Museum.

The Botanical Gardens of the University of Tokyo, a museum-related facility, originated in the medicinal plant gardens of the Tokugawa shogunate, founded in 1638. In 1684, the south garden moved to the Hakusan residence [of 5th shogun Tunayoshi], in which the present-day Botanical Gardens has roots. In 1875, it became the Koishikawa Botanical Garden attached to the Ministry of Education Museum. In January 1877, the Museum began to function as an educational museum. In the April immediately after, the Garden was placed under the University of Tokyo as an attached facility.

The Museum subsequently underwent organizational changes. Its facilities and specimens were devastated during the Great Kanto earthquake. Its full activities as a museum took hold in 1949, when it became the National Museum of Nature and Science pursuant to the Act for Establishment of the Ministry of Education.

However, compared with colleges and similar research institutions, natural history research activities conducted at the Museum remained sluggish as a whole, although it had outstanding researchers on an individual level. The lack of active achievements at Japan's representative natural science museum gave the public a strong impression that a museum was a building only containing stacks of dusty literature. Particularly for some periods in Japan, museums continued to fail to gain recognition in society as natural science research institution.

Despite an understanding that museums conduct research on cultural properties and other materials, most people do not know that museums pursue natural scientific research, a situation that persists at some museums even today. In Japan, museums began to attract some attention as research institutions in natural sciences around the time that the National Museum of Ethnology was founded (i.e., 1974 and opening to the public in 1977). It should be noted that this museum came under the control of the bureau that governed research institutions (the then Science and International Affairs Bureau) at the Ministry of Education. Therefore, its researchers were assigned educational posts, such as professor and assistant professor. The introduction of these posts to a museum was a new idea. Probably, the general public would not acknowledge a staff member as a researcher unless he or she was called a professor. Subsequently, the National Museum of Japanese History (founded in 1981) and other museums were founded in the same style. Presently, these museums are major members of the Graduate University for Advanced Studies, and under the name of museum, they nominally play the role of a college as a higher educational institution. Notwithstanding this trend, when the Graduate University for Advanced Studies was founded, the National Museum of Nature and Science did not join the member institutions.

Meanwhile, the latecomer Museum of Nature and Human Activities, Hyogo (founded in 1992), a public institution, introduced at establishment a system to treat many researchers as college teachers (4.2).

At the National Museum of Nature and Science, the museum staff at times pursued their activities as administrative officials rather than as researchers driven by scientific curiosity, although the Museum had a specified objective to conduct research. This was probably because the institution was under control of the Social Education Bureau of the Ministry of Education (restructured into the present-day Lifelong Learning Policy Bureau, the Ministry of Education, Culture, Sports, Science and Technology) rather than under a research-related bureau. This was not simply for organizational reasons but was also affected by the personal traits of the active researchers and administrative staff of the Museum. It is well known that even among college faculty members, those exist whose primary activities do not contribute to research and education.

Besides the National Museum of Nature and Science, natural history museums include the Osaka Museum of Natural History (open since 1974), which has a long history with its predecessor Natural Sciences Museum beginning exhibitions in 1950. Since its founding, this museum has produced successful research results, even under difficult conditions. In contrast, some local museums and similar facilities subsequently founded during the bubble economy did not expect their staff to pursue research, as their primary aim was to construct facility buildings. Zoos and botanical gardens in the wide sense were also built on the similar idea of their facilities being the equivalent to a museum. In this sense, supporting lifelong learning tended to be thought of as a role of special facilities, or organizations designed for adult education.

The above results from the influence of the provisioning of an intellectual training-oriented school education system, intended for wealth and military strength and the promotion of intensive educational measures for a strengthened intellectual foundation. This was adopted to catch up with and overtake Western civilization since the Meiji Restoration, when the government became clearly aware of Japan's cultural backwardness. The attitude of catching up with and overtaking the West was effective for the Japanese people, who had been slow to acquire Western knowledge, to learn Western science and technology. However, this system missed the idea of intellectual training for the development of originality and creativity. Once Japan caught up the West, it was at a loss as to what to do next.

Certainly, it is remarkable that since the Meiji period the Japanese education system has improved the intellectual foundation of the Japanese people. However, at the same time, the Japanese people forgot their traditional lifestyle of living in harmony with nature, allowing their originality and intellectual curiosity to decline. Museums and similar facilities can be effectively used to encourage lifelong learning by people, including school-age children, in a form of museum-academia collaboration that allows full educational enjoyment, while also gaining voluntary-learning benefits. Such trends are recently emerging, although somewhat limited extent, which is a hopeful situation.

Natural History Museums in Chiba and Hyogo Prefectures In the 1980s, when Japan was booming in the bubble economy, preparations for natural history museums were being made in Chiba and Hyogo prefectures on an unprecedented scale. Both prefectures had a reasonable scale of finance. As a favorable condition, the governors at the time were willing to build a natural history museum. Of course, the propositions did not directly follow the traditional style of natural history museums. Chiba aimed at building a museum that emphasized coordination with ecology. Hyogo sought to present contributions to citizens. They intended to actively provide what the era was demanding from museums, that is, enhanced support for lifelong learning and think-tank functions. 'New Approaches of Museum for People's Learning' (2012), edited by the Museum of Nature and Human Activities, Hyogo, describes the 20-year history of the Museum. In particular, Chapter 6 of the book gives background information on the Museum's founding.

Both museums were on par in that they each had more than 40 outstanding full-time researchers assembled. The museum in Chiba had more name recognition in society due to its somewhat earlier opening and proximity to the central part of Japan. The museum in Hyogo in its early period was not very successful in making its activity widely known beyond regional boundaries, although locally it began to be highly reputed at the time of its founding. The actuality of its growth as a natural history museum in line with the times was not well understood nationally.

The museum in Chiba exceeded traditional local museums [of Japan] in size. However, its research activity was restricted to prefectural staff because their status remained within the conventional framework. They were certainly active in research and providing support for lifelong learning, which nonetheless did not lead to drastically changing society's impression of museum staff.

The museum in Hyogo opted to hire the majority of its researchers as college faculty (4.2). Therefore, researchers pursued think-tank functions and support for lifelong learning in the manner of faculty staffs. This was not surprising because they were indeed the members of college Institute. In these activities, capable researchers made far more vigorous contributions than some less willing ordinary college teachers. Insiders highly assessed their achievements. However, it unfortunately took some time before these accomplishments gained nationwide reputation because their activities were primarily centered in Hyogo Prefecture and its vicinity as a prefectural institution.

At the turn of the century, the museum in Hyogo began to consider reforms to return to the very basics of museum activities by its tenth anniversary as a target date. The museum set out activities, called the new phase, through self-initiated discussions, while drawing on advice from within and outside the institution. At first, some museum staff were not willing to provide think-tank functions or support to lifelong learning. Even such researchers have now become interested in using their research skills for relevant museum activities to promote productive knowledge dissemination.

Unfortunately, in the period of growth of these activities, the museum was severely affected by personnel downsizing and reduced budgets in the overall Hyogo prefectural government during the desperately tight financial situations that local governments experienced throughout Japan, particularly due to various consequences that followed the 1995 Southern Hyogo Prefecture Earthquake. These problems were not something that could be solved by the efforts of the museum staff.

Twenty and some years ago, the museum founding was backed by the financial strength of the time. In Hyogo, the museum opened using buildings constructed for an exhibition known as "Holonpia '88." The issue that the buildings were not built as museum facilities persists. Presently, the younger museum staff sustains the museum's growth. However, to maintain the present scope of activities in the long term, fundamental research activities will naturally be restricted. Researchers not conducting research will be unable to follow the ideal way of providing think-tank functions or support for lifelong learning. This situation might lead to declines in museum activities.

[It may better be added here that in 2022 (after the issue of the original Japanese edition of this book) a Collectionarium was newly built, intending to promote lifelong learning as well as for housing collections.]

Museum-Academia Collaboration and Natural History School education accounts for an overwhelmingly large part of the educational sector, particularly in Japan. Even with dramatic enhancements, museums and similar facilities remain as a supplementary contribution to education, unless the education system undergoes a drastic change. This is unavoidable considering the quantitative difference between museum and school education facilities. Nevertheless, even if very small in quantity, museums and similar facilities play a role that is difficult to fill by school education alone, as revealed by recent museum activities [in Japan].

One feature of museums is that they store specimens and other real things as research materials. It is not easy for schools to have a section specializing in the storage of materials. To store specimens used by researchers, colleges need to have an attached museum, as a facility similar to an independent museum. Attached museums are expected to promote basic research activities suitable for a museum, rather than to provide support for lifelong learning or think-tank functions. Moreover, at a university, which should collect comprehensive knowledge, the attached museum is expected to initiate research that integrates diverse areas of study using materials in storage (cf. the following section).

The term *museum-academia collaboration* has recently come into common use, in anticipation of improving educational results by making the optimal use of the characteristics of museums and promoting collaboration between museums and schools. For museum learning in Japan, museum staff are expected to be greatly helpful in sharing their abilities. In Europe, it is common that a teacher takes school children to a museum or botanical garden and speaks for some minutes first. The children then head in all directions to learn using real things, and they return to a designated place to discuss their observation results around the teacher.

Kindergarten and elementary school teachers are obliged to play a very important role in building an intellectual foundation in each child, while stimulating their intellectual interest. Individual teachers have to be concerned with all subjects in most cases. Beyond intellectual education, they are expected to train children to adapt to group life. Moreover, in recent years, families and society tend to expect schools to cover moral education, although families should essentially provide this. It is desirable that museums and similar facilities undertake fostering curiosity about wonderful objects and phenomena in nature to supplement school education. In the class, one teacher shoulders the responsibility for all the students' intellectual activities and is expected to provide social education for children's habituation to group life. In such a school education environment, it is difficult for children with a flexible mind to achieve ideal and enjoyable learning results about nature.

Junior high and high schools have recently strongly leant towards concentrating on preparation for entrance exams. Using tacitly accepted numerical targets known as deviation values, parents hope for immediate results that their children will enroll in a prestigious school, with the aim of having a stable job in the future. Essentially, classrooms should not serve as a place for the embodiment of materialism. However, useless laments cannot solve the problem if most parents desire this situation. It should be useful to reexamine the significance of learning that took place at *terakoya*, primarily intended for practical training.

The Museum of Nature and Human Activities has been continuing the presentation meeting, Plaza for Harmonious Coexistence, for nearly 20 years. At this meeting, school students, including nursery school infants, present their results of observation. Moreover, students who participate in the meeting have achieved reasonable success in entrance exams for higher schools. These admirable achievements make me feel sorry that good students are swept in the generalizations that criticize present-day students. This is especially true in the presence of superb teachers, however few, who are able to help highly motivated children learn by fostering scientific curiosity.

For museums and college students who are supposed to move away from passive learning, think for themselves, and build up their own learning, they should not expect museums and similar facilities to provide them with instructions when visiting. Rather, they should stimulate their intellectual curiosity by accessing materials in collections and exhibitions to build their own learning. Those who think and act by themselves may often encounter researchers at a museum who lend a helping hand, if they are not visiting the museum exclusively for passive learning. For college students' museum-academia collaborations, the influence of new findings from joint research between college teachers and museum staff has an interesting impact on students.

Schools seem to approach museums, to some extent, for exchange. However, visitors to museums and similar facilities do not include many junior high schools to college students. In this regard, I am not so familiar with what is going with humanities-related museums. However, natural history and science and technology museums should understand that dramatic increases in visitors from the above age groups are vital for promoting museum-academia collaboration in its true sense. One relevant and important role of museums is to plan exhibitions and events that motivate these groups to visit museums in spite of the busyness of their studies. In this sense, natural history serves as an ideal route for people in these age groups to have a deeper interest in nature.

Some measures must be taken to change the recent trends in Japan that people associate education only with school education. As intellectual animals, humans should be living organisms learning for a lifetime. Moreover, humans should build their learning beyond simply being taught something. Having built cultures, humans are experts in shifting from passive to active learning based on knowledge passed down from those who came before. A lot of things can be learned from existing stacks of knowledge, mainly through classroom study. However, after knowing the joy of learning, humans find pleasure in constructing and creating something new, while making effective use of stacks of existing knowledge.

Museums can offer opportunities for students to have contact with nature, to evolve from classroom study to hands-on learning, and to pivot from passive to active learning. When this is done, kindergarten and elementary school children, as well as high school and college students, will be able to deepen their intellectual curiosity suitably for their respective ages. To develop science literacy, it is necessary not only to learn stacks of knowledge, but also to foster one's original ideas. Natural history study should be regarded as highly suitable for raising the awareness of issues in all age groups.

Natural Science Museums and Education The National Museum of Nature and Science has been active since the early days of the provisioning of academic systems in Japan following the Meiji Restoration. Following this, natural science museums have been gradually founded and enriched, including science and technology museums and natural history museums. That said, lifelong learning support programs and the enhancement of museums and similar facilities tended to receive a lower priority because, since the Meiji period in Japan, education and academic systems have focused on school education to catch up with and overtake advanced European and

American cultures. This momentum has basically remained unchanged even after the Second World War, when Japan discarded the wealth and military strength slogan. The reason for this unchanging momentum probably includes the attitude of those involved in natural history, even though not intentional, as well as the public's perspective about education.

In the Edo period, *terakoya* served as a place of social education, as well as a facility for gaining practical knowledge and fostering intellectual skills. At the same time, families and society shared the understanding that they should all be involved in raising children, rather than delegating the task to *terakoya*.

The school education system established in the Meiji period, although enriching schools as intellectual education facilities and placing priority on the intellectual training of teachers, has resulted in leaving everything under the name of education up to schools, imposing even tasks that families and society should essentially undertake on schools. This tendency went overboard in the post-Second World War period, for which there are several reasons. The emergence of what is known as "monster parents" is very unnatural. Children should originally be trained at home and in society. In local communities, children should grow in their own suitable social structure. If this no longer takes place, it is equivalent to the loss of the social life built by the animal species *Homo sapiens.*

Schools are supposed to pass down knowledge. Under the name of education, major efforts have been made to transfer existing knowledge from those who taught to those who were taught. The teacher demands that children study and does not expect them to do what they love to do. School education has rather prohibited students from becoming enthusiastic about a self-serving activity. One implication of the term *study* in Japanese [in Chinese characters] is to force children to work hard, precisely depicting the above reality. While inheriting the system and facilities of *terakoya* from the Edo period, many schools came to have gradually altered education policies since the Meiji period in which compulsory education was legislated. Needless to say, the change entailed beneficial improvements. At the same time, difficult problems also developed, which cannot be overlooked.

The academic framework established in the Meiji period achieved its goals. The Japanese people's basic scholastic abilities improved. The Far Eastern island country in no time developed a culture comparable to the West. However, the Japanese people, like sheep following leaders, built a history single-mindedly towards ugly wars. Dreaming of becoming a first-class power resulted in falling into an abyss. However, notwithstanding the slogan "a national confession of Japanese war guilt," the true nature of that history has been forgotten with time. When the guilty feeling has gone away, those who crave wealth from war seem to instigate public admiration of a nation with powerful military strength.

Scientific museums include the National Museum of Emerging Science and Innovation (open since 2001) and other large national museums, those which keep records on science-based technologies, and space museums which offer star-watching services. Demand has been relatively high for these museums, which have opened and are indeed popular. Museums in the domain of natural history have been built one after another since around the 1960s by prefectural governments and other municipalities. There were reasons for these developments, such as that Japan gained the necessary economic strength and that emerging serious environmental problems led to the need for places where people could access current knowledge about the earth.

Natural history museums include those which focus on dinosaurs, as in Fukui and Gunma. These museums hope that the subject fossil animals, or dinosaurs, which strongly pique children's interest, will attract many visitors. Local museums could be more locally specific. However, the characteristics of nature in a locality do not serve to attract increasing visitors, even though such characteristics should be useful for research and exhibitions by museums. If measures to utilize local resources have not been developed, consideration should promptly be given to devise effective utilization measures.

In very recent years, natural history museums in Japan have begun to contribute to local communities in a tangible manner. In addition to holding general exhibitions to attract general citizens, they set challenges to provide support for lifelong learning and desired think-tank functions. Nowadays, with sustainable use of biodiversity included in political challenges, museums help with building relevant knowledge. More specifically, they serve as a hub for compiling red data on endangered species and making efforts to develop regional strategies in line with national biodiversity strategies. For these purposes, natural history specimens are used to provide a scientific basis. However, few local museums have sufficient authenticated natural history specimens to provide scientific basis of study. In this regard, they probably need to enhance their abilities to make contributions.

General Museums Affiliated with Universities There have been increasing chances for museum staff to train successor researchers. In addition to individual research institutions that meet the needs of graduate school courses for training specialists, programs have developed for these institutions to collaborate with colleges, allowing superb researchers affiliated with small research institutions to take part in the training of successors more easily. The University of Tokyo set up an organization to receive researchers from the National Museum of Nature and Science and other institutions as dual-service teacher. The growth of this kind of program provided isolated researchers with opportunities to participate in post-graduated education. I believe this is a preferable way of using their precious abilities. The program has been in place already for many years; however, its effectiveness is not known. Certainly, it may not be possible to assess the effectiveness of programs such as this in a short period of time. Nonetheless, now may be the time to make a proper evaluation.

Against this backdrop, institutional forms of college museums developed. In 1966, the University Museum opened at the University of Tokyo to bring specimens and the like maintained in each specialized field under unified control for the purpose of convenience in research. Formerly, individual researchers kept specimens relating to their research areas at hand. However, such specimens increased in quantity, incurred maintenance costs and labor exceeding the suitable level of a laboratory, and were often disposed of because of professors coming and going. The University Museum was founded to meet the need for the University to take responsibility for specimens. In most of the universities, their Museum had been treated as a resource center, partly due to poor organizational recognition of museum activities within the University.

The Museum promoted research activities and was gradually enriched as an organization owing to insider efforts. In 1996, rules were formulated regarding the establishment of the University Museum. The organization has developed into an institution staffed by more than ten full-time researchers as of 2015, including professors and associate professors. Nonetheless, against the initial expectations, the floor area remains insufficient for the storage of the materials. More than thirty percent of the plant specimens are still stored in a building of the Botanical Gardens of the University of Tokyo, even after more than half a century since the Museum's inauguration as a resource center.

Subsequently, the Kyoto University Museum (established in 1997 and open since 2001) was officially founded as an organization encompassing natural history, cultural history, and technological history. [It is to be noted here that some plant specimens are still preserved in a space of the Department of Botany, like in the case of the Botanical Gardens, the University of Tokyo. The Museums of these universities have no enough spaces to house all the specimens they have.] The Hokkaido University Museum (established in 1999) and the Kyushu University Museum (established in 2000) also contribute to research activities.

Aside from museum insiders, most college staff strongly expect museums only to serve as a window to society for colleges engaged in individual and highly specialized research projects, which themselves are not in direct contact with society. This is probably due to the notion that museums are linked with exhibitions.

The aspect of the social contribution of the tremendous amount of research conducted at colleges, which serves as a foundation for cultural advances, cannot be grasped at a glance by exhibitions associated with research conducted at museums. Even if exhibiting the roles played by colleges openly, it is not possible to gain a correct understanding about a university unless effective public relations methods are formulated beyond the existing framework of museums to encompass all faculties and research institutes. Substituting such methods with exhibitions by existing museums would not satisfy the public. In light of the significance of support for lifelong learning by public museums, as described in the section on museums in Chiba and Hyogo, believing that museums play the intermediary role between the public and colleges and that it is possible for colleges to render a service to the public by open lectures and museum exhibitions will never enable intellectual exchange between colleges and the public. Again, researchers should have a stronger awareness of making science available for society.

Among college research institutions, general (research) museums are organizations that can piece together cross-departmental knowledge. Liberal arts departments were once expected to play a similar role. However, they did not serve as a place for knowledge integration because they were meant to train first- to second-year students. (If teachers were motivated, liberal arts departments could evolve differently as a place of higher education with a more fundamental significance. The issue that liberal arts has not taken hold in Japan cannot be discussed in this book due to space constraints.)

College museums are characterized as research institutions. Museum staff comprise researchers in natural and cultural history. Essentially, they should conduct research from an integrative perspective based on research materials. In the Japanese academic community, college museums represent the sole form of organization that enables the integration of disciplines, including the bridging of humanities and science, a process slow to take a tangible form, despite a loud voice for its necessity. Hopes rest on people who begin to pose questions to college museums as to the true integration of disciplines. That said, there is still a concern that more than a few naturalists who specialize exclusively on their particular materials (e.g. organism groups) build a wall around themselves and refuse to have a broader perspective. If naturalists continue to have such an attitude in the future, general museums attached to colleges will end with fragmented knowledge commensurate with the number of researchers and their results confined by individual research areas.

Some private universities have also founded social education centers and museums attached to related faculties. However, they do not trend towards the establishment of museums as research institutions. While private education institutions expect their museums to preserve and exhibit materials owned by them to demonstrate college assets and performance, there is no sign indicating that their museums will work on larger scientific breakthroughs.

The Makino Herbarium at Tokyo Metropolitan University is a museum-related research institution attached to a university. The Herbarium, founded featuring specimens collected by MAKINO Tomitarō, is a hub of research on plant species diversity. The institution was established in 1958, the year after Makino's death. He left a huge number of unorganized specimens, estimated at 400,000, between newspaper sheets. During some 40 years, the Herbarium has organized to grow these specimens as research materials, with type specimens computerized in a database, to serve as a center of research on plant species diversity. In 2018, an annex was built, which has a storage capacity of 1,500,000 specimens. Institutionally, it is a research facility affiliated to the university and is administered and operated regularly by the teachers and other staff of the Faculty of Science. At the Herbarium, researchers from Japan and abroad make effective use of the materials to achieve results. The institution conducts vigorous research on plants from the Bonin Islands, taking advantage that the Herbarium is a Tokyo metropolitan government-run facility.

An increasing number of large universities have come to have their own museums and similar facilities. Until very recently, researchers had research materials at their individual laboratories to use during their tenure. However, research materials were often lost after the researchers responsible for them moved or retired. It was rare that such materials were preserved and used at a suitable facility on campus. In recent years, there have been increasing cases of valuable materials transferred to museums and similar facilities for use even after the owner leaves. These relief measures are still implemented on an individual basis. In natural history no institutional guarantee has been provided for controlling research materials. It is a good sign that individual researchers have become motivated to prevent the loss of materials. While in more than a few cases it has been rather distressing to properly handle or dispose of personal materials, it is expected that information exchange and the building of collaboration frameworks will progress in the future.

The Museum of Nature and Human Activities is an organization that renders dual services as a facility attached to the University of Hyogo. In terms of clerical organization, the Museum is a subordinate organization under the Hyogo Prefectural Board of Education, separate from the University. There are high expectations for the institution to vitalize its activities as a prefectural museum and as a research institute (museum) attached to the University, given their organizational advantages, overcoming the seemingly complex organizational inconsistency.

The institution received approximately 250,000 plant specimens from the collection of Shōei Junior College in Kobe because the responsible researcher left the College at the end of the academic year 2012/13. Thus, enhancing and using the authenticated specimens in its plants department, the Museum drives forward with its activities.

Natural History Research and Education at Museum-Related Facilities

Museum-related facilities include botanical gardens, zoos, and aquariums. I am not well-versed in zoos or aquariums. Fortunately, the Natural History series includes 'Zoos in Japan' (2010) by ISHIDA Osamu and 'Aquariums in Japan' (2014) by UCHIDA Senzō, ARAI Kazutoshi, and NISHIDA Kiyonori. These books describe zoos and aquariums in detail, covering their research aspects as well. For botanical gardens, in which I have been deeply involved and have contributed to in Japan, I detailed 'Botanical Gardens in Japan' (Iwatsuki, 2004), included in the Natural History series. Therefore, this chapter provides only brief descriptions about them. In Japan, botanical gardens are regarded as facilities that grow and exhibit beautiful flowering plants to serve as a refreshing place for people. Indeed, it is essentially important for botanical gardens to play this role. In addition, botanical gardens are often compared with zoos, described in terms of amusement parks, probably due to the similarity in name in Japanese [動物園 (zoo) versus 植物園 (botanical garden)].

Botanical gardens in foreign countries include notable organizations that have historically developed as research institutions making contributions to society. Indeed, many major herbariums are attached to botanical gardens. Among an endless list of examples from both advanced and developing countries are the Royal Botanical Gardens, Kew (the global hub of research on plant species diversity), Royal Botanic Garden Edinburgh, the Copenhagen Botanical Garden, the Berlin-Dahlem Botanical Garden and Botanical Museum, the Royal Botanical Garden of Madrid, the New York Botanical Garden, the Missouri Botanical Garden, Jardim Botanico do Rio de Janeiro, Kubun Raya (the Bogor Botanical Gardens), the Singapore Botanic Gardens, the

Calcutta Botanical Garden, and Royal Botanic Gardens, Peradeniya.

Japan has many institutions that can be categorized as botanical gardens. The Japan Association of Botanical Gardens had 112 members as of 2017 (122 in 2024; in the 1990s, regular members of the Association numbered more than 140). The Association was founded in 1947 as a private organization and has been active as a corporate organization since 1966. Member institutions were largely established after the Second World War. More than a few member institutions are not particular about the term *botanical garden*, with their names containing terms such as *flower park*, *flower center*, and *botanical park*. Incidentally, the members include many physic gardens because it is compulsory for college faculties of pharmacy to have an attached physic garden.

Many botanical gardens in Japan, founded by municipalities, serve as institutions that supplement the functions of public parks and provide flowers and green foliage. Those which also work on natural history research are limited mostly to botanical gardens attached to colleges or run by the national government. However, since joining the Association, members have actively run joint projects contributing to the conservation of endangered species within their facilities to make achievements as an association of botanical gardens. The Association has had many research successes, including endangered species-related research results. Many member gardens provide support to projects that make social contributions, in addition to exerting the functions of public parks.

Nevertheless, as described later, activities at Japanese botanical gardens pale in comparison with major overseas botanical gardens, such as Kew, New York, and Missouri, which have led and are serving as hubs in natural history research. This is partly due to differences in size. In Japan, the Botanical Gardens of the University of Tokyo has the largest number of full-time researchers, four as of 2017. In comparison, Kew has 140 researchers with academic degrees, including those at the attached Jodrell Laboratory. The Missouri Botanical Garden has more than 50. In Europe and America, research assistants comparable in number to researchers are constantly hired. Differences between Japan and the West in staffing and necessary budget size are too large to cite without hesitation. These quantitative differences are naturally and clearly reflected in research results.

However, even a small organization has a chance to gain significant results if it focuses on a specialized challenge. Take the International Association of Pteridology Symposium, held in 1990 at the University of Michigan, as an example of quality achievements. At the time, the Botanical Gardens of the University of Tokyo selected ferns as a major research theme partly because I, then the director, was studying ferns. Various research projects were implemented using quality materials. Among them, multiple research projects were highly valued, such that of the slightly less than 30 total symposium oral reporters, four were members of the Botanical Gardens of the University of Tokyo. (There were two other Japanese symposium speakers from Tokyo Metropolitan University who spoke on their themes.) Even large research institutions with full teams of researchers may not always have innovative research contributions.

The International Botanical Congress is held every six years. For the 1987 Berlin congress, the Berlin-Dahlem Botanical Garden and Botanical Museum worked as the Secretariat. In 1999, the Missouri Botanical Garden played the role of the Secretariat. The Congress was held in Yokohama for the first time in Asia in 1993. For this congress, I served as the Secretary-General. The Botanical Gardens of the University of Tokyo virtually took care of the meeting as I was the Director. The staff size was not large, and they were very busy running errands during the period between preparation and cleanup. Owing to them, the international congress was successful, with nearly 5,000 participants.

Nevertheless, differences in human resources and budgets are critical. Although it is possible to produce an illustrious booming event one time, necessitating a huge amount of work, organizations with limited resources are unable to guarantee or sustain such activities over a long period of time.

Social contributions made by Japanese botanical gardens in the broad sense have been successful not only in visible aspects, such as providing flower and green foliage exhibitions as a refreshing place for the public, but also in inconspicuous aspects of helping the conservation of endangered plants. Botanical gardens can make beneficial use of their plant cultivation techniques by contributing to ex-situ conservation or propagating endangered species in a facility and reintroducing them to their natural habitat to the greatest extent possible. For this purpose, botanical gardens cultivate living endangered species. They have a well-equipped environment to make this widely known to the public, thus improving social education. Counting on this condition, the Japan Association of Botanical Gardens has set endangered species as a core of its recent programs.

Many of municipal botanical gardens founded during the rapid economic growth period following the Second World War highlighted their public park-like functions, then insufficient in Japan. The aim was to provide busy people with refreshment and pleasure by plants. This pleasure is beginning to be adorned with functions intended for enjoyable learning.

The Japanese people have long valued contact with nature. In over-developed and excessively artificial urban settings, they yearn for green places. One solution is for them to enjoy green places while increasing their science literacy and having a renewed awareness of living in harmony with nature. Towards this purpose, botanical gardens still have been slow to initiate action. However, hopes rest on strengthening future connections between people and botanical gardens and similar facilities. If in this process the perspective of natural history takes hold, there is hope that developments for humans to live in harmony with nature to ensure sustainable use of resources on earth will occur.

Concept and Analysis Techniques in Natural History To establish an integrative perspective, this book emphasizes the significance of natural history study to analyze the integrated whole of diverse living organisms. From this standpoint there come ideas difficult for many researchers in modern sciences, who have been trained to produce analytical research results.

Some researchers analyzing living organisms are aware that research solely relying on analytical techniques is limited. Developmental biologists who look at each multicellular body as an integrated whole and study the formation processes of individual bodies have developed the analysis method known as evo-devo (3.3). In a sense, this was a natural course of research development. Nonetheless, if I include these areas in the category of natural history, natural history researchers may become confused.

However, to contribute to society as a modern science, natural history researchers will increasingly need to be thoroughly familiar with the techniques of investigative science and to study individual phenomena in an analytical manner. Not every researcher needs to hold an integrative perspective firmly in mind. I would emphasize that the right path of natural history will make essential and great contributions to science. Scientists devoting themselves to investigative research and being content with the results thereof will face the risk of losing the perspective of natural history, i.e., an integrative perspective.

NAKAMURA Keiko at the JT Biohistory Research Hall advocates biohistory research, an attempt by molecular biologists to view the kingdom of organisms as an integrated whole, while focusing on the diversity of DNA. Huge expectations are placed on the research pursuits of biochemists. Unfortunately, most biochemists do not correctly understand this concept.

Evo-devo and biohistory both represent contributory efforts made by researchers not satisfied with classical natural history. In the field of natural history, researchers are also willing to adopt molecular-level analysis methods when available to analyze the state of being alive. It is desirable that researchers look at living things from an integrative perspective, making beneficial use of available analysis methods, rather than to partition their work into microscopic and macroscopic scales.

(4) Promotion of natural history supported by society

For natural history, the issue of research expenses and educational program costs is serious. Since mainstream sciences began to demand empirical results by hypothesis-testing and physicochemical analyses, it has been very difficult to gain financial support for field surveys and for the management and maintenance of research materials. Since they became corporate bodies in 2004, national universities have been facing cuts in operating expenses. It has become necessary to gain more external funds to cover not only research expenses, but also even educational costs. Faculty members complain that their schedules are too tight to conduct research due to the hours spent raising research fund.

Grants-in-Aid for Scientific Research and Other Promotion for Research
For natural history research, field surveys remain essential, no matter how important the analytic work becomes. In this regard, Japanese naturalists found great assurance in the 1960s when the Grants-in-Aid for Scientific Research added the category of overseas academic research. This category survives still today, despite a history of twists and turns. This category is rare in countries outside Japan. Moreover, even when strict single-fiscal year expenditure was required with the Grants-in-Aid for Scientific Research, which restricts large research projects, reasonable consideration was given to field research. Thus, the funds have been effectively used.

The Grant-in-Aid for Scientific Research (Overseas Academic Research) was a category once set up and noticeably effective in the fields of physical anthropology and cultural anthropology, which uniquely evolved in Japan. The grant also made steady contributions to the domain of basic research in biological species diversity, although with less remarkable topics. In this research fund category, as an aspect of field research, programs were vigorously implemented to work jointly with local researchers and render a service to the furtherance of research in developing countries. Therefore, I value it as substantially contributing to international joint research in a broad sense. Joint research programs funded by this Grant-in-Aid have noticeably led to the invitation of many students to study and receive training in Japan to become future researchers.

Private Research Foundations and Award Programs In Europe and

America, donations from the private sector play a substantial role in the management and maintenance of museums and similar facilities. For example, the British Museum began with a personal collection of Doctor H. Sloane (1660–1753), bequeathed to the British government for a sum of £20,000, paid from the Parliament to the bereaved family. The Smithsonian Institution in Washington was founded by funds bequeathed to the Government of the United States by, as the name indicates, the British scientist J. Smithson (1765–1829).

Japan has seen less of these philanthropic cases, perhaps due to differences in philosophy about donation, differences in pragmatic aspects such as tax incentives, or the absence of the true rich in Japan. Certainly, it is not rare that personal collections of specimens are entrusted from a bereaved family to museums after the death of a collector. The Yamashina Institute for Ornithology is a research institute founded based on the collection of Dr. YAMASHINA Yoshimaro. The Center for Academic Resources and Archives, Tohoku University opened the Tsuda Memorial Herbarium in 1987, which was built primarily on funds donated by TSUDA Hiroshi, who wished for the furtherance of plant taxonomy. However, donations of funds to museums for research purposes have been very limited.

Recently, environment-related NGOs and NPOs collect donations from a wide range of groups. Alongside the incorporation of national and municipal universities, schemes for accepting donations have developed.

Donations to academic activities made by corporate philanthropy programs have gained social recognition to some extent. That said, considering that Japan is the world's third-largest economy, donations for academic purposes are still limited, except in areas directly contributing to technology. This may be proof of the public's level of awareness about academic and cultural activities. Alternatively, it may indicate that the intellectual contributions of science to society remain limited in extent.

Foundations subsidizing scientific research have recently increased. However, subsidies to natural history-related research projects are few. Under these circumstances, the public-interest foundation described below incorporated to exclusively foster natural history.

Fujiwara Natural History Foundation The Fujiwara Natural History Foundation was founded in 1980 using funds bequeathed by FUJIWARA Motoo, who successfully worked at the Tsukiji Market. At first the organization, as an incorporated foundation authorized by the Tokyo Metropolitan Board of Education, aimed principally at implementing biological education in high school. In 1992, the organization restructured into an incorporated foundation under the control of the Ministry of Education and began to subsidize natural history research. In 2012, it was reorganized into a public-interest incorporated foundation, as necessitated by a reform of the corporation system. It has actively subsidized research ever since.

Fujiwara was reportedly a business proprietor dealing with fish at Tsukiji. He had strong interest in subsidizing basic research and bequeathed large funds to run a research funding foundation. Thus, the foundation that helps natural history education and research was inaugurated.

To finance research subsidies, the foundation mostly relies on interest earnings from the basic fund. When the deposit interest is low, the total sum of subsidies remains low, accordingly. The foundation primarily subsidizes individuals (or small groups of researchers), with individual sums moderately amounting to less than one million yen. Nevertheless, subsidies granted each year to researchers in the domain of natural history are dependable support specifically earmarked for young researchers, including graduate students. Researchers affiliated with museums and similar facilities account for a high percentage of those subsidized, a distinguishing characteristic among research subsidies in the field of natural history.

In addition to the research subsidies, the foundation continues to donate tools for science education to high schools, hold open symposia for the dissemination of natural history knowledge, and run other programs tailored for schools and the general public. Their website <fujiwara-nh.or.jp> provides more details about the foundation's activities.

Transformation of Administration at Museums and Similar Facilities In recent years, social recognition has substantially grown regarding the activities of naturalists at museums and similar facilities. Meanwhile, a contradictory trend against the support of natural history research has also grown of late. Public museums and similar facilities have been under economic pressures. Under the slogan of "becoming semi-public," they have been incorporated or placed under the designated administrator system to meet expectations for them to focus on profitable programs.

In Japan in particular, national and municipal institutions tend to strongly control people in an overconfident way, partly as a negative effect of reforms made during the Meiji period. Even museums and similar facilities, with increasing bureaucratic awareness, tend to do what museum staff desired, rather than to stimulate the scientific curiosity of people. Even at national universities, some disciplines are criticized for their lack of social contributions and the use of research activities as an excuse. At many institutions, reduced expenses increasingly constrain activities, with the nominal aim of ensuring the allocation of larger budgets to socially beneficial sectors. Noticeably, museums and similar facilities have little time, energy, or funds to conduct fundamental research to sustain their essential activities, such as providing support for lifelong learning and think-tank functions.

It is difficult to make an objective evaluation of the resultant implications of semi-public museums and similar facilities, which cannot be verified in terms of numbers. This is not only because of the characteristics of research and social contribution activities. In fact, organizational reforms were made precisely when staff at museums and similar facilities were beginning to have a raised awareness for promoting tangible activities for society. It is necessary to scientifically observe the real trends to understand the best organization forms, without uncertainties.

The Museum of Nature and Human Activities is producing reasonably successful results, despite strict cost and personnel cuts across the prefectural administration. As a former insider, I am proud of this. I have evaluated the Museum's 20 years of history in a book [Museum of Nature and Human Activities ed., 2012] and a museum brochure. The Museum provides an illustrious example of how organizations relating to natural history should act, as described in the following section.

4.2 Collaboration between Museums and Universities

I stated earlier that capable museum staff can train successors by collaborating with a college that has graduate doctoral courses. Some museums (not limited to natural history museums in the narrow sense) proactively take part in graduate education, as well as conduct research and disseminate knowledge. Examples include the National Museum of Ethnology and the National Museum of Japanese History, which are key members of the Graduate University for Advanced Studies, even though they proclaim that they are museums. If considerably large museums are viewed as research institutions, museums can work on graduate education to serve to train their successors, an idealized purpose for these facilities.

A large national museum can conduct research as a key purpose of its establishment. However, local museums under control of municipalities place knowledge dissemination as the principal purpose of founding, in accordance with their requirements for establishment. This is true even if the size of an institution is considerably large, and research is regarded as a basic activity to meet that purpose. Under these conditions, a new organization was desired with the aim of allowing research and knowledge dissemination to go hand-in-hand. The solution was the Museum of Nature and Human Activities.

(1) Administration of prefectural museums in Japan – Museum of Nature and Human Activities, Hyogo as an example

Curators of Museums in Japan Until recently, museums, or at least natural history museums, in Japan needed a limited number of staff members to carry out various tasks. Research staff were unable to smoothly carry out essential services, such as research and provision of support to lifelong learning, partly due to their small sizes.

Considering legally recognized purposes of establishment, the National Museum of Nature and Science alone is a research institution, which is expected to work on knowledge dissemination based on research results. (The Museum did not always have a smooth exchange with colleges on an institutional basis, except at the personal level, due to its specific characterization as an organization. When the Graduate University for Advanced Studies was founded, the Museum did not consider becoming a member.) In contrast, municipal and private museums are supposed to mainly work on knowledge dissemination and to collect and manage necessary materials and conduct research for that purpose.

The status of museum staff, whether they are researchers or curators, is categorized as government staff. Unlike college faculty (educational posts), they are not allowed to conduct self-initiated research activities. This restricted researcher status has constrained research activities and think-tank functions of museums in Japan.

For the status of college faculty, the Special Act for Educational Personnel and relevant customs ensure a substantially higher degree of freedom in research and education than for general personnel. Curators at natural history museums in Europe and America are expected to be active as researchers. They are also allowed to conduct self-initiated activities, like college faculty do in Japan. Indeed, many curators make greater scientific contributions than faculty staffs and are accordingly evaluated by academic associations. They provide think-tank functions and substantially contribute to society. In contrast, to pursue research, museum staff in Japan face many formal constraints. Additionally, social recognition of museums as research institutions has remained rather low for a long period of time.

Here follow some examples of status differences regarding museum staff in Japan. Faculty used to enjoy some degree of flexible duty hours and could do learning at home almost at will. It is said that humanities faculty once included those who seldom came to the campus and appeared at their offices only when they needed to give a lecture, attend a meeting, or use the library.

Many departments of science, for which experiments were essential, differed from departments of humanities in that faculty spent many hours in the laboratory because experiments could not be carried out at home.

For educational posts, research performance (treatises and presentations at academic association meetings among other achievements) is important for promotion and obtaining a Grant-in-Aid for Scientific Research. Therefore, faculty use some of their time to improve research achievements. Biological researchers deal with living organisms. They need to develop a plan according to the specific rhythm of the organism in question. Therefore, it is not rare that they work outside normal working hours, such as midnight, and are absent from their offices during the normal working hours if not constrained by lectures and hands-on training programs.

In comparison, to evaluate the work performance of government staff, tangible results from routine work are difficult to objectively recognize. Partly due to this reason, importance is placed on how much they conform to the specified working hours, with tardiness and absence receiving negative ratings.

Faculty staffs are generally free to go on research trips. To go for a research trip abroad, they formerly needed to obtain approval from the faculty council. As overseas research trip occasions increased, the procedure was simplified. Except for faculty staffs in a managerial position, faculty could leave the workplace for a destination (including a foreign country) for investigation purpose, if they complete the minimum required tasks. In contrast, government staff follow the directions of their managers, even if they are researchers. They are not as free as faculty staffs and not allowed to work at their own initiative and direction, although this may vary to some degree according to workplace customs and the intention of the manager.

The above difference was also true with the use of research subsidies. While government staff were subject to various constraints, college faculty could use subsidies flexibly, as long as they followed accounting rules. However, ironically, a small number of immoral researchers committed fraud, inviting mistrust towards the usage of subsidies in general. It was a very regrettable scandal. Notably, some researchers provided with ample funds engaged in improper transactions with suppliers to disguise single-fiscal year accounting practices because of their inability to use up funds within the fiscal year.

Specialists in a research field may occasionally provide expert knowledge in response to requests from outsiders (= exerting think-tank functions). Faculty staffs may receive gratuities almost without any restrictions, although they are required to report on received money. Government staff are prohibited in principle from personally receiving any gratuity. Perhaps the specialized knowledge of researchers in teaching positions is considered to belong to individual researchers, while government staff's knowledge is considered to belong to the organization.

This results in incongruities. Researchers at the Museum of Nature and Human Activities include college faculty and curators under the board of education, as described later. Two similarly capable individuals differing in terms of job status may be regularly engaged in similar activities sitting side-by-side in an office. When these two are asked to serve as members of an outside committee, they may attend meetings and perform similar services. However, the researcher affiliated with the college receives a gratuity for attending the meeting, while the curator under the board of education can receive no gratuity.

It cannot be denied that in Japan the above-described differences (or maybe *discrimination*) have left researchers and museum curators (government staff in principle) impregnated with a kind of paranoid awareness and vague feelings of envy towards college faculty. In Europe and America, society reasonably respects museum curators according to their abilities. In Japan, poor social recognition of museums as research institutions could have resulted in spurning the above-described trend.

A Trial of the Museum of Nature and Human Activities When the Museum of Nature and Human Activities was founded in Hyogo Prefecture in the early 1990s, measures were explored to eliminate the above-described status-related constraints (Museum of Nature and Human Activities ed., 2012). To solve this issue completely, the law concerning the establishment of museums needs to be changed. However, amending the law without a full social understanding will require considerable amounts of time and energy. In cases like this, it is common to develop a feasible framework, while complying with the applicable legal requirements to achieve concrete results through service. When a proven record of accomplishment meets a need recognized by society, it is not so difficult to make necessary legal amendments.

Preparations for the establishment of the new museum in Hyogo prefecture began in the late 1980s. As a natural history museum, it was planned at an unprecedented size in Japan (although the Natural History Museum and Institute, Chiba of a comparable size opened in 1989). For the museum to be comparable to those in Europe and America in terms of performance including support to lifelong learning and think-tank functions, the existing organizational conventions had too many formal constraints for the roughly 40 prospective researchers to be active.

Some of the researchers playing a pivotal role in the preparations for the establishment noted this issue. The National Museum of Ethnology, which opened in 1977, was founded as an inter-university research institute. This museum is similar to research institutes attached to colleges in terms of the status of its researchers. They used titles such as professor, assistant professor (currently associate professor) and assistant (currently assistant professor). The Special Act for Educational Personnel applies to the researchers there. Although its name contains "Museum," the researchers there are treated differently from conventional museum curators.

Following the example of the National Museum of Ethnology, the National Museum of Japanese History founded in 1981 started with a similar structure. Unlike national organizations, there were no legal grounds for prefectural institutions to follow the above-mentioned formalities. Nonetheless, they wisely devised operational measures to improve the given conditions. The people involved in the founding of the Museum of Nature and Human Activities racked their brains to build an organization in which museum researchers would be faculty staffs of the prefectural university. The aim of this solution was to establish a style that faculty of the prefectural university in Hyogo could conduct research activities at the museum, enabling them to work in the status of a faculty staff.

In practice, the devised solution was to establish a new research institute attached to the Himeji Institute of Technology (the present-day University of Hyogo) and locate the institute at the museum so that the faculty staffs affiliated with the research institute could conduct daily activities at the museum. It is desirable for a natural history museum that faculty staffs (researchers) at an institute attached to university (Research Institutes and Centers) are engaged in the same activities as the museum staff.

The devised style envisioned an unheard-of organization. To build such an organization, it was necessary to discuss the matter with the authorities and obtain their approval. Importantly, to meet these requirements, the project had the right people for the mission. In practice, the project turned out to be in line with the regulations and established a precedent to overcome the challenge, which had been a major impediment in Japan. Consequently, a nameplate for the Institute of Natural and Environmental Sciences, Himeji Institute of Technology was put side-by-side with the nameplate for the Museum of Nature and Human Activities, Hyogo. Researchers at the research institute (in the status of professor, assistant professor, or assistant) conducted their activities as museum staff by nominally commuting to the museum and performing dual services.

Regrettably, however, all prospective researchers could not be affiliated with the research institute attached to the university. To found a museum at the time, establishment required a specified number of museum curators assigned to the facility. Because the true aim was to found a museum and not to establish a new college research institute, it was necessary to assign the required number of researchers as full-time museum curators. This requirement for establishment has been altered. However, it is very difficult for administrative offices to change the status of already hired personnel. The result is that status discrimination is still present between individuals, although they are similarly working at the same workplace.

It is commonly known that one must go through many formalities to develop a new system. It seems that the development of the system at the Museum of Nature and Human Activities was very laborious. It is easy to imagine the difficulty they went through, even today. However, notwithstanding the difficulty, the new system effectively prevented researchers at the museum from developing a jaundiced view towards the status of college faculty, a favorable wind for activities at the museum.

That said, whether every researcher at the museum could conduct ideal museum activities is another story. If favorable conditions ensure that every individual will work in the ideal way, the world would be far better than it is. Conversely, even under harsh conditions, some individuals can produce activities with great results. Finally, in every part of the world, the key is the quality and motivation of the people involved.

It seems that individuals engaged in museum activities under the status of college faculty do not fully understand the advantage of their status and its magnitude of value at museums in Japan. I tell them about this repeatedly, although knowing about the poorer conditions will not necessarily (or should not) bring anything beneficial.

Narrating the above story may simply be a regretful look at the past. Recently, young researchers who desire to work as museum staff rather than college faculty have emerged. As I have had continued interest in this issue, the change appears as an impressive advance in the situation.

As it is generally understood that the activities of museum researchers are of the same nature as those of college teachers, then some degree of flexible work arrangements for museum researchers can be allowed to the extent managerially possible, even if their status is administrative service staff. Along with the revalidation of the key elements in evaluating the activities of museum staff, this is helping boost museum activities.

(2) Graduate schools in collaboration with museums – Biodiversity and Evolution Courses at the University of Tokyo

The organizational style of the Museum of Nature and Human Activities, which attached a college research institute to the museum, was developed between 1989, when the preparation for establishment actually began, and 1992, when the museum opened. At the time, as a favorable backdrop for this development, there was a burgeoning trend towards academia-museum collaboration. The term *academia-museum collaboration* has now come to be understood to some extent by the public. However, at the time, education programs implemented by collaborations between schools, including colleges and museums, were not openly drawing public attention.

A little later than the founding of the Museum of Nature and Human Activities, an initiative emerged to establish a graduate school department with the aim of training successors for biodiversity research by means of collaboration between the University of Tokyo and the National Museum of Nature and Science, among others. This graduate school initiative took the form of setting up a (then) Evolutionary and Biodiversity Biology Course under the (then) Department of Biological Sciences, Graduate School of Science, the University of Tokyo. The initiative envisioned instructing excellent graduate students by creating a collaborative team comprising faculty staffs at the University of Tokyo and a delegated group of dual-service teachers consisting of researchers at the National Museum of Nature and Science and teachers in biodiversity-related subjects from other colleges in Japan not engaged in doctoral programs guidance. This Course was inaugurated in the academic year 1995/96.

The initiative sprang from a sense of crisis that MARUYAMA Koscak (1930–2003), who held the posts of President of Chiba University and President of the National Center for University Entrance Examinations, had about the absence of an organizational foundation for training successors in animal species diversity research. He explored whether it would be possible to found an organization for training excellent successors by utilizing the abilities of researchers who were currently engaged in research at a college but had no opportunity of providing guidance at a graduate school with doctoral programs. He also considered whether collaborating with science museums that held active researchers in related fields would help with training successors. In the field of botany, the sense of crisis about organizational issues was not so serious because, at the time, doctoral programs were effective in training successors in natural history at graduate schools of several universities. Certainly, there were various non-organizational problems, however, which are unrelated in the context of the present topic.

At first, Maruyama and others explored possibilities of forming a new organization relating to animal diversity in their own way. They strived to set up the organization for some time. However, it turned out that the candidate universities they worked with all failed to form an organization that met their expectations. Before long, with the strong intention to build an organization relating to biodiversity, they approached biologists. In this process, Maruyama made the best use of a personal connection with myself, as we were colleagues at Kyoto University and, having moved to Kanto at a similar time, had collaborated many times through academic associations and at the Science Council.

After trying various measures, Maruyama told me that it was impossible to implement the innovative idea in the domain of zoology alone and that he desired to further the idea by pushing biodiversity to the front. He proposed that the University of Tokyo could manage the necessary formalities.

It was difficult for the University of Tokyo to meet his request, as it came immediately after a budget request was approved to transform from a faculty-dominated organization to a graduate school-dominated one at the Graduate School of Science, the University of Tokyo. An additional organizational change would normally be out of question. Notwithstanding this situation, the organizational change was implemented partly due to efforts of the National Museum of Nature and Science, the collaborating partner, who pushed for the change. Fortunately, at the time the Ministry of Education was also motivated to raise the degree of organizational freedom to enable more diversified graduate education.

It is useful to look at the relevant historical background for this change. In 1988, the Graduate University for Advanced Studies was founded, as a graduate school consisting of multiple institutions and providing doctoral programs only, which linked research institutes governed under the control of the Ministry of Education. The National Museum of Ethnology and the National Museum of Japanese History are key members of this graduate school. The sense of resistance to combining the term *museum* with *graduate school* should have melted away by then. However, the National Museum of Nature and Science did not join this graduate school. I asked various people about the reason and received formal explanations. However, no one could provide a scientifically grounded answer that I could comprehend.

Consortiums of colleges were also formed. As examples of allied colleges that had no doctoral programs and who were setting up graduate schools with new doctoral programs, the United Graduate School of Education and the United Graduate School of Agricultural Sciences formed.

Although its name is similar and confusing, the Kyoto Graduate Union of Religious Studies is an association of faculties and graduate schools at eight existing universities, formed to enable smooth credit transfers and other services and created as an alternative to a formal graduate school or a graduate department. The organizational formation of united graduate schools probably helped promote alliances like this.

Due to a favorable evaluation of the united graduate schools and departments, the trend gained momentum, enabling many examples of the founding of a graduate school by allied colleges. Research institutes, as well as colleges, also joined coalitions. Not only research institutes under the control of the Ministry of Education, but also research institutions under the control of other governmental departments and private institutions became able to ally with colleges. In this regard, the Evolutionary and Diversity Biology Course at the University of Tokyo developed a system which enabled researchers at municipal and private universities, as well as national institutions, to take part, a forward-thinking attempt at deregulation.

For natural history-related research organizations, topics protected under the research unit system dissolved. Although some corrections were made, situations like this would ultimately depend on the people participating in the organization, in addition to formal amendments. One can achieve reasonably successful results even under difficult circumstances in some cases, while in many cases, one can fail to achieve successful results even with dependable formal protection. Amendments (far from a perfect provision) of conditions for natural history were supported by a sense of crisis from outsiders, rather than the efforts of insiders, and they did not dramatically improve the motivation of researchers in natural history-related fields.

Currently, what is needed most for the furtherance of natural history is a wide recognition of the significance of research that uses naturalist techniques and the elucidation of accurate facts about phenomena. Furthermore, it is key that the general public understands that fulfillment of scientific curiosity by scientific inference based on the elucidated facts helps improve people's science literacy and that this resultant raised interest in natural history contributes to modern science.

4.3 Natural History Research Viewed from a Global Perspective

Examining the current state of natural history research in Japan begs the question of what position it takes in the world. Tracing the past and present research in this domain globally would produce a tremendous amount of information. This section will look at a representative portion.

(1) Natural history research in Europe and America

Research in natural sciences has greatly advanced in Europe and America, as concretely indicated by the number of Nobel Laureates. Essentially, the modernization of sciences in Europe since the Renaissance has built and fostered natural sciences into their present-day forms. Of course, people in every part of the world are working on research in natural sciences in anticipation of the benefits that their studies may bring. However, at present, without doubt, Europe and America are at the forefront in this field, both in quality and quantity.

This is true from the perspective of natural sciences. However, what of the specific context of natural history? If what matters is strict conformance to hypothesis-testing methods based on reductionist analyses, as understood in natural sciences, attempts to look at organism bodies from a total perspective in natural history are criticized as not being scientific. The reason is that science is supposed to lie solely in analyses by physicochemical methods of a single aspect of phenomena exhibited by substances that constitute an individual body. This tendency is noticeably seen in the proportional allocation of research funds, for example, in the United States.

Nonetheless, attempts to understand individual organisms by ascertaining the whole picture of their lifestyle are also vigorously promoted. Although the aim is a holistic perspective, no adequate methods for that purpose have been established. Therefore, for the time being, subjects of natural history research should be biota, from which researchers gain hints by exploring information about which organisms are living in which habitats. In the United States, evolutionary research is generally pursued at the forefront of the understanding of living things, although some states are resisting teaching evolution due to its religious implications. Moves towards understanding life from an integrative perspective are also apparent, with research groups (departments) being formed to pursue integrated biology. As another reality, research valued as excellent in this domain is the result of accumulating empirical achievements in meeting specialized challenges.

More specifically, for biota surveys, Europe and America are at the vanguard of the production of basic information, including speciation processes, partly due to moderate levels of species diversity in these areas. Their lists of diverse species complemented with regional characteristics enable one to examine the origin and development process of species diversity. European and American researchers have contributed to basic surveys of biodiversity in all parts of the globe and have produced noticeable results.

Moreover, they conduct analytic research to inquire into universal principles underlying species diversity, and at the same time, they promote integrative research on species and species groups by noting inherent phenomena that diverse organisms individually exhibit.

To investigate universal principles in the kingdom of organisms, researchers analyze the phenomena in question using model organisms and consider how to universalize the analysis results across the kingdom of organisms. Also, by selecting organisms and inquiring into the lifestyles of the selected species, their research intends to clarify the implications of diversity. Bracken is a commonplace plant. John Lawton (1943–) and others observed the ecology of various living things on individual bodies of this genus, producing research that won the Japan Prize in 2004.

The desolate autumn landscape of windy heights in northern England is lonelier when the ground is covered with dead bracken leaves. In England, where the people do not eat bracken sprouts, this genus is treated as a troublesome weed. Herbivorous animals, such as cattle, do not eat live bracken. However, bracken can be contained in hay, which is eaten by cattle. Reports on a high rate of cancer development from eating it once attracted people's interest. Bracken has nectaries on its plant bodies, which lure ants to guard against entomophagous insects. Lawton and others (1976) observed organism communities on brackens and described inter-species coexistence and competition in detail. Starting from this observation, they furthered their research by analyzing ecological diversity from a natural history-specific perspective. In doing so, they combined field observations of communities of various species with analysis of phenomena in laboratory and conducted pattern analysis of large datasets and mathematical model-based analysis.

These achievements were possible by Lawton making the best use of his personality as a naturalist interested in contacting living things. With the above research background, Lawton was later involved in science policy, working on the development of large environmental test apparatus known as the Ecotron and achieving successful results for the conservation of birds.

In Europe, research based on the naturalist tradition described above is deployed in various forms. Although they are not economic giants, countries in North Europe also play a pivotal role in basic biodiversity research. For example, the Netherlands contributes to the compilation of 'Flora Malesiana,' and Denmark works as a leader on 'Flora of Thailand.' The secretariat for the Global Biodiversity Information Facility [3.3 (2)] is located at Copenhagen, Denmark. The country's proactive contribution is a major drive to the development of this project.

In Europe, contributions by museums are a basic factor in the furtherance of natural history research. In England, the Natural History Museum in London and the Geological Museum are backed by a gorgeous history, are tremendously large, and contain incomparable numbers of collections. The same applies to the National Museum of Natural History in Paris and the Smithsonian Institution in Washington. These museums make huge contributions to basic research and store a vast range of natural history specimens that can be used to further the evolution of research.

During 1969 to 1970, I stayed and conducted research for the first time at the Natural History Museum with a scholarship awarded by the British Council. I oversaw ferns for 'Flora of Thailand' and thus investigated specimens. I examined a full collection of ferns taken out of the basement storage of unidentified specimens, which were from Thailand in the early 20th century, including specimens that A.F.G. Kerr (1877–1942) collected from throughout Thailand in the early 1900s. There were stacks of research materials like this in storage for more than half a century, neatly organized for immediate use when an interested researcher appeared.

I had a similar impression when I later visited the National Museum of Natural History in Paris several times. I used to make beneficial use of such services each time I visited an herbarium in Europe and America. While independent natural history museums such as in Paris, London, and Stockholm have herbariums, many are attached to colleges (e.g. Harvard, California, Chicago, and Copenhagen. Incidentally, in 1992, herbariums at Leiden University, Utrecht University, and others were united into the National Herbarium of the Netherlands). In addition, herbariums are often attached to botanical gardens with the aim of locating them in tandem to living plants.

In the field of botany, botanical gardens and similar research institutions play a major role in natural history research. The Royal Botanical Gardens, Kew in the United Kingdom maintains an herbarium that, contributes most vigorously out of all herbariums in the world to flora research. Kew also has the attached Jodrell Laboratory, which actively contributes to species characteristics analysis. In the 1930s when cytotaxonomy became a topic, the botanical research institution enriched laboratories for this new branch of study. In the 1960s when chemotaxonomy was a topic, it invited E.A. Bell as the Director and provided the Laboratory with more researchers in this field of study. As molecular phylogeny became available as an analysis technique, Kew recruited M.W. Chase (1951–) from the United States in 1992 and appointed him the leader of a laboratory to perform data digitization. In tandem with the promotion of visible activities intended for the general public, the Royal Botanical Gardens continues to make leading-edge contributions to basic research.

In very recent years in Europe and America, botanical gardens have been facing the need to conduct profitable activities to support management. Projects that do not directly lead to revenue are under increasing pressure. Contributions to basic research are confronting this harsh reality, both in funding and human resources.

(2) Natural history research in Asia

In China, natural history developed through herbalism, as described earlier, in connection with its introduction to Japan. China has a tradition of recognizing nature's creations in the country as valuable assets. In China, researchers conduct research on nature in line with the country's herbalism tradition by making asset lists. Even during the Cultural Revolution period, in which the government was hostile to culture and basic scientific research was virtually suspended, research on fauna and flora continued to some degree. Some exceptions I have heard are that researchers were forced to give up their research due to the Down to the Countryside Movement and that a collection of potted tropical plants was labeled as bourgeois science materials and taken out of the greenhouse for hours, resulting in devastating overnight damage to the plants in extremely cold Beijing.

Although this section focuses on natural history research, we will also look at the state of research on plant species diversity. Present-day China is using its resources on research in leading-edge life sciences. In a sense, this is natural in connection with bioindustry and medical care. At the same time, records of nature in the country, which is blessed with high species diversity, are also vigorously promoted.

Since the days when China was still poor, many researchers have been working on biota research at a research institute affiliated with the Chinese Academy of Sciences. As early as in 1959, the 'Flora of China' compilation project started. At first, Chinese researchers implemented this project to publish records of flora in the Chinese language. In this project, because China is a vast country, provincial and other regional compilations have been published. A project to publish 'Flora of China' in English jointly with American researchers was completed in 2013, in cooperation with related researchers throughout the world. Because the principal goal was to organize currently available information, the project included areas still to be investigated. Hence, the presented content is subject to some strict reviews.

Current scientific knowledge about plants in China is still insufficient. Nonetheless, it is advancing steadily. On one hand, the compilation of the records of plants is a project that elucidates plants in China; on the other hand, it provides basic information that will greatly facilitate plant species diversity research on a global scale. Moreover, it is not surprising that research results on plants in China, for which I provided some assistance, will also greatly help my research when I reexamine species of continental Southeast Asia and the Himalayan region, based on field surveys that I and others conducted.

In China, basic species diversity surveys and comparative research to produce books on flora are vigorously conducted. However, in the country, there is no sign of the development of research to ascertain the whole picture of living things based on these surveys or to analyze life from the perspective of natural history. Hope for future developments in the field rests on the recognition of facts about species diversity by traditional methods. China's economy has grown remarkably since the turn of the century. For this country, there are high expectations for substantial advances in science as well.

China has rich biota, particularly in its southwest part, including the provinces of Sichuan and Yunnan. This is because the biota there is part of the biota of the Himalayan region, which has the richest gene resources in the temperate zones on earth. As such, investigating the biota of the regions from the southwest part of China to the Himalayan region for the elucidation of biological species diversity on earth is an important challenge. Indeed, investigating the biota of the Himalayan region has been a subject of scientific curiosity by Western researchers since the Age of Exploration. In addition, it has been an important subject of surveys for securing resources. Full-fledged research on Himalayan flora began in the 1850s, when J.D. Hooker, who served the Director of Kew and is well known for his interactions with Darwin, conducted surveys. Surveys on Himalayan flora by Japanese researchers began after the 2nd World War. The early survey style was such that researchers stayed for a long period of time on an individual basis. Before long, research teams working on field surveys were sent successively. Currently, research is conducted in various manners beyond the framework of monographs.

In investigations across South Asia, including the Himalayan region, plants in India and Sri Lanka were the focus of attention. This is because these areas served as bases for Europe to dominate the Orient. Europeans conducted basic surveys on species diversity to build a foundation for securing resources, in a way typical for the time, although, for researchers conducting surveys in practice, the purpose was to fulfill their scientific curiosity. That said, books of flora and fauna across the Indian subcontinent have not been compiled in practice, although projects have been repeatedly planned.

In Southeast Asia, European explorers conducted surveys with an interest in animals and plants as natural resources in the early period of the development of the sea route to the Orient. Before long, naturalists emerging from explorers conducted surveys to build basic knowledge on species diversity. However, tropical zones have an extraordinary expanse of biological diversity. The range that a small number of naturalists can elucidate with only scientific curiosity is limited. It is easy to make full use of the resources currently in use. However, even still today, science has not been successful in collecting basic findings sufficient to make beneficial use of biodiversity as potential gene resources.

In Southeast Asia, basic species diversity surveys were conducted on the continents by willing researchers from the former suzerains because the continental part of this region was once largely colonized by the United Kingdom and France. However, surveys were slow on islands with the richest biodiversity. For Polynesia, Melanesia, Micronesia, and other similar areas containing small islands in the Pacific region, research materials were available from people traveling in these areas, despite the sluggish pace of surveys. For large islands, particularly inland areas in the geographical Malesian region (not equal to the country Malaysia), it was difficult to collect information on nature due to very low accessibility. In 1951, a project began to edit the 'Flora Malesiana' in an international collaboration framework. The achievement rate of this project has met one-third of its goal, although it is progressing gradually. The project remains at this level of progress even though its aim is merely to collect basic information about what is present and where it is located. [I am happy to note here that our contribution of Hymenophyllaceae for this Flora was issued in 2023.]

Natural History of Copeland in the Philippines

As an example of the lifestyle of a scientist, let us look at Edwin B. Copeland (1873–1964) who, born in the late 19th century in western America, studied biology in Europe and served research and research policy in the Asian tropical zone. As a biologist, he is not as renowned as Darwin or Mendel. However, a lot of things can be learned from Copeland, whose subject of study was ferns (Wagner, 1964).

First, the reason that this section looks at Copeland comes from my interest in the lifestyles-of-fern research. This was first sparked when I founded a science club at the newly established junior high school in Oku-tamba and independently strolled in the hills. The science teacher YOSHIMI Hajime, who was the advisor to the club, told me that few people knew ferns, while many people had a good knowledge of flowering plants. Encouraged by his words, I prepared fern specimens without any instruction (Iwatsuki, 2012).

When I went on to high school and joined the activities of a biology group, I collected a small quantity of fern specimens, which I thought of as my hobby. However, when I became a college student, I was invited and joined the Ferns & Mosses Study Club, where I attended authentic lectures by Professor TAGAWA Motoji. Professor Tagawa used the book 'Genera Filicum' by Copeland as a textbook of taxonomic hierarchy. Published in 1947, this book is the compilation of research on ferns conducted by Copeland. As I joined the Club in 1953, it was the newest publication in its field at the time. This Club was inaugurated in 1950. To my knowledge, the first lecture by Professor Tagawa was on the fern allies like whisk ferns and club moss. It might have been a matter of course that Professor Tagawa selected the authority's new publication as a textbook when he began to give lectures on ferns (early 1951 according to the records).

Subsequently, I decided to professionally study ferns, which had formerly been my self-declared hobby. I first spoke at the Ferns & Mosses Study Club on the history of fern taxonomy. An abstract of my speech was printed in the mimeograph magazine of the Society, 'Shida to Koke.' That piece was a youthful indiscretion. Rereading it, I wonder how I could dare to post my imperfect discussion of research history, and I have an impulse to erase the article entirely. This research history picked up Copeland's treatise on taxonomic hierarchy published in 1929. In this treatise, Copeland discussed naturalness and usefulness in the context of the basics of organizing hierarchy for ranks above species. I was a graduate student when I wrote the article, in which I reorganized Copeland's discussion of scientific inevitability and convenience for use as an individual beginning a research career. Nevertheless, the article seems to indicate signs of the philosophy underlying this book.

For my master's thesis, I studied the family Hymenophyllaceae. Although it did not develop into an academic dissertation, it was a material I later worked on seriously. This is because, in the 1930s, using this material Copeland built his taxonomy, which served as a good guide for me. I began observing the family Hymenophyllaceae to gain a better understanding of his 1929 treatise. This family later became the central part of my research.

After graduating from Stanford University, Copeland studied at Leipzig University and the University of Halle in Germany. Majoring in physiology, he earned his degree at Halle in 1896. After working as a teacher for some time in the United States, he was asked to work as an advisor to the Philippine government and took the post in Manilla in 1903. It was around the time that the Spanish-American War (1898) ended, and the Philippines was colonized by the United States. Although the anti-American struggle was suppressed, anti-American guerillas were still active then.

In the days of Copeland's studies, the major challenges in plant physiology were nutrition, internal transport, phototropism, and stomata. Indeed, he addressed these problems. He investigated tropical vegetation with the above physiological background. The first challenge he seriously met was the ecology of plants in the coastal hills of San Ramon, in the western part of Mindanao. His treatise, completed in 1906, resulted from his enthusiastic field work depicting autecological behaviors observed in vegetation while investigating the ecology of individual plants from the standpoint of a physiologist. The treatise also subtly reveals his attitude in looking into plant lifestyles using the working of stomata as a clue.

Subsequently, however, he became deeply involved in the study of fern taxonomy. The prime reason was that he became unable to concentrate on physiological experiments that required continuous observation because he was involved in the founding of the University of the Philippines College of Agriculture, which increasingly necessitated his devotion to science policy and attendance at occasional conferences and meetings. Copeland is said to have been blessed with superb memory to such an extent that, upon returning to his office from a meeting, he could continue his study where he left off, using specimens and literature laid out on the desk.

Copeland served as the founding dean at the University of the Philippines College of Agriculture during the period from 1909, the year of the inauguration of the College at Los Baños at the foot of Mount Makiling, to 1917. In this period it might have been difficult for him to work on a research challenge that would require staying in a laboratory and continuing experiments, even though he could give lectures as a professor in plant physiology.

Another compelling reason was that he keenly felt the necessity of organized basic knowledge about species diversity in order both to pursue physiological studies and to accurately recognize the plants he studied. He felt this given the reality that studied plants were difficult to identify or give species names, according to his experience in physiological surveys in San Ramon. He embarked on the study of species identification from a keen awareness of the issue that, among others, taxonomic information must increase in quantity. He selected ferns as the subject of his research because the progress in research to date was slow. The college campus was propitiously located at the foot of Mount Makiling, which had a rich fern species diversity. It was an ideal location in terms of accessibility to research fields.

I have no detailed knowledge of his achievements as the Dean of the College of Agriculture. To my knowledge, however, he published several research papers on useful plants, and based on his experience as the dean, among others, he published books titled 'The Coconut' (1921) and 'Rice' (1924). These books published from the Macmillan Company, New York were apparently used as textbooks in their field for generations. He played a substantial role in training superb agriculturists in the Philippines. However, this book refrains from providing detailed descriptions of his contributions to the study of agriculture.

Along with the contributions he made to founding the College at the University, his great service and noticeable science policy achievements include the founding and growth of 'Philippine Journal of Science,' intended to publish scientific accomplishments. Indeed, many of his treatises on ferns (including the three-part treatise on the family

Hymenophyllaceae, which has been most meaningful to me) appeared in this journal and were disseminated to the world.

Nonetheless, publication of his papers in this Philippine journal resulted in a disadvantage to Copeland. In a eulogy to C.A. Weatherby (1875–1949), who achieved successful research results at the herbarium at Harvard University, Copeland commended his error-free treatises. Certainly, Copeland's papers contain frequent inadvertent errors. However, many of them were due to deficient printing techniques and inadequate proofreading. Although Copeland held the ultimate responsibility, it resulted from the absence of competent assistants, and his insistence to publish his research results in a journal edited by a poorly staffed office.

A minimum responsibility of scientists is to ensure that published papers are error-free, as Copeland expressed his respect for Weatherby. I do highly value Weatherby's papers as data. Nevertheless, I have never found anything in them influential to my research. In comparison, I feel greater and more positive influences from Copeland's papers than from any other scientist, despite more than a few typos and inadvertent errors. This may clearly indicate a difference in greatness as a scientist. One would learn little from a precious paper if he or she continually points out superficial errors and fails to read the superb thought described in it, as treatises are meant to convey the author's personality in unfolding the true essence of science, rather than to simply record observations as data. This is true in humanities and social sciences, as well as in the realm of natural sciences.

Along with carrying out detailed observations for the identification of individual species, Copeland proposed a very bold hypothesis that ferns originated in Antarctica (Copeland, 1940). He inferred this by putting together his phytogeographical knowledge on the entirety of ferns after he had examined all plant species in the family Hymenophyllaceae from the old world and during a process in which he systematically organized the specific nature of their distributions. He attempted to show that ferns diversified in the Antarctic region during the time when the temperature was far higher than today and during the prominence of the diversification of ferns. Of course, his inference is not an empirical theory constructed on facts. It is a hypothesis that can be criticized for being a simple presentation of facts he knew to provide grounds for an idea that occurred to him when he was comparing the current distributions of many species. Notwithstanding this criticism, facts that serve as grounds for denying his idea are scarce.

Phytogeographical knowledge and analysis methods at the time were by no means comparable with present-day knowledge and methods. Accordingly, his process of argumentation and the point of view he proposed had many problems. However, in his inferences one should see his sincere effort as a researcher to have insight into the truth, though based on limited knowledge.

That said, Copeland at the time was apparently frustrated by the limitations of the analysis methods he could use. He was often unable to argue in a persuasive manner against criticism to his species identification. He even ended in saying that one could criticize his views only when one had finished as much fieldwork as he had, which was inconsistent with empirical research methods. Nonetheless, Copeland's treatises are very easy to understand after carrying out fieldwork, even if not to the same extent as he did. I enjoyed reading his papers once and again with a smile, wondering if this could be called science.

In addition to his contributions to steady advances in the study of agriculture in the Philippines, a country colonized by his home country, his services to the founding and initial development of the college for agriculture, and his publication of agricultural treatises based on his study results in plant physiology, Copeland produced additionally greater achievements in floristic research on ferns. In this regard, considering his duty as the Dean of the College of Agriculture, his research was a service not connected with his monetary compensation. Monographs on ferns in Southeast Asia were simply research challenges that he addressed to fulfill his scientific curiosity.

After Copeland returned to California in his later life, the above-described motivation he had as a researcher served to drive the outstanding fern scientist Warren H. Wagner, Jr. (1920–2000). Although Copeland was not Wagner's supervisor at the college, Wagner named Copeland as his trainer instead of his supervisor at the college. The collections that Copeland had were moved to, organized, and stored at the herbarium at the University of Michigan, at which Wagner worked for life. I also used the materials for my research.

When I was a graduate student, Copeland, who was in retirement after going blind, sent me a large volume of offprints of his treatises in response to my ungracious request. In fact, Wagner took care of this task, which is a story I heard directly from him after he and I became more familiar.

Under Copeland's influence, Wagner achieved results to usher in a new era, raising researchers who would play a key role in subsequent studies on ferns in America. Considering this, I believe that what Copeland achieved in his deployment of research methods as a natural historian in the first half of the 20th century will facilitate research growth in the next era.

---

(3) Natural history of earth

It is very difficult to explain concisely the whole picture of biological species diversity on earth. Making a list of known biological species, numbering well over one million, to understand the whole story is simply a tremendously laborious task.

To have a comprehensive view of all living organisms is what naturalists expect to or dream of accomplishing. As a biologist in the 18th century, Linnaeus attempted to make a list encompassing all animals and plants then known in the world. After him, no one willing to make a similar effort has emerged because the scale of such a project cannot be achieved by an individual or institution. The amount of information is too large, and new findings constantly increase.

Vascular plants constitute a large taxonomic group, which can nonetheless be treated in a relatively well-organized manner. Consequently, attempts to make a comprehensive list of vascular plants have been made several times. After Linnaeus, scientists have repeatedly tried to list all species in line with an inherent taxonomic hierarchy. Although none of these have been conclusive, a comprehensive view of the then leading-edge findings was expected. When the Berlin-Dahlem Botanical Garden was very active, projects such as 'Die natürlichen Pflanzenfamilien' (The Natural Plant Families) and 'Das Pflanzenreich' (The World of Plants) were promoted. Also, a comprehensive list (D.C., 1824–41) by A.P. de Candolle (1778–1841) and 'Genera Plantarum' by G. Bentham (1800–1884) and J.D. Hooker (1817–1911) were published. However, in the second half of the 20th century, it became difficult to compile these types of comprehensive lists, although monographs of families and genera have been consecutively published, and books of regional flora have been printed in various forms.

Since around the mid-20th century, knowledge has not only been printed on paper and published, but also been compiled into databases that make optimal use of electronic media. Thereafter, the compilation of lists in an electronic format was the focus of attention, including constant information updating for subjects such as biodiversity with a tremendous amount of information. As detailed in the section on GBIF (3.3), attempts to digitize and compile biodiversity information in various forms began.

The fauna of earth, encompassing all the diverse animal groups, exceeds the scope that present-day science can handle in terms of the size of the subject and its diversity. It is apparently impossible to quickly formulate a compilation project under current conditions, which is understandable. Making a list of endangered species has the implication of depicting the whole picture by models. It is inevitable that doing so should be based on information segmented by taxonomic groups and regions. This is unavoidable due to the current state of available knowledge. For fungi and microorganisms, the lack of species diversity information is additionally serious. Although it is possible to point this out, it is not easy to produce further relevant information.

Advances of electronic information processing methods are expected to enable biodiversity-related information to be put together in an easier-to-use form. However, biologists are particular about the compilation of books of earth's fauna and flora, despite the present state of scientific knowledge. This is not only due to the motivation for the production of useful data but also for hopes of an approach, established during the compilation process, that could illuminate the integrated whole of biological species diversity. To ascertain facts accurately and reliably, it is essential to subdivide the event to be studied and analyze and understand the details. This approach is good at clearly analyzing specific events. A taxonomic hierarchy of organisms is formed by making the best use of findings at the time. This is, of course, intended to serve as a reference for understanding to what extent segmented events can be used to explore universal principles. Nonetheless, to know in practice the universal principles underlying the whole, a perspective for grasping the whole is still indispensable. This is a characteristic of the perspective of natural history. One needs to face up to the fact that physicochemical analysis alone cannot lead to the elucidation of what it is to be alive.

In this regard, an important element of the perspective of natural history is whether it is possible to have a bird's eye view of the entire kingdom of organisms. By getting proactively involved in 'Species Plantarum: Flora of the World,' I endeavor to look at the entirety of the organism group of vascular plants. However, this effort is far from having a bird's eye view of the integrated whole of the kingdom of organisms, although it enables a comprehensive view of a specific taxonomic group. The 'Species Plantarum: Flora of the World' project may be an attempt to gain a whole picture using vascular plants. However, to approach the integrated whole of the kingdom of organisms, information on the entire kingdom of organisms must be organized by covering the entire Earth (as attempted by GBIF and described in Chapter 3).

On one hand, the volume of information is too large for the human brain to process; on the other hand, one can correctly recognize the subject by having a bird's eye view of the whole. However, in the context of information processing, a question arises as to whether it is possible to gain an integrative understanding without processing the information by a single human brain. The use of relevant technology is expected to enable one to organize and process information to reach an understanding of the whole. To examine this tremendous volume of information, one can seek the truth of the laws that dominate it. For this, a full effort to organize the necessary information is key to discovering truth by using the existing information. To know truth, an integrative perspective is required, until information is perfectly organized.

Chapter
5

Developing Natural History – What Are We Faced With?

When discussing the science of natural history, it helps to examine the present to depict what we can expect for the future. What should natural historians, given the rapid advance of science, seek out and contribute, considering contributions of the past and explorations of the present? Based on the preceding discussion, Chapter 5 returns to natural history's fundamental perspective to clarify the challenges for future exploration by elucidating the past: what have natural historians sought; how have goals changed with the times; and what do natural historians expect to discover in natural history.

5.1 Science and Intellectual Curiosity

At its ultimate, science is believed to delineate the whole picture of nature knowable to humans. Different people may have substantially different predictions as to when and in what form science will ultimately achieve this purpose. Some may say it will happen on the last day of science, probably centuries in the future. While based on agnosticism, some say that such a day will never come, others apprehend that humanity will be extinct before the end of science. This chapter explores arguments from the perspective that science will elucidate the whole picture of nature some day in the future, leaving that question of what exactly is meant by elucidation unanswered.

Human intellectual curiosity generally drives people to understand nature in its entirety, although, for some, examination of natural phenomena by discrete facts is satisfactory. Intellectual curiosity motivates people to know what it means to live and how the earth and the universe are composed. For humans, these are among the most attractive questions. A person may be fulfilled to some degree by knowing one aspect of life or one property of earth or the universe and thereby finding intellectual satisfaction (it is true the interest of many is limited to only understanding portions of a phenomenon). However, curiosity is not completely satisfied unless the whole story becomes clear. It is certain that no scientific elucidation of the whole picture will be made in our lifetime. Accordingly, my speculation may be no different from the conclusion of agnostics.

Centuries from now, everything may become clear and people may know the whole story. Imagining this outcome, however, does not satisfy the present-day. Even if our knowledge of facts is very limited, we, driven by curiosity, anticipate having accurate knowledge (or inference) of the whole picture.

Present-day science, particularly natural science, is expected to produce research results backed up by demonstrated evidence. Since they adopt the reductionist way of addressing challenges, scientists necessarily work in fragmented domains and embrace the analytical pursuit of individual challenges. Therefore, empirically clarified discrete phenomena of the natural world are often studied to develop fundamental technologies used in relevant production activities that meet the expectations of people aspiring to safer and more affluent living. Strongly material- and energy-oriented, present-day society values scientific findings as the foundation of technologies that enable material wealth and safety. As a natural result, science that serves as technological foundation is vigorously promoted. Nowadays, advanced technologies are mostly based on findings achieved by science, and as such, society places high significance on science as a technological foundation.

The above form of research promotion deviates somewhat from the pursuit of science motivated by purely scientific curiosity. In many cases, the evolution of science is directed towards analysis of phenomena for use in technology, rather than driven solely by scientific curiosity. Nowadays, when you hear the phrase "promotion of science," it mostly implies the promotion of science for use in technology (or science for society). This is manifest most strongly when it is said that science develops greatly during war. Military research, conducted urgently, nourishes technology and helps the science underlying it grow.

From a long-term perspective, the growth of science motivated by human scientific curiosity will enrich human society, make it safer, and lead to greater stability, although it may not help the immediate affluence and safety of human society. History has proved this assertion, which remains a common notion.

Despite this, scientists place priority on being responsive to the immediate demands of society, and their first principle is to contribute to urgent material affluence rather than to improve society's science literacy for the growth of pure science. Why is that so? Petty politicians are only interested in votes of the next election. Business executives focus their attention solely on the figures in the next annual report. Likewise, the concern for communities of scientists is that they may be too busy responding strategically to short-term science policy and have lost their long-range strategies for fulfilling intellectual curiosity. To understand scandals in scientific community, it is necessary to look at their backdrop, rather than to view them as individual incidents.

When I have discussions on the above topics with groups of outstanding scientists, I have the impression that most of them share an understanding regarding the promotion of science. However, reality does not align with this understanding, just like the worlds of politics and economics tend to readily neglect their principles. This might result from the irrationality of how human beings conduct themselves.

Since the science backing technology became extremely important as a means for building a safe and affluent society, scientific contributions to society have overwhelmingly been measured in terms of their contribution to the foundations of technology. Stemming from this, in my view, science has been confined to its own arena and understood separately from the arts, and the notion that the arts and science should exist in the same field as intellectual activities has been forgotten. Therefore, lists of policy issues are apt to neglect purely intellectual science, although science as a foundation of technology is powerfully promoted in policy, while plans for the furtherance of pure arts are also implemented. Aside from policymakers, more than a few scientists are immersed in the world of pure science, spurring the above trend. This situation is of concern as a cultural issue, particularly that the above trend is observed specifically in Japan and corresponds to the fact that people in Japan, unlike most other countries, make a clear distinction between natural science and humanities.

Scientific curiosity motivates natural historians to engage in purely intellectual activities, as natural history responds directly to human intellectual curiosity, regardless of the distinction between humanities and science. Advances in research in this discipline, or the results thereof, are not expected to immediately lead to technological applications directly useful for human life. On the contrary, even if the issues posed by naturalists are made for the sake of future generations, natural history research results might negatively influence production activities, as in the case of environmental conservation. The fundamental significance of the advancement of natural history lies in that it helps people return to the essentials of science to take another look at improving the science literacy of present-day society.

In the days when the Second World War shadowed Japanese society, some scholars worked on refining existing knowledge of the biota of the regions that Japan invaded. They might have not been able to conduct thorough field research, but they could conduct extensive literature research. It is likely that they had interest in regions they weren't familiar with to extend their research and study. For this purpose, they could easily receive grants in aid for research in the name of producing literature valuable in the search for food and other resources. Rumors even exist that they were under the order of the authorities to conduct resource surveys, although I am not familiar with these specific facts. Of course, this chapter refrains from touching such dubious topics of past events. Described herein are the disciplines of science pursued as intellectual activities and driven by scientific curiosity that will lead to long-term contributions to society, even today, when science is primarily applied to technology.

In their respective times, humans have attempted to understand nature using all the information they had available. In other words, in every period of history, humans strived to gain understanding by drawing on all current knowledge of the times. Aristotle's findings, Leonardo da Vinci's cognition, and Darwin and Mendel's insights were all state of the art at the time. However, from the perspective of today's people, with access to a far greater amount of knowledge, their understanding of the facts was archaic. Likewise, present-day scientists continue making strenuous efforts to elucidate facts that might be considered archaic by scientists in the near future.

In an era of rapid advances in science, a question arises as to for what natural history should aim and contribute. When discussing the science of natural history, examining the present helps to create expectations for the future from a fundamental standpoint, based on the past contributions and present explorations in the field. This chapter returns to the fundamental perspective of natural history, including what topics naturalists sought, how their aims have changed in response to the times, and what the present-day people expect from, or are trying to discover in, natural history.

Before this chapter's detailed discussion, the issue of scientific curiosity needs to be reexamined in the context of this book. Among groups of individuals who currently actively contribute to natural sciences, some consider that scientific curiosity is the motivation by which a person, after clarifying their present state of knowledge, decides what to study next and commences the relevant analyses. This may certainly describe the scientific curiosity that natural scientists see in real life. Scientific curiosity, as described repeatedly in this book is somewhat more fundamental, describing the truth-seeking curiosity seen in human intellectual activities. It is curiosity for attaining a holistic view, for example, questing for an understanding of the universe or what it is to be alive. It is not curiosity about challenges in a fragmented or specialized area. The primordial scientific curiosity that drives human intellectual activities is on the verge of being left in obscurity. The issue here is whether such a situation is satisfactory for scientists.

5.2 Life Viewed from a Natural History Perspective

(1) Scientific analysis of life

Integration in Science Aristotle grouped thought into *physica* (natural philosophy) and *metaphysica* (metaphysics). His concepts (if even not directly proposed by him but systemized by later scholars during the process of organizing his work) have served as the foundation for the growth of human knowledge, or culture and academia, for the 2300 years since his time. Before long, *physica* and *metaphysica* evolved clearly differentiated from each other. It has been forgotten, however, that in Aristotle's mind they constituted integrated thought. For example, Aristotle, although known as a philosopher, also observed the organs of animals. This is explicitly symbolized by the morphologic term "Aristotle's lantern," which refers to the chewing organ at the mouth of sea urchins.

Nevertheless, the question remains of whether humanity can understand the entirety of intellectual information in an integrated framework, given the present-day level of academic growth and complexity achieved by humanity. If Aristotle were alive today, would he be able to make additional academic contributions to, for instance, biological textbooks, while being a leader in mainstream philosophy?

Although I have spoken with a limited number of such individuals, I have had discussions with ARIMA Akito (1930–2020), an authority in particle physics and also a haiku poet who leads a poetry group in this genre, and NAGATA Kazuhiro (1947–), a well-known *tanka* poet who has also made remarkable achievements in cytology. Both have been successful in playing a leading role in their domains in humanities and the natural sciences. However, even they could not attain the same range of contributions achieved by Aristotle. This constitutes an important cultural and academic difference between the times of ancient Greece and the 21st century.

That said, in the context of this section, discussions regarding the ability to cross fields do not hold much significance. Separate pieces of human knowledge about individual phenomena, even if they can be utilized individually according to their attributes, cannot fulfill the fundamental scientific curiosity of humanity. The necessity of promoting empirical and investigative research to clarify individual phenomena need not bear repeating. Its importance is understandable based on the phenomenal growth of the present-day world of science. Nonetheless, the questions remain as to what extent of effort present-day scientists make to develop science that sees nature as a whole, as driven by human scientific curiosity, and whether or not making such efforts will lead to satisfactory outcomes.

Natural history aims to have a comprehensive view of individual facts and, at the same time, to know the true nature of the world through them. Given that it is essential to describe individual facts, naturalists must theoretically and practically pursue and record the scientific meaning of the investigation of individual facts.

Establishing an integrative perspective in science remains a challenge pointed out on various occasions. In the domain of biology, universities and research institutes advocating "integrated biology" have been long-standing, even incorporating the term in their names. However, superb scientific achievements have been more intensively realized in reductive and analytical research. This is probably because the expectations for the application of this type of research to technology are high. This reality exists notwithstanding a high regard for research motivated by pure scientific curiosity. The question arises as to what this situation signifies.

As described earlier (in Tea Time 2), naturalists are naturally expected to have an integrative viewpoint, although it is not easy to explain what integrative science is. Natural history strives to recognize fundamental phenomena in the natural world by producing detailed descriptions of them. Descriptions are not expected to spontaneously provide any straightforward answer about the raison d'être of the described phenomena. From the perspective of scientific curiosity, describing phenomena is an operation by which naturalists are awed, recognizing the myriad diversity of phenomena.

In the natural world, phenomena do not occur individually or separately. All things form, develop, and evolve interconnected with each other in the unending flow of nature. For those who have interest in life, it must be profoundly moving to recognize and observe the diversity of real living things.

Although diverse forms of living things exist, life on earth originally appeared in a single form. This present diversity was created through evolution spanning nearly four billion years. This historical fact persuasively conveys that, whether at the gene, species, or ecosystem level, various phenomena exhibited by apparently diverse living things formed through the evolutionary process nearly four billion years and are interconnected with each other.

When it comes to diverse phenomena, humans have fun even looking at the simple fact of diversity. More than a few of them are fully intellectually satisfied by simply closely observing diverse phenomena and listing as many events as possible. Meanwhile, when describing diverse events, the writer not only knows the facts about the diversity, but also becomes curious about the backdrop to the development of this diversity. When the relationship between the development of diversity and the resultant forms is revealed, the writer feels the mystery of nature anew. While fulfilling curiosity by simply listing diverse facts is beneficial, integrating individual events by researching their true forms has even greater value.

The task is, of course, more complicated than expressed by several lines of text here. As described in this book, humans have diligently recorded the forms of diverse living things throughout history. However, even with today's advanced science, less than 10 percent of the postulated number of species of living things on earth, as an optimistic estimate, have been recognized by humans. Moreover, this result comes simply from the identification of species. When it comes to describing diverse facts in depth at various levels and understanding diversity manifest in various phenomena – let alone describing the relationships among them – the end is nowhere in sight.

It is a common knowledge that some relationships always exist between diverse events. However, defining the specific relationships between specific events is not so simple. The amount of information regarding relationships thus elucidated so far is too small to view the whole picture of diversity, although humans have a pile of fragmentary knowledge about a very small portion of diversely combined relationships.

Regarding diversity, individual events differ in their respective ways; moreover, interrelationships also differ in their respective ways. This is one essential part of the nature of diversity. Keeping track of each of these would take unimaginable amounts of time and energy. Naturally, unless we integrate known events, have a bird's eye view of them, and depict what becomes visible through this process, we will fail to clearly recognize even those things that we believe we know.

Since Aristotle's 'History of Animals' (4th century BC), numerous records of fauna and flora (biota) have been compiled in various styles. Nonetheless, no work clarifying all kinds of events has been compiled. No one expects the production of such a work, because even ordinary people vaguely know how much knowledge we lack about biodiversity. That is why, still today, the discovery of a new species and concerns over an endangered species are newsworthy events.

Even understanding the significance of the analysis of specific events conducted by scientific methods, and the present-day challenges implicit in the task, it may not be possible at present to integrate all pieces of our fragmentary knowledge, have a bird's eye view of the whole, and accurately organize this by scientific methods to promote general human knowledge. Even so, without insight into the reality of nature, we cannot ensure a life of stability or affluence. It is essential, particularly in the domains of life sciences and environmental sciences, to address challenges according to logically constructed facts construed by modern scientific techniques. Additionally, it is also essential to deal with events on a trial-and-error basis, based on truth inferred from an integrative perspective. In this way, indeed, humanity can utilize its knowledge.

Species Biology and Asexual Species The term *biology of species diversity* came into use to refer to taxonomy [in Japan] in the 1990s. It was in 1993 when I first titled a textbook using the term (Iwatsuki, 1993). At that time, the term was not readily accepted. The textbook attempted to identify universal principles in the phenomena of species diversity, as well as to describe the taxonomic hierarchy in the world of organisms.

Textbooks and commentaries emphasize solidly proven facts. It is even said that school textbooks should avoid facts that have not been an established theory for at least 30 years. However, every taxonomic system is a hypothesis. Although it compiles existing findings in a systematic form, it still expresses inconclusive matters. This hypothesis is frequently described in textbooks. When new facts are elucidated, and another system is to be proposed in the future, the previous system (= hypothetical conclusion) will be understood as incorrect. While in other fields hypothetical theories are not taken as conclusive, taxonomic system has been treated as conclusive to the extent that it is used in exam questions, and students must learn it by heart. For science education, it is necessary to clearly state what is known and what is hypothetical; however, for years, taxonomy has failed to do this. The same can be said for the definitions of species.

In 1942, Ernst Mayr (1904–2005) clearly defined the biological species concept (Mayr, 1942). The title of his book explicitly states that his premise is, "from the viewpoint of a zoologist." Although Mayr claimed that his definition developed from a limited standpoint, his definition is criticized as incapable of universalization beyond sexually reproductive species because the ability to interbreed is used as grounds for species delimitation (Iwatsuki, 2012 and others).

There are a priori expectations that the species concept should cover all organisms because the goal of science is to look for universal principles. Despite the need for a definition universally applicable to the world of organisms, when defining species in taxonomy, scholars always present the excuse that such a definition is difficult given the current extent of scientific findings.

When living organisms first appeared, they reproduced asexually. At an evolutionary stage (estimated at more than one billion years ago), they began sexual reproduction, by which they boosted the speed of evolution and accelerated diversification. Through subsequent evolution, metazoa and terrestrial plants appeared on earth. For most of the highly evolved and diversified living things we see, it is possible to use the ability to interbreed as the basis of defining species, as Mayr did.

As stated earlier in this book, approximately 10% of pteridophyte species (approximately 13% for Japanese pteridophyte species) have abandoned sexual reproduction in their life cycles (2.3; Iwatsuki, 1997). Mayr's definition is not directly applicable to these species. Consequently, the search for a definition that is universally applicable to the world of organisms makes the biological species concept incomplete. (It may be superfluous to say that, with the indirect aid of Mayr's definition, it is possible to universalize the sexually reproductive species definition of vascular plants to the pteridophyte groups. Microorganisms and other originally asexually reproduced species require a completely different logic for differentiation than that used in discussing vascular plants that have abandoned sexual reproduction.)

Because various phenomena seen in the world of organisms originate through diversification of one type or another, taxonomy attempts to organize them into one system. According to this trial, an incomplete taxonomic definition is the most up-to-date. From the standpoint of science, predicated on the provision of logically impeccable proof, one can criticize perspectives with poor logical proof. However, from the perspective of the biology of species diversity, it is a must to consider diverse phenomena as impossible to cover with a single definition. Thus, phenomena that are universal to all diversified living things become increasingly differentiated from phenomena that exhibit diversity.

Of course, this chapter refrains from verifying whether it is biologically correct to organize, under a single concept, species of microorganisms and those established in metazoa and vascular plants groups. Suffice it to say that species as fundamental units of biodiversity are, in precise fact, fluidic in the logic of science.

In the history of life, sexual reproduction is a trait developed when diversification speed up after the emergence of diversified lineages. However, the most diverse metazoa, terrestrial plants, and true fungi (eumycetes) are basically sexually reproductive species. Consequently, the apparent dynamics of species readily seen on the earth is sexual reproduction. Therefore, a definition of species structure commiserate with sexually reproductive species appears to define species of all organisms. One may certainly infer and apply a definition of species in the domain of asexually reproductive species based on the definition applied to sexually reproductive species.

However, the existence of currently thriving organisms due to the development of sexual reproduction does not necessarily imply that organisms are viable owing to sexual reproduction. To adapt to their environment, some organisms acquired sexual reproduction and developed into diverse sexual organisms. Nevertheless, sexual reproduction is not a fundamental trait indispensable for life. It is a mechanism that accelerates and simplifies alternation of generations and facilitates the introduction of diversity for survival.

Notwithstanding this, to discuss all currently extant organism species, the dynamics exhibited by sexually reproductive species are applied to the fundamental definition of organisms. In exploring what is life, scholars are aware that living things acquiring the extraordinary life cycle of sexual reproduction exhibit an atypically evolved way of living. Despite this knowledge, when describing the true nature of species, scientists refer to sexually reproductive species. They are, in a sense, preoccupied by the present approach of living things. This may result from a reality which stresses the importance of science for humans by dealing with currently living organisms. In this way, scientific curiosity that asks what it is to be alive takes a back seat, even in the world of science. (In addition, the trend in human culture towards neglecting history and addressing the present alone is also widely observed.)

When examining what it is to be alive, it is unreasonable to interpret the undiscovered life of bacteria directly as my life. Life is a universal phenomenon in living things, commonly observed in bacteria and *Homo sapiens*. However, the lifestyle of bacteria differs radically from that of *Homo sapiens*. These differences have resulted from the process of evolution during nearly four billion years. DNA science has proven that bacteria are not different from *Homo sapiens* in life's fundamental building blocks. However, understanding the biology of bacteria does not sufficiently explain the principle behind the motion of an elephant's nose.

Extending the biological species concept to measure the distance between species by the degree of difference in DNA sequences is of no use. Species differ substantially with taxonomic group, to which it is impossible to apply a single figure. This naturally results from the fact that biodiversity is not a phenomenon by which organisms diversify by following a specific rule, but a phenomenon of diversification occurring in diverse manners.

The question then arises: Is there not a universal principle that applies across extant living things that have diversified in diverse ways? This question in turn reminds me of the success of molecular genetics, which allowed

20th century biology to scientifically track genetic phenomena across all life. All living things maintain a structure of four bases comprising DNA, properly transmitting the propensities of parents to offspring at both cellular and individual levels.

Therefore, the answer to the above question is that a universal principle is present behind life. Given that living organisms survive as a species, it then must be possible to provide a universal definition of species. Meanwhile, it is far-fetched to apply the concept of species recognized with sexually reproductive species to such organisms that have not evolved to be sexually reproductive. Within sexually reproductive species, there are apomictic fern species that have, for their convenience, abandoned sexual reproduction. They should be subject to discussion of species structure, using the biological species concept that is applied to sexually reproductive species.

Species is a concept devised to organize phenomena observed in organism groups, such as metazoa, vascular plants, and fungi, that have diversified to increase the speed of species differentiation by means of sexual reproduction. This concept has been successful in explaining the world of organisms to some degree. Subsequently, the existence and life mode of more primitive creatures has been elucidated. More correctly speaking, scholars have become aware of the necessity to review the fundamental significance of living things, sorted based on the species diversity concept by type of basic life form. We should firmly understand this historical consequence before proceeding to an analysis of species diversity in various organism groups.

Although it may sound superfluous to readers, the author would like to add more commentary on agamosporous fern species. Fern species that produce offspring by apomixis are a specialized type of fern, which chose to abandon sexual reproduction for some reason of convenience, although they had once evolved to be sexually reproductive. They appear to be like many microorganisms, in the sense that they lack sexual reproduction; however, there are substantial lifestyle differences between the two organism groups. Their specialized way of life does not fundamentally differ from unspecialized fern species.

Taxonomic Systems and a Bird's Eye View of the Kingdom of Life Whether it is called taxonomy or the science of biological species diversity, this domain of science attempts to elucidate the true forms of diverse individual species. At the same time, it necessarily investigates related species of the target species. Consequently, scientists' thought extends to the ranks of genus and family. Eventually, examining taxonomic system leads to a growing interest in all living organisms. Even if the starting point is an individual species within a limited area, the scholar's interest in the species will naturally lead to the entire distributional area of the species and extend further to the distributional areas of related species. Eventually, in this domain of research, the entire earth's surface (= biosphere) will inevitably be the subject of research.

Natural history studies of life inevitably cover the entire biosphere on earth and investigate the whole process of the history of life, spanning more than three billion years. It is impossible for every individual researcher to discuss life across the entire biosphere throughout the history of life. Individual research projects are implemented concerning a specific area and a specific span of history. However, it should be remembered that researchers in this field are strongly concerned with the entire biosphere and all lineages.

While following everyday research processes, researchers can limit the subject to fit the narrow range of their senses, as if they are in a well. Even regarding natural history studies in Japan, Japanese natural historians must admit that the criticism of their hobby-like pursuit of observation based on a preference is fair to some extent. However, as this trend is not unique to the field of natural history, although it is quite apparent in this field, this book refrains from discussing it further.

Natural history researchers basically explore individual phenomena in a reductive manner and analyze, in an empirical way, the entity from which the phenomena emanate. However, simply clarifying the facts is not success. Natural history studies can become systematic only when the acquired data is compared with other related facts, and the facts are integrated. It is not a study completed by investigation alone. Integration of investigation results is a prerequisite for natural history.

This framework of research methodology allows researchers properly pursuing natural history to spontaneously merge investigative science with integrative science. In present-day mainstream science, individual phenomena are understood by means of reductive techniques, and the results are applied to technology. The predominance of that method blurs the idea of properly understanding the reality of a subject from a bird's eye view of discovered facts for fulfillment of scientific curiosity. Natural history methodology helps make this a clear focus of attention.

In no domain of scientific research is a question analyzed in an isolated manner, with the researcher having, so to speak, the mentality of a frog in a well. It is rare that such an analysis would achieve successful results. Large research groups might include some member researchers who, like engineers, work highly efficiently by only solving assigned tasks in a single-minded way using given techniques. This is, however, not a matter to be examined here. To analyze a challenge, associated scientific findings and the techniques used to obtain them provide valuable useful information. Moreover, known facts about universal principles underlying all events, or at least similar events, help the analysis. Therefore, it is commonly known that individuals reputed to be outstanding scientists, without exception, have profound knowledge about a wide range of scientific accomplishments.

It should be remembered that in natural history, in many cases of exploring individual challenges, specific species are picked up from diverse species. Therefore, in the worst case, researchers are apt to be narrow-minded, only considering the specific material. It is regrettable that some (more than a few) researchers are satisfied with this situation. Nonetheless, it is not rare that an individual immersed in a specific challenge and committed to efficiently describing diverse species contributes to piling up accurate information.

However, diversity research, when rolled out, will inevitably require placement in a system of organisms. For this reason, researchers are constantly required to have a bird's eye view over the entire system of the world of organisms. This leads to the necessity for natural historians to have an integrative perspective. They need to review all existing information about the organism group to be analyzed. This is the only way to enable the researcher to make the best inference of positioning in the system based on the current knowledge.

Moreover, while natural history is *physica* (natural philosophy), it is a domain in which naturalists view the quintessence of science such that natural science cannot exist separate from *metaphysica* (metaphysics). This important fact has not been widely conveyed in the world of science. In simpler terms, it is not fully understood [only] in the world of [natural] science. This is a profoundly regrettable reality for the healthy promotion of science. Scientific accomplishments that serve as the foundation of technology are very valuable and necessary for human society. Nevertheless, for humans as intellectual animals, sincerely working with scientific curiosity is a truly human activity.

(2) Natural history of the *spherophylon*

To empirically clarify universal principles, science pursues analytical and reductive research, elucidating accurate cause-and-effect relations between individual events. Another aspect of science is that it has grown by human intellectual curiosity. If science asks the question "why?" analytical and reductive research may be good at elucidating the reality of individual events. However, it is difficult for such research to capture, by itself, the integrated whole that ignites scientific curiosity, based solely on current findings.

Twentieth-century biology became known as life sciences and made steady steps in the world of science using DNA as a keyword. Likewise, in biology, analytical and reductive research has attained and is attaining rapid-paced advances. A question arises as to what achievements natural history made or attempted to make during these periods.

Spherophylon When talking about the fact of living, one thinks of one's own life. That life involves the intellectual activities of a multicellular individual. However, biology attempts to examine life at the level of individual cells and clarify it at the molecular level. The idea is to elucidate universal principles underlying living things by reducing them to matter because, being based on matter, these are viable simplifications. Biologists have come to know various aspects of life based on facts clarified in this way. Their findings applied to technology, or the development of biotechnology, have made profound contributions to human society in terms of affluence and safety.

Natural history aims to understand living things at the level of individuals or higher and continues to seek the implications of the concept of species, that is, a group of individuals. The discipline addresses the fundamental challenge to biology: what it is to be alive. It attempts to grasp life, not only in terms of matter, but also as the integrated whole composed of facts from living things that have survived for nearly four billion years, are currently extant on earth's surface, and are interconnected with each other in diverse differentiated and evolved forms. In sum, while bearing in mind that organisms are living at the level of cells and at the level of individuals, natural history looks at the fact that they diversified from one form and that, throughout their history, diverse forms have survived while maintaining interconnections with each other. I propose that this model of life should be understood as *spherophylon*, the concept of being alive in an integrated manner at a level higher than individuals (Iwatsuki, 1999, 2023).

The essence of natural history research is summarized below from the perspective of *spherophylon*. The state of living can be viewed as individual cells that make up an individual body, as well as multicellular individuals. This is clearly indicated by the fact that one cell constituting a multicellular body may grow into another multicellular body. Genetic information based on which a multicellular body develops, as a rule, is uniformly stored in each cell. While the somatic cells of an adult animal do not make direct use of the genetic information, those of plants maintain an embryonic state on a permanent basis. Plants can readily reproduce (clone induction) by, for example, layering and grafting. Accordingly, the life of an individual multicellular body is covertly stored in each somatic cell constituting the multicellular body. The state of living is represented by the cell. (Individual bodies solely indicate the lifestyle of an organism species. Individual cells do not demonstrate the lifestyle of the species, although they do render the state of living.)

This fact can be investigated further, as follows: Simply given cellular DNA that carries genetic information, one can, at least in theory, produce a cell (even if difficult under natural conditions). It has become a common knowledge that it is possible to explore living things in a reductive manner down to the molecular level, or even the level of constituent atoms. This is similar to contemplating the level of cells down from the level of individual bodies.

To view this from another perspective, since earth's beginning, no living thing has completed its life as a single individual body. Living beings have always lived interconnected with others (whether the same species or, after species differentiation, with other species). Similarly, they have always depended on the inorganic environment, rather than existing as an independent individual body. In sum, a single cell does not constitute a life form as perfect as, for example, the animal species known as *Homo sapiens* (except for germ cells and fertilized eggs). Likewise, one individual body alone, cannot be said to live as a living thing.

The first single-type organism is estimated to have appeared on the earth more than three billion years ago. The organism began diversification as soon as it emerged on earth (or rather, it could live because it had the ability to diversify as an elemental characteristic). Furthermore, through a long evolutionary process, the present-day animal species composed of multicellular bodies that are capable of performing intellectual activities, such as *Homo sapiens*, have come into existence. In addition, at every moment throughout the process, organisms diversified repeatedly and have maintained or strengthened direct or indirect interconnections between diversely differentiated organisms. In their present form, living things exhibit life as an integrated whole.

Since emergence on earth nearly four billion years ago, all living things have evolved and survived on the earth's surface (biosphere) through being interconnected with each other. Through this, first we recognize the integrated whole as one body of living matter, calling that entity *spherophylon*. Then, contemplating the science of natural history, we can understand it as science that explores the structure and function of the *spherophylon*. By having a clear understanding of the *spherophylon* as a research subject, it becomes possible to concretely understand life in an integrated manner and contemplate it from a bird's eye view. Of course, while the challenges addressed in this section are limited to living things on earth, one should remember that natural history research deals with the integrated whole of the earth and of the universe.

The idea of viewing all of biodiversity as one body of life is commonly shared in biology. The specific use of the term *spherophylon* probably stems from my fussiness about biodiversity. To explore differences between diverse organisms and to link the magnitude of each difference to taxonomic system, one may attempt to examine the historical background for clues to these differences. It is natural that such an attempt leads to knowing that taxonomic system links to the life structure of the *spherophylon*.

In contrast, with strict adherence to modern biological methodology, taxonomists stick to ascertaining accurate (= scientific) differences between organisms, whether at the level of, below, or above individual bodies. Accordingly, they have become apt to fail at assessing detected differences from a comprehensive systematic perspective. To sound the alarm about this situation, the need to analyze the life of the *spherophylon* should be emphasized. It is regrettable that researchers in the domain requiring the greatest attention to the above reality fail to properly understand the true intentions behind the above-mentioned concept.

The Science of *Spherophylon* It can be said that the *spherophylon* refers to the real entity of biodiversity. The term *biodiversity* tends to give an impression that it is an abstract concept. In contrast, the term *spherophylon* concretely indicates the reality of living existence, although the term itself does not enable a view of the integrated whole.

Moreover, the term *biodiversity* gives the impression that organisms are randomly diverse because of the implications of the word *diversity*. When you hear the term *spherophylon*, on the other hand, the concept of the whole as one body comes to mind, facilitating the recognition of hierarchy among diverse living things having evolved from a single origin.

Here a question arises as to why the phrase "the study of the life of the *spherophylon*" is more likely to fit the concept of present-day science than "biodiversity analysis." The following examines the mode of research on the life of the *spherophylon*.

Living things exhibit diverse forms. This is interpreted as resulting from diversification through the history of evolution of a single form since its appearance on the earth more than three billion years ago. Because of the long period of time for diversification, currently observed differences in diversity exhibit a hierarchy. To make it manifest, taxonomic charts assign ranks to organized taxonomic groups. Nonetheless, even at today's advanced level, scientific knowledge is only a fraction of the reality of diverse organisms. Related researchers know that current taxonomic charts are still far from adequate to be viewed objectively.

Incidentally, the constituent cells of an individual multicellular body will never exhibit a phenomenon such as a muscle cell of the left arm eating a skin cell of the chest. The relationships between humans and natural resources were built through the historical process of evolution. Those who wanted to avoid imagining eating cattle meat, began to use alternative wording, such as *beef*, rather than a cow or ox. As a more recent example, I have heard that a girl visiting her parent's hometown could not eat chicken cooked and served on the table after seeing her grandfather slaughter it.

Diverse elements that make up the *spherophylon* exhibit mutual energy metabolism, in other words, the relationships between eating and be eaten. Because it is nature, scientists calmly depict these relationships. However, when it comes to human intelligence, the relationship of killing and being killed goes beyond situations resulting from evolution. It should be remembered that adding humanity's emotional aspect to the relationship between these elements could disturb the order that exists in nature.

Regarding the *spherophylon*, the following introduction to similar concepts further explains why the concept of *spherophylon* is necessary.

One similar concept is Gaia hypothesis, expounded by James Lovelock (1919–2022; Lovelock, 1972). He proposed the idea of viewing and respecting all living things by regarding the integrated whole of all materials on earth as a living existence, while also paying attention to homeostatic exchanges between the living and nonliving. Gaia is a concept intended to enable an understanding of earth by assimilating it to living matter.

Take a living human body, for example. There is no definite difference between living and non-living matter. Hair growing on my body is part of me. As soon as I have my hair cut at a barber shop, it becomes trash. Likewise, when clipped, my nails will become trash. When rubbed in the bath, the surface of the skin becomes pieces of dead skin and discarded. Meanwhile, chewed food, when digested in the stomach, seems to be part of me. Notwithstanding this, billions of intestinal bacteria are regarded as separate living things, which are not part of me, even if they spend their entire lives in my intestine.

Viewing the earth as living matter and including its atmosphere leads directly to other celestial bodies, making it unsuitable to understand the earth separately from the universe. This becomes incongruous unless the entire universe is viewed as one living matter.

The concept of Gaia is emotionally accessible and works fine for raising awareness of respecting the earth. Nevertheless, to ensure scientific soundness, the concept is not completely comprehensible.

In contrast, the *spherophylon* focuses on the fact that since the emergence of life more than three billion years ago, living things on the earth have expressed a common life in diverse forms and have maintained or evolved for this. Material exchanges between the living and nonliving do continue ceaselessly. Scientifically, there is a clear difference between them due to control of the living by genes.

Furthermore, although unknown by Lovelock, Japanese people have held the belief in so-called 8 million deities and have recognized life residing in all things in the natural world. Due to this recognition, Japanese people's concept of harmonious coexistence between humankind and nature, although not penned in a cohesive written work, has helped build a history specific to Japan (cf. the postscript). To discuss this further, it is necessary to write a separate large volume of work. In this chapter, the author would like to point out that the Gaia hypothesis and the belief in myriad deities are concepts with commonalities.

Another similar idea is NAKAMURA Keiko's concept of biohistory. This philosophy about diverse living organisms was proposed by a molecular biologist. Biohistory suggests important challenges that need to be addressed by science in the future. It may sound somewhat bold, but biohistory is natural history from a molecular biologist perspective in that it attempts to describe diverse organism forms using a single keyword, DNA. One difference from the *spherophylon* is that the relationships between diverse organisms beautifully illustrated by sectors, lack the perspective of lineage. This can potentially and mistakenly give an impression that diverse organisms are randomly diversified. The idea of the *spherophylon* suggests that life can be understood in an integrated manner, focusing on the fact that all living things on earth live as one whole, similar to the life of an individual body or species. Consequently, this should be construed as a concept evolving natural history techniques, rather than assimilation by the philosophy of natural history.

These earlier concepts are commonly oriented towards viewing the integrated whole of biodiversity. However, living things should be defined as vehicles of life more precisely. It should also be strongly recognized that all current living things are a kind of integrated existence emerging from one primordium comparable to ontogenesis.

Death Viewed through Natural History

All living things must die. However, this statement comes from the perspective of the lives of individual bodies. In theory, the *spherophylon* continues to live eternally. Living things exist for eternal life to continue, with frequent replacement of the vehicles of life. The vehicle of life (= things that carry life) takes the form of a cell, an individual body, species, or other ranks of taxonomic hierarchy. One vehicle of life is replaced with another, or it is discarded. Hence all vehicles of life (= organisms) will die.

A death, according to the commonly held notion, is the death of an individual body. It is very unlikely that the public will talk about death at the rank of species, death of constituent cells of an individual body, or needless say, replacement of molecules or atoms, although these may be subjects noted by researchers.

The reader may think that the death of an individual body is not difficult to understand. However, regarding this issue, it is impossible to exactly define what constitutes an individual body in the world of organisms. Therefore, at present, scientists are unable to provide a definition of death. Even with the death of a person (an individual body), no common definition has been established, as exemplified by various brain death criteria.

Using death, the opposite of life, as a clue of what it is to be alive, let us examine the death of individual bodies and the life of the *spherophylon* from a natural history perspective.

To examine death, first, consider the death of one *Homo sapiens* in terms of zoology and that of a human in the context of culture. For me, what is my death? 'Pensées' by Blaise Pascal (1623–1662) says the following (translated by W.F. Trotter):

Man is but a reed, the most feeble thing in nature, but he is a thinking reed. The entire universe need not arm itself to crush him. A vapor, a drop of water, suffices to kill him. But, if the universe were to crush him, man would still be more noble than that which killed him, because he knows that he dies and the advantage which the universe has over him; the universe knows nothing of this.

When they were a wild animal species, *Homo sapiens* did not think about their own death. After evolution (specialization) to an intellectual existence, humans began to be aware of their own death. Questions arise as to what is the death of which humans are aware and what is the difference from biological death?

Death, or the End of Life for an Individual Organism We know that the lifespan of an individual organism is limited and that it will inevitably come to the end (= die).

When discussing death, one commonly thinks of the death of an individual (multicellular) body. However, death is not a phenomenon only observed with individual multicellular bodies.

An individual multicellular body is made up of cells, which die individually. However, a person does not view the death of the person's individual constituent cells as the person's death.

Death at levels other than the individual body is examined in biology. In other contexts, it rarely comes to mind. None of various forms of living things on earth, such as species, individual bodies, cells, molecules, and atoms, will live eternally. They all have a limited lifespan, at the end of which they will cease to live. Nevertheless, in no instance is the existence of living things terminated by death. They continue to live in principle, replaced by other related species, individual bodies, cells, molecules or atoms.

It is said that molecules and atoms that make up cells maintain dynamic equilibrium (Baldwin, 1952). They are replaced relatively frequently with other molecules and atoms of the same elements. Cells keep the very same molecules or atoms for a relatively short period of time.

Likewise, most individual cells that make up a multicellular body are replaced with new cells when they become old, with exceptions including neurons, which sustain individuality throughout life. Similarly, individual bodies outlive their usefulness by successfully transferring their genes to individual next-generation bodies. Then they experience the phenomenon of the end of life (= death). This also applies to species, which serve as foundation for constantly creating new species without indefinitely retaining the same forms, whether they undergo differentiation or transmutation.

However, in this process in the strict sense, cells transform into daughter cells, just like a unicellular body is divided into two individuals. In the case of multicellular bodies, when replaced with an individual next-generation body, the mass of substances that carried life are no longer alive. Nonetheless, in the sense that life is transferred over generations, death does not occur with living organisms of this rank.

Cognition of the Life of the *Spherophylon* As an integrated whole (= *spherophylon*), living things are immortal, despite the death of cells or individual bodies.

Individual humans have a limited lifespan. Likewise, the organism species known as *Homo sapiens* will eventually experience the end of its species (= death), as with other organism species. However, the integrated whole of living things (= *spherophylon*) does not have a limited lifespan. It will never die a natural death, although it can die by committing suicide. Although individual bodies and species, as constituent elements of the *spherophylon*, are repeatedly replaced at the end of their respective lives, the life of living things (= the life of the *spherophylon*) will never come to an end (= death) in the course of natural providence.

The *spherophylon* can be killed only by suicide or accident, which the activity of a specific constituent species of the *spherophylon* (e.g. *Homo sapiens*) can cause.

Biology analyzes life in a scientific way. It exclusively deals with the mechanisms of life to clarify the properties of living bodies (= vehicles of life) as special collections of matter. That which is alive is a living body – an object. Life is a set of living body-specific phenomena that occur in the structure of the living body.

Living things have a limited lifespan. Whether at the level of individual bodies; constituent cells, molecules, or atoms of an individual body; species as a collection of individual bodies; or other taxonomic group levels, they cease to live and die at the end of their lifespan. After the termination of a living state, the life-carrying object (= lump of matter if at a level below individual body) is called a dead body, which will never again exhibit a living state.

Death refers to the state of life coming to a full stop. Scientifically, death is the end of the life of a research subject, a special collection of matter. Death of an individual multicellular body is easy to recognize, because it is a body of closely collected matter. A species is a group of individual bodies, comprising a set of individual discrete bodies. Individual cells that make up an individual multicellular body are recognized as part of the individual body. It is difficult to consider them independent beings. Recognizing the end of an individual body or species is accomplished by scientific examination. Before the death of the individual, the life of a person's body involves the death of a vast number of cells.

The length of life of an individual organism body (or a cell or species) is limited. It dies when its life expires. It is commonly understood that no living things recognize the death of other living things, except for *Homo sapiens*, although it has been occasionally reported that mammals recognize death of other individual bodies, such as blood relatives.

It is said that humans alone particularly and clearly recognize their own death. Humans alone recognize themselves in such a manner that "I think; therefore, I am" (Descartes, 1637). They think of their being. Hence, they know the actuality of their death. No living thing other than humans recognizes itself. Therefore, they never recognize their own death.

That which thinks is human intelligence. Humans alone have evolved (become specialized) to have intelligence. The term *human* refers to a living thing with intelligence. The term *Homo sapiens*, italicized as with the species names of other living things, refers to the special assembly of life-carrying materials, in itself human absent the propensity known as intelligence. This leads to an understanding that *Homo sapiens* do not recognize their own death, while humans do. Individual bodies of all living organisms die when their life expires. Recognition of one's own death is not a universal propensity of living things. It is an outcome of culture developed exclusively by humans. (Needless to say, humans and *Homo sapiens* cannot be distinguished from each other. The evolution of Primates is indicative of the difficulty in discerning this issue, like ambiguous differences between life and death or differences associated with diversity. However, this book refrains from discussing it further.)

Living things, whose individual bodies are made up as an assembly of materials, have differentiated in diverse ways on earth through the history of evolution spanning more than three billion years. *Homo sapiens* are a species of these diverse living things. Like other living things, individual bodies (= vehicles of life) of *Homo sapiens* will die eventually. The species *Homo sapiens* will also likely cease to exist someday, as a rule of evolution. One form of its end is extinction of the species. Alternatively, its present form may disappear by differentiation into multiple species or evolution into another species. Individual humans alone are aware of themselves and recognize their own death. Individual bodies of species other than humans are not aware of themselves and do not recognize their death.

Humans are aware of their own death as the death of individual bodies (although having the foreknowledge of one's own death does not mean an individual can confirm it after it occurs). However, none of the constituent cells of a *Homo sapiens* body are aware of their own death. Humans have developed intelligence, but individual cells that make up a *Homo sapiens* body do not have intelligence. One *Homo sapiens* dies when most constituent cells of its individual body die.

Human Life: Involvement of Genes and Their Contribution to Culture An individual *Homo sapiens* body has a limited life span and inevitably dies after a certain period. Meanwhile, individual *Homo sapiens* bodies, as constituent elements of the *spherophylon*, live a fraction of the eternal life of the *spherophylon*. It is as if each constituent cell of an individual *Homo sapiens* body lives its limited lifespan, and they are replaced with new cells when they become too old to sustain the life of the individual *Homo sapiens* body.

Homo sapiens, as an organism species similar to other species, inherit genes from their parents and pass them to their offspring. Individual *Homo sapiens* living as individual bodies play the role of relaying the stream of life to sustain the survival of the species *Homo sapiens*. In this sense, individual bodies make up their species and live as an element of the *spherophylon*. This is not limited to *Homo sapiens*. It is the reality of life common to all living things.

The death of an individual body is the cessation of life of the living thing made up of materials. Individual *Homo sapiens* bodies die at the end of their life. However, while *Homo sapiens* as an assembly of materials (body) dies at the end of its lifespan, the knowledge that a human contributes to culture is conserved in society and continues to live even after the body dies. Given that "I think; therefore, I am," my being is grasped through thought. Thoughts lead to culture and therefore do not vanish, even when the body dies. Indeed, the influences of the lives of past people are alive in various fashions in the mind of every person.

From a zoological perspective, the life history of *Homo sapiens* spans from the cradle to the grave. However, in a sense, the life history of a human with culture extends from the fertilized egg to the ghost. By impressing its being on society, the life of a human with culture becomes a constituent part of the life of society, beyond the boundaries of the individual body. Consequently, a human life continues until the end of the organism species *Homo sapiens*. By taking part in culture, individual bodies have a responsibility to ensure the continuation of the life of the species *Homo sapiens*, as well as the personal life of individual bodies.

A single atom that momentarily takes part in forming my body is an indispensable constituent element of my individual body. Likewise, all individual living bodies are alive as an indispensable piece of the eternally living *spherophylon*.

One should recognize the significance of one's life, living as a piece of the *spherophylon* that has eternal life. This goes beyond simply living an individual life alone, expecting the death of the body at the end of a limited lifespan.

Life recognized in culture is not only that of an individual body as an organism species, but also that of individual bodies and species as foundation for the cultivation of culture. It also represents the life of the *spherophylon*, composed of the assembly of all living organisms. From a perspective that focuses on individual bodies, the life history of an individual human body spans from the fertilized egg to the ghost, even though the organism species of humanity that recognizes this existence has a culture. Like within a human body, neurons have cognitive function, while neither skin nor muscle cells are aware of life.

Instead of viewing the death of one's individual body as one's death, one may have the idea that his or her life will be retained in the human cultural sphere, even after the body dies. With this idea, one would extend the responsibility to one's activities from the body's lifetime (the period ending at the cessation of the life of the individual body) to the lifetime of the *spherophylon*, or in other words, to the period of the eternal life of the *spherophylon*.

The above reasoning equals examining the significance of life and death from the perspective of natural history. It is similar to declaring that it is not possible to know the reality of life exclusively by studying the nature of life through a reductionist analysis of materials, simply viewing living matter as a mass that carries life.

In analyzing a living thing, naturalists should lean towards looking at the individual body, the species, and the *spherophylon* together, because they are concerned with the integrated whole. Considering that death is an event when various living phenomena end, then analyzing the varieties of phenomena in a hypothesis-testing manner is an essential operation in biology. At the same time, it is indispensable to understand, from a general perspective, what phenomena are present in life and death to define the cessation of life.

(3) Science of natural history

Natural History and Science Firstly, this paragraph examines terms. In English, the term *natural-historology* does not hold, while its counterpart in Japanese may be usable. Although the term *historical science* is occasionally used, the term *history* apparently does not go well with '-logy' used to refer to the logic of natural science that involves empirical verification. The neo-Kantian Wilhelm Windelband (1848–1915) defined that *Geschichtswissenschaft* (historical science) differs from natural science in that it is a science that describes nonrecurring transitory events. (To be doubly sure, *histology* is a term used to refer to a discipline of biology and is unrelated to history. The term *histology* originates from the Greek word *histos* [tissue].) *Botany* is interchangeable with *plant science* and *phytology*, although subtle differences exist. The philosophy of science is pursued, while the science of philosophy is not. Philosophy is not something that is expected to provide solutions through analysis of individual phenomena, based on mathematical logic.

The above examination is equivalent to asking what natural history is and, at the same time, asking what science is. Natural science analyzes phenomena in the natural world for underlying universal principles on a material basis. Of course, some scientists present discussions, such as holism and the science of complex systems, beyond simple mechanistic theories, to avoid being readily entrapped by reductionism. Notwithstanding this, nowadays an outstanding scientific treatise equates to successful results gained through reductionist analysis.

As this book has repeatedly discussed, simply describing and integrating complex and diverse phenomena does not elucidate events, even in the domain of natural history. One will never approach the truth of individual phenomena unless one analyzes and empirically explores them using modern scientific methods. Furthermore, in no discipline of science is it possible to immediately determine how facts gathered simply by analyzing individual phenomena are related with the universal principles of the natural world.

Notwithstanding the above, the need to use the methodology or approach of natural history to know the integrated whole of a concrete subject, such as the universe, earth, or life, is not simply because the issue is concerned with the integrated whole or is expansive. It is also not a mere effort to explore a methodology to deduce the form of the integrated whole based on limited findings, simply because current scientific findings about individual phenomena are less than adequate.

The relationship between natural history and science is like that of science and philosophy. They are not mutually contradictory; rather, they complement each other. Present-day natural science focuses on individual phenomena and analyzes them meticulously. In contrast, natural history intends to inquire what the subject is as an integrated whole, such as life, the earth, and the universe. As such, the methodology of natural history makes manifest the existence of those that cannot be elucidated by analytical research alone, and it pursues further study from an integrative perspective. In sum, natural history is still necessarily defined as a branch of modern science, although the broadening of its methodology is recognized. In this way, it deviates from the narrow definition of natural science by being positioned in the broader field in academia.

Natural science empirically clarifies the significance of individual facts by reductionist analysis. Findings, in many cases, are applied to technology and functionally effective in augmenting the material- and energy-oriented wealth and ensuring the safety of society. (Some modern science-based technologies are profoundly efficient at destruction, as noted by the cynical view that war benefits rapid advances in science. There should certainly be knowledge to control that destructive power. However, knowledge that causes wars to break out or controls them is not what the logic of natural science is meant to construct.) The domain of natural history can in no way be a sacrosanct research-first discipline. It is also expected to contribute to society on a short-term basis. Present-day researchers should neither lack the perspective of science for society nor neglect efforts to contribute to science for science.

Natural History and Organism Species As mentioned earlier, when discussing the concept of species these days, the biological species concept is frequently referenced. The central idea of this definition of species is sexual reproduction, which is a phenomenon seen with living organisms currently thriving on the earth, including metazoa, terrestrial plants, and fungi. As also noted earlier, from the comprehensive perspective of life, sexual reproduction is a feature earned in the second half of the history of life. While remarkable diversification of living organisms occurred after they acquired sexual reproduction, this reproductive method is not common to all living organisms, although it is certainly the reproductive mode for the dominant majority of currently living organisms. To find out what constitutes the life of earth's extant organisms and, based on these findings, to derive scientific laws applicable to technology that guarantees humanity's wealth and safety, analyzing species using this reproductive mode as a key is an important challenge for present-day science.

Biology began to deal with the concept of species at around the time of the English naturalist John Ray (1627–1705). While records of biota unfolded to organize diverse metazoa and terrestrial plants, species-level difference provided the most easy-to-understand basic unit above individual bodies. Indeed, historically, species differences were the basic means used to recognize biodiversity. It became a common practice to follow this common notion when giving names to animals and plants, and giving other names at the rank of species. Diverse species were identified. Before long, scholars began to gradually notice that species differences were not uniform and that different ranks were present. Different taxonomic ranks, such as genus and family (the term *order* came into use earlier), grouped similar species for organization convenience, while *species* was gradually established as the basic unit used to recognize diversity. *Species*, as proposed by Ray, was adopted by Linnaeus as the basic unit for biota systematization to recognize morphospecies. Among others subjected to scientific argumentation, the unit *species* came into widest use due to the biological species concept (Mayr, 1942).

To grasp the whole picture of diverse organisms, it is necessary to use some measure to recognize and ascertain them. For this purpose, *species* is suitable as the basic rank in species diversity. Recognizing of species is a convenient tool for processing an enormous amount of biodiversity information. Moreover, *species* has a long history, through which a set of common international nomenclatural rules have been established. It is a concept that has conveniently helped scholars to develop and improve their research.

The question of what it is to be alive is the fundamental challenge of biological sciences. Considering this challenge, the issue of what a species is does not question the true nature of living, although species as a measure of biodiversity has been well established. To answer the question, scholars must derive laws for the whole world of living organisms, must not use a specific reproductive mode to streamline the concept, and must keep track of facts based on universal principles.

Nonetheless, in actually dealing with the aforementioned challenge, present-day science cannot be construed as approaching the true form, since it only knows a fraction of species of all current living organisms. Furthermore, if knowledge of what it is to be alive serves as the basis for developing guiding principles for everyday life, then knowing the truth about what it is to be alive is essential for those living today. That said, scientific curiosity does not lean back until necessary conditions are fulfilled for a solution. Those who are currently alive wish to know the principles of life. Regarding the issue of species diversity, at present, it is inevitable for scientists to ask, as a challenge related to the true nature of life phenomena, what a species is. This quest differs somewhat from the issue of what definitions of species can be used to understand biodiversity.

Biodiversity is a basic trait, fundamental to the essence of life.

Establishment of a mechanism to ensure diversity was the foundation that guaranteed the emergence and sustenance of life. Life is considered to have appeared on the earth as a single type. The emergence of life, at the same time, meant diversification of life phenomena. Nonetheless, even if at the time of emergence of life on earth one of the first actions of primordial organisms was to vary genetic material (this book refrains from discussing whether it was RNA or DNA), then the variation must have been initially intended to create individual differences. It is impossible to assert that the variation immediately led to the creation of species. It must have taken a reasonable length of time for primordial organisms to achieve the evolution known as species differentiation. In this way of thinking, questions arise as to when and in what manner diversification of the rank known as species was introduced to the world of organisms, and what features species differences exhibit in comparison with individual differences.

When the living matter of primordial organisms developed independently into individual bodies is not definitely known. Naturally, differences between individual bodies that had the same genetic properties should have reflected some degree of variation (those considered as the same species according to the present-day knowledge). At first, living matter might have been living on a cell-by-cell or, at best, a colony-by-colony basis. To infer their early state, one can use present-day unicellular asexual organisms for comparison purposes. With microorganisms as well, species differences in taxonomic terms are considered a basic property that indicates diversity.

Currently living species of asexually reproductive unicellular organisms are identified with the aid of the biological species concept. Although it is known that different definitions of species are used with different phyletic groups, in many instances the numbers of species are compared between different phyletic groups. Even with prokaryotes that frequently produce mutants, differences between the original and mutant are distinguished from differences between individual bodies (= cells) of different species. The identification of species is predicated on the establishment of sufficient genetic differences. It is understood that such genetic differences should equate to species differences of sexually reproductive species, as recognized in the biological species concept. The recognition of such species differences has led to an understanding based on a common (though ambiguous) measure as to what it means that differences occur at a rank above individual bodies and how these interpretations avoid bias by differences in phenotypic characters produced by simple genetic mutation. Of course, as indicated using the adjective *ambiguous*, it is not possible to establish a scientifically clear definition of species difference. For that reason alone, it is not rare that different researchers differently identify the species of an individual body found in the natural world.

Since the emergence of multicellular bodies, sexually reproductive types evolved from asexually reproductive types separately in their respective lineages of unicellular and multicellular organisms. The rate of gradually growing genetic differences between individual bodies due to the accumulation of variations accelerated with sexually-reproducing organisms. However, although experiencing a relatively faster rate of evolution, these organisms did not develop a fundamental change in their mode of being as living organisms.

Species defined and distinguished by the biological species concept differ from each other to the degree that their crossbred type, if produced, cannot reproduce. As different types, they have established isolation of taxonomic characteristics. Cryptic species are those which are not readily identified based on phenotypic characters, but can only be genetically determined as definitely different.

Genetic mutation can occur at some rate with any individual bodies, even with organisms that produce next generations solely by asexual reproduction. Various individual variant bodies accumulate randomly within a population. Non-adaptive variants die early and become extinct through generations because they produce offspring at a low rate. Variants that bring about no inconvenience in the living environment accumulate within the population. Variants accumulate differently over time between different populations. Through this process, differences between groups of individual bodies develop, which present-day science recognizes as species differences.

With organism groups that produce next generations solely by asexual reproduction, it is difficult to find a universal principle which accounts for delineating species among different ranks. From a bird's eye view of the whole world of organisms, differences at the species rank are not essential for organisms to live. In fact, living organisms have increasingly thrived by evolving to sexual reproduction and acquiring the ability of rapid diversification. Along with the evolution of sexual reproduction, living organisms began to exhibit definite rank-to-rank differences in diversity as ascertained by the biological species concept. Evolution and diversification are the most basic properties of living organisms. One such property, although hidden, is the tendency to inevitably augment diversity. Biological evolution leading to currently living species has consisted of a series of changes enabling diversity to increase.

To have a universal bird's eye view of the world of organisms, one should be aware that reliance on the number of species as a representative measure of diversity throughout the world of organisms is only understood in the context of the above-mentioned ambiguity. In the effort to recognize diversity by species diversity, counting the numbers of species throughout the world of living organisms is the easiest method to understand and, for everyone, the most accessible way of doing it.

5.3 Human Society Viewed from a Natural History Perspective: Contributions to Society

(1) Culture based on natural history: How do we associate with nature?

The question of to what degree science is useful to society is constantly posed to fundamental scientific researchers. In contrast, the economic usefulness of arts and religion to society is seldom questioned. Aside from the usefulness of science, the seriousness of the question has intensified since natural scientific research began to necessarily expend a large amount of money, directly or indirectly cause environmental issues, and connect with military research. Until the early 20th century, the general public viewed scientists as honorable and pure, living in poverty similar to artists. Nowadays, the societal image of scientists has altered substantially. Probably in line with that trend, scientists have generally begun to adopt an altered attitude.

Upon examination, religion and arts (aside from studies on the accomplishments and history of religion and arts) still today emphasize original creative work, speculation, and missionary work, among other endeavors, from standpoint less constrained by founding a college or requiring large budgets to perform work or train successors. In contrast, modern sciences – not only natural sciences, but also social sciences and humanities – have a considerable operating budget for college study, as well as subject-specific research expenses. The raison d'être of science is occasionally questioned due to economic issues. In present-day society, which is oriented towards materials and energy, it may be inevitable that returns on investments are scrutinized, and the raison d'être of science is subject to examination as a matter of course.

Nevertheless, it is perilous to argue the issue of whether science is useful to society or not solely on the basis of natural science contributing to the wealth and safety of society by applying its achievements to technology. This is particularly true when viewing science as an attribute of the intellectual organism known as humanity.

Economic Significance of Biodiversity The section discussing bioinformatics (3.3) examined weather forecasting as having similar economic effects as biodiversity. It is widely known that the accuracy of weather forecasting has a tremendous influence on the economy. Present-day people, even with no understanding of concrete facts, readily accept the idea that megascience, or supercomputers, will have a huge effect on the wealth of society. However, biodiversity, although discussed as a megascience, has a conspicuous conservation aspect, evidenced by efforts towards the preservation of species. As such, biodiversity projects are even subject to occasional biased views, being thought of as inhibiting economic development.

Conservation, or keeping things as they are at an immense cost, is not very appealing to society. Present-day society is heavily material- and energy-oriented and generally assumes that material wealth equals human happiness. In such times, investing in a project with little economic impact is postponed until economic conditions largely improve.

To explain the conservation of biodiversity in relation to economic effects, the topic of ecosystem services comes into play. The argument, which clarifies the roles that biodiversity plays in certain situations, stresses the magnitude of potential economic losses incurred by disturbances to biodiversity and points out the need to formulate complete measures to avoid such consequences.

The effects thus explained are maintained with a conservative attitude. In this way, the issue is simply prevented from deteriorating. Consequently, the rationale for conservation can never assert that investing into biodiversity will positively build up wealth. Discussing biodiversity in this way simply offers a warning that sitting back and doing nothing will result in loss. To avoid simply presenting the threats, it is desirable to also advocate for utilizing biodiversity as resources.

At one time when the implications of endangered species were not properly understood, scholars emphasized the value of biodiversity as genetic resources and the importance of biodiversity in terms of environmental conservation to explain why it was important to think about and protect endangered species from extinction. As expected, at the time, the explanation tended to be conservation-oriented. Nevertheless, in their heart, they thought that sustainable use of biodiversity was a minimum goal for the survival of the species *Homo sapiens* and that known endangered species could serve as the best models for scientifically describing the dynamics of biodiversity. Moreover, they explained that the increase in endangered species was indicative of the crisis of biodiversity and that endangering biodiversity equaled bringing humanity to extinction. However, they themselves did not believe that there were sufficient scientific grounds for the explanation to be persuasive.

Rather than emphasizing conservation, pointing out how human society's way of life has increased the number of endangered species can illustrate the proper utilization of natural history accomplishments in a scientific context. The important point is to develop, by enhancing scientific thinking, a lifestyle that enjoyably aspires to live in harmonious coexistence with biodiversity, or nature, rather than conservation. This should emphasize the happiness that can be reached through natural history studies in a constructive manner.

The issue of endangered species is often depicted in the media. However, it is not understood as a critical public concern. Media insiders suggested that I "threaten" the general public with more stringent warnings as a researcher with knowledge of this fragile reality. Otherwise, according to them, the entire society would not understand the reality. Of course, I am not fond of *threatening* society overtly. Pointing out problems to a scientifically sound extent and stating the current science is the limit of what can be communicated. I believe that every specialist in this field should adopt this attitude, although whether it is commensurate with a specialist's responsibility to society is a difficult matter.

Biodiversity and Culture The reader may question what benefits biodiversity, whose loss implies a huge loss, has for human society. When tackling this question, if you live in Japan it is vitally important to intimately think about biodiversity on the Japanese archipelago. The Japanese people live with rich biodiversity on the archipelago. They have developed a culture suitable for it in their unique way.

Blessed with a rich biodiversity and many products from nature, the Japanese people could live a life without the necessity of exterminating others (e.g. other tribes) in a scramble for resources. They developed the wisdom for living a life by appropriately sharing resources, which works well unless a person aims for extraordinary wealth beyond what is necessary. They have never had reason for blood feuds with neighbors in resource-poor locations.

Therefore, the Japanese could avoid silly ideas such as developing the whole archipelago. Since the beginning of agricultural life, farmland development was mostly limited to narrow lowlands, valleys between mountains, and intermountain basins. In addition, after the development of rural areas, the Japanese retained part of the cleared forest in the rural area, which they respected as a place where myriad deities resided. They have maintained back hilly terrains as *satoyama*, or tamed forest areas, in which they inherit part of their ancestors' foraging lifestyle by periodically cutting trees for firewood as an energy source, hunting small animals, and gathering edible wild plants. Deep mountains, spanning about a half of the archipelago, exist in almost natural conditions, preserved as the territory of a myriad of deities, which control all creation. The Japanese have attained this spiritual richness by finding beauty in nature.

People on the Japanese archipelago have been friendly to nature and have avoided development that would completely destroy nature. Their culture has incorporated nature's complex and diverse composition. The Japanese people have appreciated wild plants and vegetation at the level of art, as exemplified by the 'Man'yoshu,' compiled as early as in the seventh and eighth centuries and having no parallel in the literature of any other country. The Japanese have never known the idea of conquering nature and have proficiently developed a lifestyle in harmony with nature.

Since the Meiji period, during which the Japanese people began to contact Western civilization in a quasi-religious attitude, they converted their lifestyle to become material- and energy-oriented. Since the defeat in the Second World War and increasingly affected by American materialism, they affirmed the view of nature as resources for humanity, the noblest of all animals. Because technology based on modern science bestowed some degree of power, the Japanese people began to develop their land to acquire material wealth in a simplistic manner, believing that they would be able to conquer nature. Somewhere along the line they were infected by the Western concept that developing forests, the places where old witches live, is a beneficial act, and they began to utilize them as resources for humans, the noblest of all animals.

As a result, the Japanese people have now fully realized how nature has responded to haphazard land development pursued in the second half of the 20th century. They have now learned to regulate land development to some degree. However, despite understanding the significance of environmental conservation as a vision, quick profits blind some people to apply undesirable pressure on nature. Various surveys have revealed that there are many such instances even today.

The organism species *Homo sapiens* is undoubtedly one member of the earth's biodiversity. Embracing this fundamental understanding, one can live with biodiversity, while recognizing there are various benefits from biodiversity. In doing so, nature on the Japanese archipelago will surely guarantee an affluent life for *Homo sapiens.* It is also true that the Japanese archipelago is bound to experience frequent natural disasters. Due to its location, terrain, and structure, it is even known as the "disaster archipelago." These events have been fundamental in forging Japanese people's view of nature. The culture created by the Japanese people has clearly delineated the harmonious coexistence between nature on the Japanese archipelago, including disasters, and the people living there.

Present-day Japanese people should correctly understand that the archipelago they live in is embroidered in diverse elements of nature, rich and beautiful. They should know its value. They should know that applying useless stress on nature and biodiversity is harmful to the life of the *spherophylon* on earth, from which their life is not separable. To do so is a kind of self-injury. The best shortcut to aid this understanding is to elucidate the natural history of the Japanese archipelago and learn its dynamic state. Natural history-related researchers who have come to know the facts about the archipelago's natural history, even if only a fraction of them, should publicize what they have learned widely to society. Furthermore, they should disseminate the value of natural history by using correct expressions that allow the largest possible number of people to become interested in natural history.

The above-stated logic can directly and infallibly be applied worldwide, although this section has focused on the Japanese archipelago.

Contemporary Harmonious Co-existence with Biodiversity The preceding argument leads to the need to ascertain that the life you live is an element of the *spherophylon*.

Upon encountering works of art, you are inspired by the artist's exquisite sense and moved by the beauty of the work. Human artwork gives fresh sensations about the beauty of natural or artificial objects. Outstanding artwork induces or transmits the excitement arising from beauty, and you newly discover the happiness of living.

Science does have similar aspects, which present-day people might have forgotten. Humans have evolved to have extraordinary intellectual abilities and have come to experience intellectual pleasures that other organism species may not know. As materials indispensable for their lives, humans value things that are useful for their inner life, as well as those useful for their physiological activities. An example is the cultivation of mind-soothing flowering plants in their yards. This may relate to the fact that *Homo sapiens* alone has developed the propensity for scientific curiosity, which other animals appear to lack.

Scientific curiosity has helped science emerge and develop, and by applying resultant knowledge to technology, has led to the development of human resource and energy-oriented lifestyles. However, science originally developed driven by human scientific curiosity. Many scientists motivated by this curiosity still today conduct their research irrespective of whether or not the outcome contributes to material wealth or safety.

To illustrate a concrete example, the following describes my experience. When I became a graduate school student, a mid-career researcher active in the then stellar field wondered why I chose to study such a (then) archaic subject as taxonomy. Ten and several years later, when I was promoted to professor, government officials from more than a few ministries and agencies asked me to impart my knowledge to them. Regarding the greatest benefits of science to human society, we should be more confident that our contributions to science for science, motivated by the fun of science, are conducive to creating science for society. Alternatively, it is important to at least endeavor to promote science in attaining this type of confidence by addressing idealized challenges, such as science for science, rather than being stuck in one's narrow range of curiosity.

(2) Natural history in our society

Despite their status as non-professional scientists, naturalists have played a major role in natural history analyses in Japan. This might imply that the study of natural history was taking hold in Japanese society, separately from the professional course of natural history as science.

Science That Seeks Truth and That Aims for Profit In principle, genetic information is accurately transmitted. However, mutations occurring at a very low rate cause dramatic developments in the world of living organisms, a process known as evolution. Because of evolution, life can exist on the earth. This is a course of natural providence free from anyone's control. Naturally, evolution is not a phenomenon that manufactured science can control, even though science can deliver a blow for the destruction of nature that substantially impedes the unending flow of evolution.

An individual stores their findings in the brain to develop, verbalize, and share knowledge with other individuals. Information is organized by means of a spoken or written language and stored in society to use for the common knowledge of society. These processes have served as motivating powers to create and foster culture. The direction of the growth of information is determined by the thought of people who create culture, rather than controlled by natural providence.

The direction of the growth of information accumulation motivated by pure intellectual curiosity may be in line with natural providence. However, having come to have culture, humans demand their own or their nation's rights. Furthermore, they are greedy to make profits solely for themselves, whether as individuals or as a nation. Increasingly leaning towards materials and energy, they are strongly interested in knowledge that is effective for safety and wealth and in science applicable to technology. In this situation, the direction of the growth of knowledge, or culture, is controlled at the discretion of people who make up society. If the direction of the growth is wrong, the extinction of the species *Homo sapiens* will become a possibility.

The term *industry-academia cooperation* symbolizes contributions made by science to society. Science is evaluated by the measure of how much it contributes to industrial growth. This might be related to the attitude of society that assesses a person by the size of their income, as if economic affluence determines human happiness. I have heard that according to statistics, the small group of people in the richest class are not those who have the highest degree of happiness. It is said that even in the United States, while people's sense of happiness is proportional to their income up to the medium level, the level of happiness of people at higher income levels shows a less than proportional increase (D. Kahneman, 2011 and others).

It is difficult to select a measure of affluence; on the other hand, suffering in dire poverty is the opposite of happiness. Because the measure of poverty is relative, it is not easy to determine what level of poverty will lead to a miserable outcome. Being relatively poor due to disparity has a different implication from being absolutely poor in terms of survival. The material needs of the poorest segment of Japanese people in the 21st century are not in the same league as those of Japan immediately after the defeat in the Second World War.

José Alberto Mujica Cordano (1935–), former president of Uruguay, relayed a saying of the native South Americans, "poor people are not those who have little, but are those who have an endless greed and always want more and more (Mujica, 2012)." The *kanji* characters "吾唯足知" (I know only satisfaction) are written on a basin at Ryōan-ji temple, which is said to represent the philosophy of Zen and is also attributed to Lao-tze. In either case, this idea underlies the traditional thought of the Japanese people. Mujica's lifestyle would be congenial to the Japanese in old days.

Some people, although suffering poverty from the material- and energy-oriented perspective, are not necessarily at a low level of spiritual happiness. Frequently, some sincere clergy are filled with spiritual richness without aiming for material affluence. Nevertheless, except for a very small number of geniuses, the large majority of average citizens look to a more affluent and safe life as a measure of happiness. Ordinary people's sense of happiness may be such that food is available without excessive hunger or illness; delicious dishes are occasionally enjoyed, rather than existing for long periods of living on very little for survival, in addition to the availability of every meal; some of their time can be spent for learning activities; and they have the luxury to travel for recreation to see more of life. They may become envious of enormous fortune and dream of grabbing it. However, they are aware that it is no more than a dream and live a healthy life with satisfaction. This may describe the great majority of the Japanese people. If this topic is extended to a global scale, the figures would be somewhat different.

The contributions of science to society are mostly assessed on the degree at which more people are provided with the benefits of material- and energy-oriented affluence and safety. Consciously or subconsciously, whether or not scientific achievements are useful for people's real life in society is viewed as an important measure of the degree of contributions of science to society. The potential usefulness of science, which has become apparent in later years through its application to technology, does not seem to be a relevant issue for the time being.

In the reality of science contributing to society, achievements with a large contribution to the economy are highly rewarded from a material- and energy-oriented perspective. However, accomplishments highly esteemed in terms of scientific growth do not always immediately gain economic compensation. Indeed, there are challenges that are fundamental for the growth of science but that are not necessarily immediately useful for society. Those who are motivated by scientific curiosity and study said challenges to contribute to the growth of science are not always active because of the hope for economic returns. Scientists gain satisfaction from fulfilling their scientific curiosity when their research produces fruitful results. More than a few researchers who contribute to natural history studies sincerely follow this kind of curiosity.

The above is true not only for those who contribute to scientific creation and are called scientists. Enjoying the benefits of science should not solely mean receiving the wealth and safety brought about by scientific discoveries applied to technology for technical improvements. Those who learn about scientific achievements have the joy of learning solutions that fulfill their scientific curiosity. Most of all, improved science literacy adds stability and a sense of fulfillment to one's life. Humanity, the noblest of all animals was given its position by God and allowed to use nature as a resource. Moreover, human existence realizes the value of grabbing happiness only knowable to humans through intellectual activities. Longing for peace to ensure safety may be unique to humans with the capability of conducting intellectual activities.

This topic is related to the learning that individuals pursue on a lifelong basis, often referred to as "lifelong learning" in recent years. In Japan most people assume that learning is confined to schools. Certainly, what is learned in school plays a major role in developing people's edifice of knowledge. However, human beings are intellectual animals; therefore, learning is a natural human act. Moreover, the happiness felt through learning is, in magnitude and quality, similar to the satisfaction of a mind soothed by a superb work of art. That said, present-day people might have almost forgotten the significance of such learning.

Among retirees with plenty of leisure time, those who are interested in new opportunities for learning at museums and other facilities are increasing. However, particularly in Japan with an increasing number of time-rich people, the framework to respond to their demand has not yet fully developed. These are my strong feelings about the current state of natural history, and I wish that (truly literally) lifelong learning opportunities would be provided to both the old and young in Japan. Improving science literacy seems to be one of the most urgent contributions that science can make to society. I expect that for the Japanese people and Japan, further dissemination of the scientific way of thinking will contribute to making a healthier society.

Science for Society: From the Joy of Science to Environmental Conservation There is a recent tendency in society to often consider that science for society is science for materials and energy profits. Meanwhile, promoting science for science, driven by scientific curiosity, and widely disseminating scientific achievements help science take hold in society as part of cultural enrichment. This leads to people's enhanced skill in the scientific way of thinking. Living a life built on the foundation of intellectual activities leads to establishing relationships in human society through mature consideration. However, this is hardly valued in reality.

To be a science for society in the context of this book, natural history needs to meet several requirements. Typically, as a science it should contribute to establishing a scientific way of thinking in culture, serve to solve environmental issues, and function to heal people's psychological scars.

Maintaining the environment surrounding humans in good condition is a bare minimum challenge for a fulfilled human life. However, science is not yet fully capable of providing guidelines for the protection of the environment because science has only a fraction of information on the environment.

The provisioning of a system designed to collect and compile extensive meteorological information used to forecast weather enables the weather forecaster to predict very near future meteorological changes with high accuracy based on the collected and compiled data. However, when it comes to the environment surrounding human beings, a wide range of information is required, and the impact of the environment on humans ranges across diverse aspects. Consequently, even with full use of all currently available information, it is not possible to accurately predict what will happen in the very near future or to suitably design what responses to prepare for the predicted outcome.

Regarding the prediction of earthquakes and volcanic eruptions, the phenomena are very simple to see. However, scientists have still been unsuccessful in acquiring the whole body of accurate information to understand the mechanisms or the variations taking place underground in the preliminary accumulation of energy before an earthquake or eruption, despite the expenditure of a huge amount of research funds.

It is natural that people have a strong interest in the forecast of weather and natural disasters that may occur in the environment surrounding human beings. Preparations have been reasonably made in response to as many knowable forecasts as possible. Nonetheless, those who are seriously interested in the future of biodiversity, which is none other than one's life itself (in other words, the *spherophylon*), are only a limited group of individuals, even among learned intellectuals.

Despite the actual difficulty of ascertaining the whole picture of the dynamic state of biodiversity, surveys have been conducted on the current state of endangered species, which is used as a model to capture the dynamic state. How to respond to the current situation is currently under consideration (3.2–3). However, details of the current state of endangered species are known only in few advanced countries. In most areas of the globe, even basic surveys, such as where and what organism species are living, have been conducted only on a limited scale. Moreover, this applies exclusively to organism groups that are easy to recognize, such as vascular plants, vertebrate animals, and large fungi. The reasons for the poor accumulation of this kind of information have been pointed out here and there in this book.

In preparing the red list of vascular plants in Japan, non-professional naturalists made cooperative efforts, as described earlier (Tea Time 8). The findings obtained by those who began observation motivated by purely scientific interest in the dynamic state of regional biota made important contributions to the project, which required a vast amount of information. This is a good example revealing what contributions natural history makes to society.

It is a matter of course for natural history researchers to play the role of think tank, based on their experience in analyzing their respective research challenges and making full use of their trained scientific abilities. An ultimate answer to the question does not address what science can do today or tomorrow. However, the responsibility of researchers is to let society know to what degree science at the present-day level can dependably respond to the situation in question and to what extent science can reliably predict necessary measures.

There is joy in the pursuit of science as a means of healing people's psychological scars. As this book has described repeatedly, the growth of science was initially driven by human intellectual curiosity about diverse phenomena occurring in the immediate environment. It did not originate from a profit-and-reward mentality, which aims to gain something by solving problems. Solving a challenge simply brings the pleasure of fulfilling curiosity. Nevertheless, this joy in science offers the greatest sense of fulfillment in human intellectual activities, brings about spiritual enrichment, and leads to intellectual delight.

It is understood that humanity began to call itself the noblest of all animals after it had begun to perform intellectual activities. As the term *noblest* implies, intellectual activities and the creation of culture are assessed a step higher than the activities of other animals. It is a Western way of thinking. Evolution is an event, the value of the resulting outcome which humans should not assess. Rather, humans should understand that the organisms currently living are in the best adapted condition as a result of evolution. Organisms best adapted to different environments are subject to constant variation, along with environmental changes. During natural providence, the species dynamics does move towards the eternal survival of only those highly valued by humans. It is an objective fact that humans have grown intellectually and come to have culture. However, it is incorrect to say that humanity has evolved to the highest stage because it has knowledge. The inappropriate application of a high level of technology can lead to the extinction of humanity. This truth must be remembered.

At the same time, it is also an objective fact that the sense of intellectual fulfillment gives great pleasure to humans. The sense of intellectual fulfillment demonstrates the value of living, similar to the joy of living felt when enjoying exquisite dishes or being awed by beautiful things. The greatest benefit of science lies in such kinds of fulfillment. It is beyond question that natural history brings the delight of scientific pursuits to a wide range of people. Busy looking for material- and energy-oriented pleasures, many present-day people are not currently aware of these very simple joys.

Of course, the intellectual delight brought by science is more readily recognized if one is engaged in an investigative science. Certainly, many impressive things are discovered, even in the process of observation alone. However, new phases of intellectual excitement arise by analyzing problems or, in scientific terms, by performing experimental exploration.

For art meant to induce aesthetic excitement, the segment that readily leads to profits by pandering to popular tastes has grown tremendously. This may be a phenomenon like the biased interest towards sciences that can be applied to technology, although this book refrains from discussing the matter further.

Co-existence of Natural History with Society: Lifelong Learning I was born and raised in a mountain village in the Oku-tamba area. When I was a child, I enjoyed picking horsetails, hunting fireflies, and gathering edible wild plants and firewood. It was a life in *satoyama*. In fact, more accurately, in a rural area richly endowed with nature, my childhood was in close contact with the diversity of nature.

When my children were growing, my family lived at the foot of the Inari-yama hill, Kyoto, where my children had fun playing outside without a care about becoming dirty. When my grandchildren were growing, my house and theirs were in a neighborhood near Jike-furusatomura in Yokohama. I and my grandchildren used to go out for walks. Although I was fairly quick-eyed, they were better than me; we had a good time finding four-leaf clovers. The experience I describe above demonstrates that on the Japanese archipelago, it is not difficult to find opportunities to come into contact with wild plants and insects, although people do appear to live in a concrete jungle.

Some people grieve that children have few chances of playing with nature. However, there are few arguments about why they do not strive to provide more chances of coming into contact with wilderness, despite their impression that having few chances of playing with nature is a regrettable situation. In the process of developing a school education system since the Meiji period, the government has skillfully provided a system for pupils and students to learn the existing system of knowledge, motivated by catching up with and overtaking Western countries. Nonetheless, they have disregarded building a system to support lifelong learning for people voluntarily motivated to learn. It might not be desirable for totalitarian leaders to develop individuals who stand on their own feet.

Even today, lifelong learning is often confused with lifelong education, with many people misreading it as adult education. Rather, a minority might think it important to develop a custom of voluntarily learning throughout life, including learning in society and at home, as well as provisioning intellectual education at school. In this book (e.g. 4.1 (3) and 4.2), I have repeated that it is vitally important that museums and other facilities function effectively to support lifelong learning because such institutions are meant to play that role. Notwithstanding their significance, this idea is understood by only a limited sector of society.

Museums are suitable institutions for people to learn natural history. Many on a global scale function as hub institutions contributing to natural history research and owning a rich collection of literature and specimens. In Japan, indeed, several institutions are successfully performing such activities, although on a small scale. As such institutions, museums can provide support for lifelong learning in a way different from schools. In Japan, however, they have been slower than in Europe or America in exerting their functions efficiently. Finally, museum staff have recently begun to work effectively to support lifelong learning. Hopes rest with their future development.

Learning should be a lifelong delight for humans as intellectual animals. Based on existing scientific findings, humans should seriously think about both what is known and what is not known about living things and nature as the backdrop to life – from wildlife abundantly seen in the immediate environment to artificially-bred animals and cultivated plants. Efforts motivated by intellectual curiosity to familiarize oneself with the scientific way of thinking will potentially improve the human ability to think logically and enable individuals to properly organize their lives.

Museums and other facilities should endeavor to help everyone pursue lifelong learning. In the domain of lifelong learning, natural history will potentially be deeply tied with, and make tangible contributions to, society. Contribution in this context does not necessarily augment material- and energy-oriented affluence. Instead, it will cultivate spiritual richness, comparable to the value of arts and religion to the general public.

5.4 The Present-Day Role of Natural History

In the above discussion, the current challenges posed to natural history should be clear. As a conclusion, the following examines what has been clarified, while avoiding repeating previous discussions.

Since great advances were made based on modern science and technology, expectations for science seem to have substantially changed in comparison with Aristotle's time. Nevertheless, people's expectations for the goals of natural history and other sciences have not seemed to have changed as much as the fundamental changes made to the form of developed society. This is probably because humans have had basically the same question, although their scientific curiosity has focused towards more detailed subjects.

In this case, the question arises as to whether natural history is functioning as effectively as it was in the times of ancient Greece. The challenges below are clarified from the perspective of an individual involved in natural history as to whether naturalists have clear insight when responding to the demands of society on science.

(1) Natural history research and studies

Natural History Pursued by Individual Researchers Firstly, natural history researchers are required to promote their studies. Many researchers interested in challenges in this field are motivated by very strong scientific curiosity. The magnitude of their enthusiasm for research is not of much concern. Rather, the problem may be that they are driven by too much frantic curiosity.

Research and studies conducted on one's favorite subjects and in one's favorite way do lead to advances in research. However, if all naturalists work in this way, accumulation of data based solely on curiosity would be the outcome, which could impede the healthy growth of the discipline. Indeed, as far as I can tell, unbalanced research-for-research ideology in some areas of specialization has led to fragmented and peculiar research. At some point, fragmented competition may have increased the efficiency of building fundamental information. Nonetheless, excessive competition in the skills of individuals has certainly been detrimental to the growth of science. (To be fair, fundamental information produced during the time of competition has actually been utilized in various ways. The problem was that at some point, research that grew in an unbalanced manner in the aforementioned areas was poorly valued by outsiders and forced into an extremely unsafe research policy situation.)

It is desirable for research promotion to be balanced in its relevant domain. For this reason, individual researchers should maintain the perspective of promoting research across the domain, without being confined to their respective worlds. This is universally desirable in the field of science. Meanwhile, in the process of deepening research in a particular subject, an extremely narrow range of knowledge in the relevant field would enable efficient short-term benefits for the researcher. However, this can lead to immersion solely in segmented subjects, particularly dangerous for natural history and other domains in which researchers are expected to take a broad view of nature, while investigating individual subjects. Lamarck successfully established a new field of biology known as invertebrate zoology after he took the post of manager responsible for handling organism groups, which he had not previously studied, despite his love for natural history.

Natural history research is constantly bestowed with discoveries, although they may not be major. Natural history subjects are interesting enough that researchers can find deep satisfaction in a limited area of research and study. As such, one must be very careful to not be trapped by the depth of one subject. Of course, geniuses, with their frantic activity, may serve as major drivers in promoting deep studies. In addition, the field expects the appearance of an extraordinary genius. Nevertheless, to promote natural history overall, many average researchers also make sound contributions. This is no different than in all other fields of science.

It may be an inevitable trend that more information is produced in accessible research domains than other domains. It cannot be denied that, for fundamental biodiversity information, advanced research has been conducted on vascular plants and vertebrate animals, while research has been slow on microorganisms and marine invertebrate animals. Moreover, compared with biota research in advanced countries, research in developing countries is generally slow. From the view of elucidating the life of the *spherophylon* on earth as an integrated whole, research is hardly promoted in a balanced manner.

To discuss natural history research on an individual basis, the role played by non-professional naturalists, especially notable in Japan, is indispensable, along with the contributions made by full-time researchers. In reality, for producing the voluminous basic literature on natural history, full-time researchers are somewhat limited in number. More specifically, for example, non-professional naturalists have achieved tremendous results in producing observation records and collecting literature and specimens representing nature on the Japanese archipelago. Despite the constraint of having a very limited number of full-time researchers, natural history research has excelled to a world-leading level on the Japanese archipelago, home to a diverse and rich nature. This may result from the surety of valuable human resources contributing to fundamental research and studies from an overall perspective. Moreover, this is only possible because of the lifestyle and the way of thinking of the Japanese people. I am concerned with the tendency that this is gradually forgotten as valuable and good Japanese traditions disappear.

Furthermore, it is easy to cultivate scientific curiosity in the domain of natural history. Challenges in the domain are waiting everywhere for solutions for anyone to find when driven by curiosity. While various research challenges are met by full-time researchers, enjoyable challenges for non-professional specialists are inexhaustible.

Natural history research not only produces research results, but also provides substantial rewards to personal and organizational contributors. While research results that contribute to the accumulation of human knowledge principally gain the spotlight, the aspect of research that enables researchers to become better human beings is not very much the focus of attention.

When pursuing natural history research, non-professional naturalists, unlike full-time researchers, cannot expect their contributions to serve as a foundation for them to earn research funds or be promoted to a higher position in an institution. They contribute to natural history motivated in large part by pure scientific curiosity. Indeed, those who work on natural history appear to develop a spiritual richness through their serious efforts motivated by curiosity. The spiritual growth of individuals learning natural history is a beautiful example that demonstrates the benefits of learning pursued by the intellectual animal, *Homo sapiens.*

In fact, intellectual learning can underlie humanity's joy of living. It is worth revalidating that natural history enables individuals, far outside the group of full-time researchers, to recognize this far more deeply than any other discipline of science.

While observing nature in a subject area, closely examining living things of a specific taxonomic group, or witnessing nature's creation in a certain range, interesting discoveries can develop in numerous directions, commensurate with the amount of knowledge of the discoverer. For the observer with curiosity, the degree of satisfaction experienced each time a small discovery is made is very valuable and beneficial.

Just like a person who is not a full-time researcher delights in making a discovery, even if not intending to publish it as a treatise, full-time researchers may often become besotted with a new discovery before writing a paper. Upon discovery of a new organism species or star, the discoverer is intoxicated by the results of his or her enthusiastic efforts, whether the discoverer is a full-time researcher or non-professional naturalist. What is more, the discovery is frequently covered by the media, whereby the discoverer feels honored.

Humans are intellectually fulfilled in the most desirable manner when they experience the joy of observing nature, motivated by scientific curiosity. This is exemplified manifestly during natural history research and study, an effect which should be more widely known.

Meanwhile, there are several pitfalls of note in which full-time researchers tend to fall. Non-professional naturalists delight in presentations by full-time researchers. Therefore, while speaking at various gatherings, some professional researchers join in the fun, are satisfied by the fun, and forget their responsibility to perform the proper duty for which they are paid. This sort of conduct establishes the reason that natural history is mistaken for an amateurish domain or, in extreme cases, is viewed as a low-grade science. Unfortunately, these instances are occasionally observed.

Even during their studies, some researchers, while tackling a specific challenge, disregard related problems and continue to work exclusively on their material. Frequently, within their confined scope, they continue to produce treatises in such numbers that they could be counted by weight. While writing numerous papers, they have the illusion that they are doing a good job. Indeed, in some instances, they do contribute to the production of fundamental literature within their specific range. However, such work is useful for building a scientific foundation, but not for research. Such efforts lack the bird's eye view of issues needed to flourish into an outstanding study, tending instead to slip into a personal hobby. Sometimes accumulation of records produced by accurate observations does build a valuable research foundation. However, a stack of literature produced from the viewpoint of technical assistance can hardly be said to result from original scientific research.

A risk for full-time natural history researchers is the state of self-complacency. Indeed, researchers experience delight when their scientific curiosity is fulfilled by a fragmented aspect of research, for which they easily misread the nature of their study as science. That said, continued production of fundamental literature within a limited range of research challenges does not occur exclusively in the domain of natural history. Such activities are routinely conducted everywhere in the wide domain of science. It is merely that, in the domain of natural history, conditions exist that make it easy for researchers to fall in the pit. Needless to say, not every researcher in the field becomes trapped.

While individual researchers stumble into problems, as lengthily described above, there are a number of research challenges that can be individually explored in the domain of natural history. Expectations are high for such efforts in this field. More than a few researchers are steadily, though inconspicuously, producing accurate literature. In recent years, several research aid foundations have provided modest support to these researchers. Although modest, many of these prizes have worked effectively, as demonstrated by the results of research projects promoted by this practice.

Group Research Activities at the Laboratory Level Research produces results ultimately judged by the quality and efforts of individual researchers conducting the research. However, for a large part of present-day scientific research, the scope of an individual's work is limited. A mainstream trend is that groups at a laboratory level produce results. This is also true in the domain of natural history. The trend towards larger research projects, which require considerable research funds, is apparent.

At Japanese universities, natural science laboratories used to follow a system built in the Meiji period, operating what is known as a small unit course system, consisting of one professor, one assistant professor, and two assistants (plus two clerical members and two technical assistants). Research and education were conducted on this unit. This is certainly a desirable research system to pursue research in, for instance, biology. It is appropriate that in terms of age structure that a senior leads research, a mid-career person actively promotes research, several junior faculty members work within their abilities to furnish fresh ideas and adapt to new technologies, and graduate and undergraduate students help in producing information while acquiring knowledge and analysis techniques. It is an historical fact that the small unit course system, which operated in a closed manner, became, in a sense, problematic due to the morality of researchers. They too exhibited the downsides of general human beings. Various human relationship problems became apparent, which needed to be addressed by systemic measures. In recent years, therefore, research systems at universities have been substantially altered. While research groups continue to pursue their activities, additional deliberations are needed to develop what system is suitable for future research promotion.

For natural history studies, it seems that there is no universally viable form of research system, notwithstanding the specific, common characteristics of the domain. Of course, with natural history as a modern science, the basic form of a research group is desirable. However, in natural history a firmly organized laboratory is not always a prerequisite because there are inexhaustible research challenges that researchers can explore individually.

When I was posted in the office at the University of Tokyo, the quorum of the Botanical Gardens in which I was in charge consisted of one professor, one assistant professor, and two assistants. During my tenure (in the Botanical Gardens, the University of Tokyo), it was authorized to add one assistant professor (with the Nikko Botanical Garden in mind). Nonetheless, to cover all research and education programs on the phylogeny of plants, we could not specialize in a specific research challenge defining the overall research theme of the institution. Moreover, this was during the burgeoning period of molecular phylogeny. To promote molecular phylogeny research, including the provisioning of relevant facilities, while pursuing traditional biota research and studies, the size of the above-mentioned group of researchers was inadequate.

It was fortunate that at the University of Tokyo around the 1980s, the University Museum in Hongo and the College of Arts and Sciences (present Graduate School of Arts and Sciences) in Komaba had related researchers, who gave guidance to graduate students. As such, staff and graduate students in related disciplines at these institutions jointly took part with us in holding seminars and implementing research and education programs. Organizational barriers were quite low in the domain of the natural history of plants.

At one time, researchers from outside the university frequently came to our Botanical Gardens, as if this institute was a nationwide joint research facility, although, in fact, it was an educational facility annexed to the University. Extramural researchers often learned the usage of our equipment, such as DNA sequencers, which were then in the early phase of wide use, and they participated in our seminars. When it came to the royal road of study, research cannot be expected to grow at a laboratory operating in a tight-knit way. Specifically, in the domain of natural history, promotion and exchange with related researchers stimulates researchers at the main research institution and provides visiting researchers with extensive learning opportunities.

While claiming a portion of natural history on terrestrial plants, the institution of this size even conducted research exploring the universal principles underlying diversity. In doing so, it used all conceivable analysis techniques and ways to promote research in an intertwined manner, from molecular-level analysis to biota research. Of course, some of the individual researchers who took part in these research efforts failed to step beyond the boundaries of their own research. Besides, some grieved that researchers motivated to have an overall view were a minority. Nonetheless, research was promoted through friendly rivalry at the laboratory, and efforts were made for individual research projects to go beyond their assumed boundaries.

The promotion of research by laboratory-level groups may still be uncommon in natural history in Japan. In Europe and America, many research institutions staffed by 50 to 100 individuals at museums and botanical gardens make healthy and fruitful contributions to the production of information. Nonetheless, they still have a way to go to successfully analyze diversity from an integrative perspective.

Cross-Institutional Joint Research The most typical form of research pursuit at universities in Japan is through a specific institution, with the aim of fostering new generations of researchers. However, research conducted at a university or research institute level in a closed manner has difficulty in flexibly forming a group of researchers when the need arises. Therefore, when forming the ideal team of researchers, researchers collaborate through cross-institutional research.

Around the time when the term *biodiversity* was still used to a very limited extent even in the community of biology, a joint research project, funded by the Grant-in-Aid for Scientific Research in the category of Minor Priority Areas, was implemented under the title of "Analysis of Plant Diversity." In this project, a joint effort was made for three years of research from academic years 1981 to 1983 and for the compilation of results in the academic year 1984. Before this began, the Botanical Society of Japan held meetings and discussed how to form an organization to pursue a large research project in the Priority Areas category and to study the major issues emerging in plant science. To explore one of the issues at that time, the joint research project formed as a joint research organization of researchers in taxonomy and ecology.

Regrettably, the project was not adopted in the category of Priority Areas, probably due to the problematic implication of the term *biodiversity* as used to tie taxonomy and ecology. However, the intent associated with the issue was understood, the project was adopted in the category of Minor Priority Areas and was launched as a joint research. The results of the project were presented in several treatises, with the overall project published under the title, 'Origin and Evolution of Diversity in Plants and Plant Communities' (Hara, 1985).

I worked hard as the project's Secretary-General from planning to completion. In my mind, the project intended to attempt to produce a modern edition of natural history, although the participant researchers, with the exception of some individuals, may not have clearly had the same awareness. Nonetheless, shortly thereafter, the Convention on Biological Diversity (adopted at the 1992 Rio de Janeiro Earth Summit) became an international topic. Since then, related ministries and agencies have submitted inquiries to me. They seem to have taken note of the project, which featured the then uncommon term *biodiversity* (Iwatsuki, 2017).

Of course, aside from this project, many cross-institutional joint research projects were organized funded by the Grant-in-Aid for Scientific Research in the category of multidisciplinary research. When I was a graduate student, I joined a joint research project on the taxonomy of ferns during 1957–1959, organized under the leadership of three professors: ITO Hirosi (Tokyo University of Education), MOMOSE Sizuo (Chiba University), and TAGAWA Motozi (Kyoto University). At that time, graduate students were still counted as sub-members, took part in research teams, and could use research funds for travel expenses. I made an expedition to the Amami Islands and other places for longer than one month, partly owing to the research funds. Since then, I have had numerous opportunities to participate in joint research with researchers from other fields. I am grateful for these opportunities, which not only enabled me to write several papers, but also helped me broaden my horizons as a researcher.

As described in earlier sections discussing GBIF [3.3 (2)], I was able to contribute to building a solid foundation in the domain of biodiversity by supporting individuals who were making inconspicuous efforts in the digitization of literature and specimens at museums and other institutions in every part of Japan and by incorporating data of exhumed literature into an international database. These achievements were related to the international GBIF and owed much to, among others, the National Museum of Nature and Science and the National Institute of Genetics, which served as hubs for these activities. As cross-institutional joint operations, these efforts were highly valuable.

In addition to the significance of the availability of the emerging database, an additional spillover effect became manifest during this process. Before the project, researchers at museums and other institutions steadily made valuable efforts to build databases on an individual basis. The significance of their efforts as museum projects was not readily recognized by their office supervisors or competent authorities (non-scientists in many cases). On the contrary, their efforts were often regarded as a personal hobby. However, this project made it clear that their endeavors, made at the nation's expense, contributed valuable service to the international scientific community. By obtaining project funding, recipients felt satisfied that their efforts were freshly appreciated by their supervisors, although the subsidy amount was not so large. This was a profound outcome of the cross-institutional joint project, in addition to the tangible contribution to the database construction.

In the 1960s, an expenditure item named Overseas Scientific Research was included in the Grants-in-Aid for Scientific Research (the category Overseas Scientific Research has since been reclassified under the category Scientific Research, as of 2017). When conducting overseas research, in many cases, a joint research project is organized in collaboration with a local institution of the host country (countries to be surveyed), as an individual can cover only a limited scope. To select suitable researchers to dispatch for this purpose, it is necessary to set up a cross-institutional research organization. For these types of Grants-in-Aid for Scientific Research, screening was formerly carried out with sufficient lead time, assuming preliminary negotiations with the host country would be time-consuming. At one time, for unprecedented research, project planning between preliminary research and the commencement of full-fledged research took about three years. While the type of research activities funded by the Grants-in-Aid for Scientific Research was unique to Japan, such activities served as a foundation for achieving great results for the research promotion in the domain of natural history.

Natural History Pursued in a Group Framework Even today, unexplored challenges in natural history research remain. Moreover, every such challenge can hold the interest of researchers enthusiastic to conduct research. For this reason, individual researchers can fall insidiously into the deep well of self-complacency. To be free of this risk and stay in the world of universal science, researchers need to make conscious effort. In addition, within a group of researchers, exchange can help a researcher to avoid this risk. While expending all their energy on individual research subjects, researchers also need to mutually share information within groups of researchers to improve their efforts through friendly rivalry. This prerequisite is applicable not only to researchers in the domain of natural history, but also, as a matter of course, to members of society. Nevertheless, I feel it is necessary to mention this when concluding discussions about natural history, as it may point to a characteristic in this domain.

Societies and Associations in the Field of Natural History The activities of natural history-related societies do not specifically differ from those of other academic organizations. In Japan, many researchers participate in general academic society activities that hold public interest incorporated association status, such as the Zoological Society of Japan and the Botanical Society of Japan. They have animated discussions at annual and other meetings. Academic societies specializing in natural history are also noticeably active, dealing with their respective taxonomic groups. However, although they hold meetings for researchers with subjects in common, desirable in-depth research discussions do not always take place.

One recent trend is the founding of unions, such as the Union of Japanese Societies for Natural History (founded in 1995), which 40 related academic societies have joined (including the Zoological Society of Japan, the Botanical Society of Japan, and other general academic societies) and the Union of the Japanese Societies for Systematic Biology, which assembles academic societies established for different taxonomic groups. These unions function reasonably usefully as liaison organizations. However, these organizations are not expected to facilitate in-depth discussions of research topics.

For fields in which large-scale research projects are promoted, researchers frequently have fulfilling discussions on their research topics. In contrast, in recent years in the domain of natural history, few successful research teams have formed. I have heard that at universities, classroom seminars held beyond the laboratory level lack liveliness. The prevalence of efficiency-oriented activities focused on fragmented research challenges might be one reason for the absence of cross-disciplinary discussions.

Clubs-Related Activities Natural history-related clubs have long operated as voluntary associations of local non-professional naturalists. While the principal members of an academic society are full-time researchers, clubs in many cases are comprised principally of non-professional naturalists. For plants, as described in 4.1, clubs at prefectural level once could ascertain the accurate dynamic state of local plants, partly due to the sponsorship of pioneers in the field, including MAKINO Tomitarō and TASHIRO Zentarō. In the 1980s, when the red list of vascular plants was prepared, the liaison networks made full use of these clubs or their spin-offs, and these personal connections worked effectively.

Until recently, these clubs regularly held local plant collecting meetings to exchange information and help club members nourish friendship, thereby fostering the next generation. Recently, these meetings have been held entirely under the title of observation meeting because the term *collecting meeting* implies a destruction of nature. Whether at an indoor meeting or a field observation meeting, in the case of a plant club, full-time researchers, as well as non-professional naturalists, often join to exchange information. Full-time researchers encourage the scientific curiosity of club participants and, at the same time, contribute to building a foundation for the production of information through intellectual exchange. In comparison, the Society for Plant Taxonomy and Geobotany and the Society for the Study of Phytogeography and Taxonomy expected non-professional naturalists to join when they were founded. In the case of these societies, it was necessary to have a reasonable number of members to publish an academic journal as the society's in-house journal. For this reason, non-professional naturalists were provided opportunities to easily access the information in possession of full-time researchers. Signs of change in the way of running academic associations and clubs have emerged, as exemplified by the merger of the above two associations into the Japanese Society for Plant Systematics, despite them being individually significant players in the second half of the 20th century.

The following discussion provides more tangible examples. The organism group I specifically study is ferns. There are many organizations for researchers and plant lovers who study ferns. In my case, I finally decided to professionally study ferns when, after I became a student of Kyoto University, I participated in the Fern and Moss Club (Kansai Meeting, Nippon Fernist Club since 1952), which was active principally in the Kansai (western Honshu) area. At this club, professional and non-professional botanists alike, including those more competent than full-time researchers and those with little knowledge, talked about existing findings on the diversity of ferns and mosses. We occasionally exchanged information at field observation meetings, discussing real-life plants. I became deeply curious about the mysteriousness of diverse living things. This experience helped ignite my curiosity to follow the path to a full-time researcher. I reported what I had learned for discussion at this club, which in turn motivated my learning. This was a way of learning similar to summarizing literature and reporting the summary at a graduate school seminar, yet I experienced it as early as in my second year of college.

Soon after I joined the Kansai Meeting, fern lovers in the Kantō area and the Kansai organization collaborated to inaugurate a nationwide organization, the Nippon Fernist Club. In the Kansai area, the Kansai Meeting continued with their previous way of holding meetings. In practice, by holding a meeting roughly every month and learning together, activities of the organization will inevitably be localized. Since its founding, the Nippon Fernist Club has annually held nationwide observation and other meetings. However, more frequent meetings are held separately in Kantō and Kansai, considering the convenience of actual participants. Additional regional meetings have been voluntarily held in many other parts of Japan. These meetings communicate with the nationwide organization, although they do not necessarily represent the club as formal branches.

While I conducted my fern research project at my laboratory at the Botanical Gardens of the University of Tokyo, I received invaluable assistance from Nippon Fernist Club members throughout Japan particularly with, among others, the collection of research materials. In treatises arising from this, I expressed my acknowledgement.

When I edited 'Ferns and Fern Allies of Japan' (ed. Iwatsuki, 1992), a professional photographer provided photographs, while, as a researcher, I prepared the text. Members of the Nippon Fernist Club helped in guiding the photographer to show exactly where desired pictures could be taken. This kind of collaboration was possible solely owing to the presence of organizations such as the Nippon Fernist Club and the volunteers that helped to run them. A quarter-century after the publication, the color-illustrated manual of Japanese ferns has been fully revised and a new edition published by EBIHARA Atsushi of the National Museum of Nature and Science (Ebihara, 2016/2017). The Nippon Fernist Club also provided full cooperation to the production of this edition.

The intent to create a society principally for full-time researchers, in addition to clubs, further developed, and in 1959, the Japan Pteridological Society was founded. This society was formed in anticipation of the participation of all individuals conducting research of ferns, rather than being simply taxonomy-oriented. Alongside the convention held by the Botanical Society of Japan, the Japan Pteridological Society has regularly held a related meeting since its founding. Trailblazers, including Japan's first female Doctor of Science YASUI Kono (1880–1971) and the world-renowned comparative anatomy researcher in ferns OGURA Yuzuru (1895–1981), were among the founding members. During the early days, the Society asked Professor Ogura to give a special lecture. In recent years, although still a society, its tangible activities are limited to an annual meeting held in relation to the annual conference of the Botanical Society of Japan and information exchange via newsletters issued several times a year.

Since around the 1980s, several university laboratories pursuing fern research have emerged in the Kantō area. At their seminars, each laboratory conducted in-depth examinations of their respective immediate challenges. In parallel with these activities, cross-organizational meetings began to be held to exchange leading-edge research information, with participants including graduate students. Thereafter, for many years this research exchange has pleasantly continued with variable frequency, although on a non-periodic basis.

As shown by the above example of ferns, some people, whether non-professional naturalists or full-time researchers, are strongly driven by scientific curiosity about various aspects of the same subject, conducting research and study on this specific taxonomic group. For these people interested in research on the specific taxonomic group known as ferns, various information exchange venues and learning opportunities are provided in a spontaneous manner. These venues and opportunities, in their respective ways, help promote research and study, grow scientific curiosity effectively, and produce successful results in lifelong learning.

Natural history research and study is not something that can be deepened by one established method. Personally, I believe that analysis and integration methods should be viable. To deepen one's learning, one should explore personally suited forms by oneself, which may include information exchange with one's fellows for research facilitation. Moreover, one should be aware of one's standpoint and have a far-reaching bird's eye view over the thorough expanse of the issue in question. Otherwise, looking at ferns might lead to only seeing one aspect of them. If one only looks at one aspect, a life of research might end without knowing what one wishes to know most, despite the desire to fulfill curiosity.

Related International Organizations The International Association of Pteridologists is a liaison organization founded in 1987, assembling societies for ferns active in several countries (long-established academic societies with published journals operate in the United Kingdom and the United States). Presently, the American Fern Society takes the initiative in Association activities, such as editing newsletters. The Association is involved in holding non-periodic symposia.

For international organizations for botany, the International Botanical Congress (IBC) is held every six years and periodically checks and manages the nomenclature for plants. The Congress is organized by the responsibility of the International Union of Biological Sciences (IUBS) under the umbrella of International Council for Science (ICUS). In interactions with international organizations for science, the Science Council of Japan is responsible for responses. When the XV International Botanical Congress was held in Yokohama in 1993, the Science Council of Japan played the role of co-organizer, and related academic societies in Japan jointly set up a committee to run the congress.

These organizations are conspicuously and variously active in international organizations, involved in think tank and broad dissemination activities that are not directly related with natural history studies.

For wildlife conservation, the World-Wide Fund for Nature (WWF) was founded in 1961 as a nongovernmental organization with the aim of protecting wildlife. Natural history-related research results and information have been utilized in WWF activities. In Japan, WWF Japan was established as a public interest incorporated foundation to support WWF activities in Japan.

In 1948, the International Union for Conservation of Nature and Natural Resources (IUCN) was founded as an international organization with the aim of conserving nature. Members of IUCN are nations, governmental agencies, and NGOs. Japan joined IUCN in 1995 as a state member. Earlier than that, in 1978, the Environment Agency (the present Ministry of the Environment) joined IUCN as a governmental agency. In addition, IUCN has many NGO members. In Japan, the Japan Committee for IUCN was set up as a liaison organization, and the IUCN Japan Project Office was also established. The Office is an international site for preparing red books that has made substantial contributions to information collection for that purpose and has established red data categories. Nature conservation requires various kinds of information about the natural world. Productive activities for nature conservation are not possible without the progress made in natural history study. Considerable efforts have been made to contribute to natural history-related information in an object-oriented manner.

The United Nations Educational, Scientific and Cultural Organization (UNESCO) was formed in 1946 under the United Nations. With the aim of achieving lasting peace in the world, the United Nations is necessarily a place for discussion centered on politics and economics. However, recognizing that politics and economics alone are not sufficient for achieving world peace, UNESCO declared in its charter that it would promote international activities oriented towards education, science, and culture to bring people peace of mind. Naturally, contributing to science is one of the pillars of UNESCO. In this context, the expedient domain of natural history is a key challenge. Of course, UNESCO is not an organization that takes initiative in promoting science. However, it plans diverse activities aiming to bring peace to the earth through science.

Compared with the above-mentioned various international organizations, GBIF described in 3.3 is the correct international organization to provide fundamental information for and promote natural history. It exemplifies that this kind of organization is essential for joint research projects in the field.

Although the number of organizations introduced here is small, the above-mentioned organizations are some examples in which I myself served as an executive and was deeply involved.

(2) Natural history and society

For those who are involved in natural history research and study, it is essential to have a bird's eye view of the current situation, to search for a desirable structure to address the situation, and to take steps in that direction. Moreover, whatever is necessary for further research and study inevitably needs to be clarified.

The above discussion has provided an overview of the breadth of the field of natural history, first from the aspect of research and then from the aspect of science and society. Clearly, for research and dissemination activities to roll out smoothly, a sound supporting framework must be provided. The domain of natural history can never thrive in Japan, as pointed out repeatedly in this book, without collaborations with non-professional naturalists, as well as cooperation between full-time researchers. In addition, a permanent structure that maintains or enhances collaborations needs to be established.

Universities: Research and Higher Education In humanities faculties at universities, research is conducted on an individual basis in many cases still today. In science faculties, in many cases, joint research involving several researchers was once the norm. However, the closed unit course operation in this format drew criticism for being a hotbed of professors' dogmatic conduct. Consequently, many national universities have abandoned the small unit course system and adopted a large course system, enabling the formation of flexibly arranged research teams. Nonetheless, this is a matter of form. It still seems common to form a research team of several members, including one professor or associate professor and junior faculty members. Empirical observations demonstrate that as a research organization, this size is convenient for facilitating research with flexibility and efficiency.

To promote research and education in individual areas, each university should first deliberate on laboratory organization structure and then determine for which specialized areas to hire researchers. Universities are responsible for screening researchers. From the perspective of academic freedom, this must not be compromised. It is natural that universities cannot be expected to always work out the best solution through this type of deliberation. They should have discussions for the sake of learning. However, frequently, their discussions develop in outrageous directions. As a faculty member, I experienced this in various ways.

Although extreme cases give a poor impression, universities in practice continue to run as a venue of academic research and education. I am sure that the large majority involved in sustaining universities are excellent people.

Regarding the promotion of natural history at Japanese universities, a breakthrough in higher education to enable fostering the next generation has occurred. Before this development, particularly in the field of systematic zoology at universities, faculty members in charge of graduate school doctoral programs had declined to a minimum level. Due in part to this, a sense of crisis arose. Therefore, the United Graduate Schools program was developed, in which outstanding researchers working for an outside institution are appointed to also work as a professor at graduate schools operating doctoral programs.

Since the academic year 1995, the graduate school at the University of Tokyo has run the Evolution and Diversity Grand Course in cooperation with the institution National Museum of Nature and Science. During this time, it has invited excellent researchers to concurrently serve as professors from national, prefectural, municipal, and private universities throughout Japan. Since the full development of this program, other universities have formed united graduate schools in various combinations. Under this program, even in fields not supported by the doctoral programs of a graduate school, outstanding researchers, if present in a university or institute, can take part in education activities to foster the next generation. Clearly, this type of education is practically provided largely by way of collaborative research. Graduate students are expected to learn by participating in concrete research activities.

Another example of this is the establishment of university museums (4.1). The University Museum was established at the University of Tokyo in 1966. In 1996, it became a research organization under the governance of the university.

Other universities have also established their respective university museums: Kyoto University Museum in 1997, Tohoku University Museum in 1998, Hokkaido University Museum, in 1999, Kyushu University Museum and Nagoya University Museum in 2000, Museum of Osaka University in 2002, Kagoshima University Museum in 2004, and Hiroshima University Museum in 2006. Each of these museums is run by less than 10 full-time staff members. Museum staff stay busy collecting, managing, and exhibiting materials specific to each university.

While these museums are characterized as a contact point between the university and society, through activities such as open lectures, it is desirable that they pursue activities specific to a university museum as a research institute.

Regarding prefectural universities, the Institute of Natural and Environmental Sciences, University of Hyogo is an organization operating in close cooperation with related institutions, such as the Museum of Nature and Human Activities, Hyogo, as described in 4.2.

Museums: Providing Support for Lifelong Learning and Serving as a Think Tank Among facilities whose names include the word "museum," Japan officially recognizes registered museums and museum-equivalent facilities. According to statistics, the number of these facilities and those in the category of museum-like facilities far exceeds 5,000 altogether. Of these facilities, the number that contribute to natural history-related research and dissemination is not known.

In recent years, many natural history-related museums, including equivalent botanical gardens, zoos, and aquariums, have been active, commensurate with their respective sizes, in providing support for lifelong learning and think tank capabilities. In these capacities their fundamental activity remains as research. In Europe, museums and related facilities play a major role for the development of natural history. A similar situation is also emerging in Japan. However, regarding contributions to fundamental research, the significance of the role played by universities is still obvious. This might be due to the structure of the Japanese academic system, in addition partly to the size of museums and related facilities.

Necessary Facilities in Japan The Science Council of Japan formed a natural history-related committee during the Council's 20th term. The committee has deliberated on the necessity of another national museum of natural history to supplement the existing natural history museums. In this regard, several forums have been held and a search for a specific location, such as in the Okinawa district, is said to have commenced as of 2016.

This kind of facility needs to cross a high barrier before its realization and may have a long road ahead before it opens. A similar concept to create a facility named BoSSCo once emerged. A private deliberation meeting body was set up and drafted a plan. The basic framework of BoSSCo is described in 'Botanical Gardens in Japan' (2004; pp. 91–98) as an ideal botanical garden-related facility.

The BoSSCo plan was conceived as one of the projects succeeding the International Garden and Greenery Exposition held in Osaka, Japan 1990. The aim was to build a botanical garden-like facility, focusing on flowers and greenery. Schematics of an ideal facility were even drawn. If this project was extended to generally cover the world of all organisms as research subjects and further extended to all natural phenomena in the domain of natural history, the concept would be basically the same as that of the second natural history museum currently under consideration.

From the perspective of natural history, the following points summarize the above concept:

1) Lead research promotion by collecting specimens and literature for natural history research and study, organizing electronic forms of related information for convenience in use, and cooperating with other similar research institutions.
2) Broadly provide society with cutting-edge information by storing, managing, and maintaining the latest specimens, literature, and derived information.
3) Develop and implement dissemination and exchange technologies

designed to enable sharing information on the current state of research, pass on excitement experienced in this research, and showcase present knowledge on natural history in an accessible and intimate manner.

4) Deepen public understanding of the deep involvement of natural history in the wealth and safety of the real world by consistently sharing research and study results with society and demonstrating the relationships between achievements in the field of natural history and the environment and natural resources.
5) Build a foundation for sharing the academic importance of natural history in the academic world by presenting tangible research results.

I am rather optimistic for the realization of these visions.

In the thriving period of the economic bubble, some local governments demonstrated strategic efforts towards early realization of BoSSCo. However, their hopes did not come true likely because the time was not ripe for such a facility, neither in society nor the academic world. It is understood that to establish a new facility, a strategic approach is required to persuade the people concerned. Moreover, an expedient foundation is required in society and the academic world. If these are inadequate, it is difficult to obtain the support and assistance of the majority, making the path to its realization long. For realization, researchers should be aware that their minimum responsibility is to aim to uplift the academic stature of the field of natural history and ensure that the uplift is influential in the academic world and society.

The United Kingdom has realized an idea like BoSSCo as a facility that publicizes the diversity of plants to society and successfully raises public awareness. The facility is the Eden Project, which has been open to the public since 2001. The Eden Project is structured similarly to the BoSSCo plan, but without research functions, predicated on the presence of the Royal Botanical Gardens, Kew and the Natural History Museum as natural history research institutions in the United Kingdom.

Regarding the enrichment of information on biodiversity, which is one of the aims of BoSSCo, GBIF (3.3) has successfully taken an international approach. In Japan, researchers who are making tangible contributions to information enrichment in this field are still limited. However, relevant information has been gradually produced due to the contributions made by some researchers with raised awareness. Expectations are high for outstanding research results achieved in the domain of biodiversity informatics by utilizing the produced information.

Existing related facilities need to be revitalized for the enrichment of specific research institutions, museums, and university research facilities, among others, to promote natural history research and study. For these facilities to be enlarged, enhanced in their contents, and further established, there should be a ground swell of anticipation for realization among the academic world and society. It is desirable that the concerned researchers boost their everyday activities for this purpose.

International Contributions This book has been translated from a Japanese manuscript written with Japanese readers in mind. When this book discusses contributions to society, it considers Japanese society. Therefore, in this section, the topic of international contributions comes from the perspective of contributions by Japan.

Growth of science should be promoted on a global scale. Japan is not poised to interfere with this. However, when human intelligence is involved in creation, individuality comes into play, both in inspiration and promotion process. The characteristics of ethnic groups also have considerable effects. This individuality and unique character should be excluded to the greatest extent possible for science to verify objective facts. While this is applicable for objectivity in conclusion-forming processes, taking the advantage of superior individuality in the method used to elucidate facts should also be encouraged. In Japan, non-professional naturalists make remarkable contributions to research and study in the domain of natural history. Their contributions may be an unusual example, yet they can be proud of their efforts.

A local culture develops under the specific effects of the environment, or nature, of the local area. Meanwhile, modern science inquires about the principles of the natural world on a global or cosmic scale, excluding regional characteristics to the greatest extent possible. The domain of natural history is predicated on regional diversity, although it is a discipline of science. For this reason, inquiring into cultures rooted in local communities also becomes meaningful.

Japanese naturalists mostly start by surveying the Japanese archipelago and closely related regions. Of course, some materials subject to their analysis can extend to the global scale. Moreover, even when analyzing materials on the Japanese archipelago, one cannot disregard nature on the planet earth. This is a matter of course on the grounds that the lifestyles of materials are vast, unless the research is limited to analyzing a specific event.

Furthermore, since their growth, Japanese researchers have observed nature in the Japanese archipelago and they have looked at the earth and the universe from the archipelago, often the viewpoint of their research and study. Accordingly, it is a general trend among them to extend the area of their research first to the immediate areas neighboring the Japanese archipelago because of the connections with materials in the Japanese archipelago and the ease of conducting field research.

My first report on ferns appeared in 'Natura,' the bulletin of the Biology Club at the Hyogo Prefectural Kaibara Senior High School. My article was on ferns in my hometown. Diverse species of currently living ferns are observed in temperate regions. For this reason, as a researcher, I participated in surveys in southern Kyushu during summer vacation of my last year as an undergraduate, on Yakushima in my first year of my master's course, and on the Amami Islands in my second year of master's course and my first year of a doctoral program. Moreover, I also joined surveys conducted in Okinawa when it was under post-war administration of the United States. After I commenced surveys in Thailand in 1965, the scope of my surveys gradually expanded into Southeast Asia. In 1980, I visited China when an academic exchange agreement was concluded between Japan and China. I waited for a while for official permission for botanical surveys in China. Since 1984, I have conducted field surveys repeatedly in China. In the second half of the 1990s, it became possible to carry out field surveys in Vietnam, after the Vietnamese war. Of course, conducting surveys depends not only on research necessity, but also on the political situation in the host country. Nonetheless, it is natural that following necessary research procedures will lead to broadening research subject areas because of broadening research objectives.

Field research subject areas should ultimately extend across the globe. That aside, when it is necessary to view related specimens and other various kinds of related information, in addition to the need to conduct field surveys, researchers request relevant facilities for permission to use materials in their collection. During the early days of my career, I used services available at major museums and similar facilities in Europe and America. In 1969 when I was in the United Kingdom, I was invited to speak at the Linnean Society meeting. The title of my speech was "Hymenophyllaceae in the British Islands under a global conspectus."

My experience offers a typical example of how to pursue natural history study. The common research practice is to begin with one's birthplace, extend to a global scale, and analyze research materials from a global perspective, whether with or without field research. This is not a simple matter and greatly concerns the sources of research subject materials and research evaluation.

This style of research should be attributed to the way of thinking of the Japanese natural historians raised in the Japanese archipelago, which hosts a rich biodiversity. There is not much plant diversity in Europe. Many botanists raised in Europe establish, in their youth, their research sites at a former colony and develop their research through a long period of repeated field surveys. In the domain of physical anthropology, Japanese researchers are at the global forefront. Several of them have achieved outstanding research results by staying for a long period of time in Africa or other regions since their early years of postgraduate schooling. Accordingly, research subject-dependent differences in the style of research may be a scientific necessity.

Field research often relies on the hospitality of the host country. Contributions to the research and study of natural objects in a specific area is necessarily concerned with the earth and human knowledge. At the same time, it intellectually helps expand the knowledge of nature in the area in question. In this sense, knowledge dissemination from Japan, although occasionally only results from materials in Japan, will contribute to knowledge construction on a global basis if the research is promoted on a global scale. For natural history research and study, as is with all other scientific studies, results should be evaluated on a global scale and, at the same time, should make contributions on a global scale. This should be even more clearly understood with research and studies deeply concerned with natural objects. In this respect, the concept underlying natural history research is in line with the basic principle of recognizing individual facts elucidated through research and study placed in the correct position within the integrated whole of nature.

In this sense, compiling a book of local flora for an area in the Japanese archipelago is conducive to compiling important findings into a basic Japanese flora literature, which provides valuable information about the earth's flora; this is very clear to see. Furthermore, with findings obtained by elucidating a specific constitutive species, it is vital, in light of the species position on the earth, to recognize that it is one constitutive element of the biodiversity on the earth (*spherophylon*) concurrent with understanding how it survives in a specific area.

Natural history research and studies should lead to integrating obtained findings constantly via a bird's eye view of the whole, while critically producing information for individual challenges with the greatest accuracy. Otherwise, the research will not fulfill scientific curiosity. Although all human knowledge constructed thus far cannot allow humans to have a scientific bird's eye view or integrate all pieces of knowledge, no healthy advance of science will be possible unless constructing logic in that direction is pursued. While applicable to all scientific quests for truth, this seems to be more clearly manifest with immediate challenges faced by natural history, which, in a sense, offers a superb model for scientific studies.

Chapter 6

The Joy of Learning Natural History: A Summary

This chapter attempts to summarize what I have stated in this book: to define how natural history should be understood, how everyone should access it, how to make use of it to build human knowledge, and how to allow it to contribute to human society. Scientific research develops to hold interrelationships between two aspects: science that evolves as a foundation for technology enriching human lives, and science expected to train human intellect to improve humanity's scientific literacy. As the conclusion to this book, Chapter 6 reexamines the most basic challenges, such as how natural history should be pursued as a science that contributes to civilizations and cultures and how one should help fulfill that pursuit.

6.1 Natural History and Science

Where should natural history be positioned in the domain of science occasionally becomes a topic of discussion. However, the term *natural history* does not refer to any specific area of scientific research. It is not possible to find any area named natural history in natural science. Rather, natural history refers to research on the integrated whole of nature. Because it refers to the integrated whole, natural history is not limited to research that uses modern scientific methods for analytically proving facts. It implies the use of scientific methods that require integrative interpretations to reach solutions about the integrated whole. The perspective of integrating facts clarified through investigation is important for natural history.

As such, the expression *natural history research* is improper. The phrase *learning natural history* is also inappropriate. The correct expression should be: conducting research with the methods of natural history and learning what the methods of natural history elucidate.

Despite the title 'Natural History,' this book has failed to discuss the integrated whole of nature within the scope of my ability. Most examples subjected to discussion have centered around living things, specifically plants. Consequently, although this book presents the topic of what is natural history, Chapter 6 focuses on the natural history of living organisms, rather than the integrated whole of nature. Nonetheless, the discussion presented here, in my view, is made like the perspective of looking at the integrated whole of nature.

(1) Accessing natural history

Natural historians attempt to understand the integrated whole, including its cause and effect, in addition to capturing the three-dimensional reality of nature in its actual state. The integrated whole is made up of an aggregation of parts. Hence, it is essential to clarify all of its constitutive elements. However, I believe that whether the integrated whole can be understood as a simple set or aggregation of parts remains unknown until all elements are elucidated.

Modern scientists use hypothesis-testing methods to prove the reality of objects and the resultant phenomena to ascertain the truth. Indeed, based on the scientific understanding gained in this way, human affluence and safety have risen to a remarkably profound level, although this comes at the cost of the increasing tendency to assign lower priority to the promotion of science in the domain that would fulfill intellectual curiosity about knowing an individual or the earth as an integrated whole.

Demonstrative, accurate verification of facts about individual phenomena greatly contributes to establishing safe technologies. In comparison, taking steps towards the correct understanding of the integrated whole of nature will make no direct contribution to immediate safety or affluence. The implication today's artificial deeds made manifest on the earth in generations to come is very distant from the success story told today.

When it comes solely to living organisms, to clarify what it is to be alive, all roles played by every individual part of matter should be elucidated, as living things exist as a mass of substances. Indeed, many aspects of the roles played by the constituents of living organisms have been rapidly clarified. Nevertheless, it is generally understood that all phenomena exhibited by parts will be elucidated only in the far-distant future.

Not every part of nature is known. Nonetheless, currently living individuals have passionate intellectual curiosity about what it is to be alive. They want to know how the universe started and what the meaning of the emergence of life is. Therefore, they extrapolate the unknown parts from the greatest existing possible knowledge. However, extrapolation is no more than extrapolation. Repeating inferences would not enable one to clarify the reality of the subject that he or she wants to know. One may interpret it, but not be able to clarify it if some parts remain unknown.

Even if all parts are known, the integrated whole cannot show itself spontaneously. To know it, it is essential to use a method of integrating all the aggregate parts. Given this fact, science faces the challenge of establishing a method of integrating existing findings without waiting until the elucidation of all parts is completed.

Modern scientists analyze and understand the subject of their research in an empirical manner. In this course of action, they reduce parts to elements. To elucidate the individual body of a living organism, they make clear what cells make up the individual body, and they understand the various cell organs that constitute cells. Moreover, to throw light on the structures and functions of cell organs, they attempt to know the behaviors of constitutive molecules and even atoms. Ultimately, living matter is made up of an aggregation of substances. Therefore, scientists need at least to know what part each constituent substance plays in living matter.

Based on the partial findings they have obtained, scientists hope to understand the integrated whole of the subject. To know the integrated whole of the subject with limited knowledge of limited parts, it becomes necessary to understand the nature of the integrated whole. In other words, there are many things that can be known only when one has the correct recognition of the integrated whole.

Remarkable advances have been made in recent years clarifying living things by disassembling them into parts in a reductionist manner. On the one hand, biology of individual bodies and groups has advanced; on the other hand, the assessment of whether the discipline is in a state to solve the question of what it is to be alive varies with the different people and ways of pursuing the question.

For example, in the study of species diversity, researchers try to throw light on diversity using species as a reference, or unit. However, species used as a unit is interpreted according to a hypothesis. Consequently, in the course of research, the study of species diversity may not be empirical. It is expected that species used as a unit will not be understood until species diversity is completely clarified. Notwithstanding this, it is impossible to analyze species diversity without the use of species as a unit. Thus, the discussion becomes a somewhat circular argument.

In the field of research with hypothetically established species used as a unit, it is not only necessary to achieve empirical results, but also to use a research method that looks for the form of the integrated whole, though based on a hypothesis. All questions raised in this book lead to asking how this research is evaluated by the methods of modern science.

Natural history should be understood as a research method for studying the integrated whole of the individual body of a living organism or the earth to throw light on the reality of nature. The subject is the universe, the earth, living organisms, and the individual body of the living thing known as myself. The living thing that exists as an individual body and what the integrated whole of the earth or the universe is are the things needing clarification.

If research based on an empirically unproven hypothetical unit, like species, is not a science, then the study of species diversity does not hold. However, without the study of species diversity, it is impossible to know the structure of the kingdom of organisms. The study of biodiversity is meant to verify the hypothesis of species. For this purpose, the reality of species should be clarified using concrete subjects. In this process, research should necessarily be conducted using hypothesis-testing methods by decomposing model species into parts.

Some say that even without the capacity for research about species differentiation it would be possible to establish healing techniques to counter human diseases and to produce food to counter hunger. Is this true? It may be true that in many aspects of food production or medicine today, the contributions of species diversity research are limited. However, without biodiversity knowledge, it is difficult to ensure human affluence and maintain safety. Contributions made based on existing findings obtained to date result merely from hypotheses formulated somewhat blindly. They are still far from perfect, as indicated by constantly discovered blemishes (which also affect affluence and safety).

As much as anyone, I place expectations on the promotion of science for immediate affluence and safety, and I value the growth of technology for that purpose. It is a fundamental human desire to eat delicious food and achieve a healthy long life. However, is that all there is to be fulfilled? Individuals that belong to the species *Homo sapiens* who have evolved intellectually have their respective intellectual curiosity. An intellectual sense of fulfillment is at the base of the human sense of happiness. In addition, it is a widely accepted notion that science based on intellectual curiosity has become the base of technology that contributes to the material- and energy-oriented sense of fulfillment. If any blemishes are present on safety and security, they exist because of a shortage of scientific understanding.

Researchers involved in natural history have pursued research methods designed to elucidate the integrated whole of the subject. While placing expectations on the evolution of reductionist research, they, based on findings about known elements, attempt to understand the integrated whole of the subject and fulfill their intellectual curiosity.

To establish methods of elucidating the integrated whole of the subject through scientific discussion, natural history research has been promoted throughout history. Whether the goals and methods of this research are truly functioning today is an acute question for natural history researchers. Truly, this question poses a consistent challenge for all sciences in asking what science should be.

It is simply fun to study natural history, probably because it fulfills individual intellectual curiosity. However, due to this fun, natural historians tend to immerse themselves in studying individual phenomena, satisfied with individual specific challenges and forgetting the fundamental attitude towards looking at the integrated whole.

(2) Natural history versus history

In this book, I tracked the lineage of living organisms somewhat in contrast to historical studies [2.1 (1)]. Nonetheless, regarding history, there have been many studies of the philosophy of history. It is not appropriate to compare them all on an equal footing with biological lineage studies that have developed within a limited research area.

The most important point of natural history is that it constantly has an eye on historical development in an attempt to understand the current moment. Life's true nature is manifest in that, immediately after the moment it emerged on earth, life began to diversify and still presents an embodiment of diversity. Without regarding historical development, biodiversity does not hold. Alternatively, narrating life in defiance of biodiversity is meaningless. In this sense, natural history should basically be approached on four-dimensional consideration.

History intends to clarify the whole temporal picture from the days when *Homo sapiens* became humans to when they began to organize social life to today. How to read this temporal passage based on records challenges the study of history, requiring historians to use their full capacity to throw light on the challenge. The challenge of tracking temporal passage is also posed to natural history. There are many things that natural historians can learn from history, which has a better and longer chronicle of methodology.

We look at nature through the facet of the present. Nature has its own historical background and is promised to have an eternal future. We are living in the present and desire to know the reality of nature in pursuit of today's affluence and safety. To this end, it is necessary to understand the integrated whole of nature. While faced with this facet of the present, we expect to understand the four-dimensional reality of nature. For this purpose, it is essential to use research methods that allow us to empirically know what facts are present before us, to learn their historical background, and to predict the future.

To clarify human history, documented records serve as reliable proof. However, for evolution in nature, records are only traces engraved in nature. For the most part, these records have been extinct along with their history. To prevent the past from becoming unknowable, it has become widely accepted that making full use of information is a helpful approach. Provided with a vast amount of information, it is possible to predict the future with improved accuracy. Information can do much to help reproduce the extinct past. Hence, for this purpose, it is essential to accurately ascertain reality and organize and use information correctly.

In natural history, basic efforts are directed towards building information. Frequently, materials commonly considered as insignificant at the time of acquisition make extremely significant contributions later. One must recognize that, in more than a few cases, making light of some materials or information because they do not have immediate utility is a huge loss of an opportunity to know the truth.

(3) **Research versus learning**

It is inappropriate to say that one is in a quest for natural history, as pointed out above. Researchers are in quest for what nature is, using the methods of natural history.

When it comes to biodiversity, the characteristic features of individual bodies of diverse living organisms are analyzed in detail, and the results are compared with those of other individual bodies. Modern scientists readily understand this technique. The same sentiment can be expressed by saying that one is in a quest for the life of the current *spherophylon*, which emerged as a single form on the earth more than three billion years ago and has differentiated into diverse forms through the long history of evolution. This expression makes clear the methods of natural history are essential for this quest. In sum, it is necessary to return to a common notion about living things, which is that life exists as an individual body, as a cell, and as the *spherophylon*.

Natural history researchers tend to focus on individual phenomena and have bias towards a specific species or genus, as repeatedly pointed out in this book. While under the flag of research and study in natural history, they are in fact striving to fulfill their individual curiosity without thinking about an integrative perspective. To see the true form of a species, it is necessary to know the history of the formation of the species. For this purpose, comparative studies with related species are essential. However, if obsessed with the interesting aspects of individual topics, the researcher will lose sight of the true nature of his or her curiosity. In this state, the researcher becomes besotted with interest in documenting specific events, rather than conducting natural history research.

Thinking about the process of learning may be more illustrative. School is a place for learning and is expected to provide intellectual training. Intellectual training is the process through which the trainee acquires information that mankind has stored in society as a result of culture. Learned local culture that children spontaneously absorb at home or in society from their parents plays a pivotal role in transmitting information accumulated in society.

School education systems were originally developed to transmit inherited knowledge in society efficiently from generation to generation. Primary and secondary education were expected to transmit knowledge efficiently. Certainly, school education has boosted the benefits of knowledge transmission. It is a historical fact that in Japan the thoroughgoing compulsory education system worked effectively in raising and maintaining the high intellectual level of the Japanese people, which served as a foundation for the nation's wealth and military strength. That said, the knowledge that I spontaneously acquired in transactions with my parents, brothers, and kind neighbors, as well as when playing with naughty lads, although inefficient, also became my blood and flesh more than piles of dry and tasteless facts. Naturally, acquiring knowledge in this way was fun for me.

Knowledge transmitted through school education has forced children to learn. Schools build curricula and establish timetables for knowledge acquisition. Learning at *terakoya* of the Edo period in Japan was promoted according to the self-initiative of the learners. However, compulsory education at school set targets for knowledge acquisition that rated pupils according to their level of proficiency. It became necessary for children to study hard to get good marks.

Forced learning is not an act pursued for love. Even if not fond of it, one cannot give it up. It is still better if learning is concerned simply with grades on term-end report cards, versus with the admission examination results to gain entrance to better schools. A recent trend is that schools are ranked according to how difficult it is to pass the admission examination, whereby entry in the better junior high schools and the better high schools means admission to a prestigious college. Graduation from a prestigious college is advantageous for employment and expected to lead to the better life. This narrative allows parents to force their children to study for examinations in consideration of their future. Consequently, schoolmates all compete for good marks. For most children, acquiring knowledge in competition with other schoolmates is simply drudgery and never pleasure. Fundamentally speaking, the acquisition of knowledge should be an intellectual pleasure that only humans can enjoy, and learning should be fun. However, when forced, the enjoyable aspect of learning is about to be lost.

The acquisition of knowledge once went beyond the confines of school. It once was expected to lead to lifelong learning, as a human characteristic. The act of learning once emanated in one's life, which is how lifelong learning should be. School education was intended to specialize in intellectual training for efficiency. Today, however, learning seems to be construed as equal to going to school. (Ironically, I hear that some say that the private preparatory school is more important than the school.) Moreover, not only intellectual training, but also character education and physical training should be learned. Since the term *forced learning* [here, the usual counterpart of original Japanese is *study* in English, though the two Chinese characters may be read *endeavor + force*] is not used to describe character education [Chinese characters mean teach + nurture], the use of the term *forced learning* becomes limited to intellectual training. Subsequently, due to the interchangeable use of the terms *forced learning* and *learning*, character education conversely became something that society demanded that school provide. The origin of school was the gymnasium in ancient Greece, an educational establishment for physical training. The Platonic Academy, Aristotle's Lyceum, and the Cynosarges, where Antisthenes taught, were three major gymnasia in Athens. These educational establishments for physical training served as places for the activities of philosophers (i.e., the spring of Sophia).

As school began to be viewed as a venue for education, the tendency to make light of home and social education grew. If the school is the place where teachers concentrate on intellectual training, they focus their energy on intellectual training for higher efficiency. However, for some time, schools have begun to take responsibility for character education and the scope of teachers' responsibility has expanded. In addition, physical training through club activities has also been entrusted to schools. Without doubt, school, where children come together in a group setting, is a suitable place for promoting character education for children as a group. Nevertheless, demanding that schools be responsible for education that only home life or society can provide is an abdication of responsibility on the part of the home and society.

From the fundamental standpoint that humans are learning-oriented animals, learning is an act that only humans can enjoy throughout life. A desirable form of education is for school education to promote advanced intellectual training efficiently and for people to receive basic character education in society and at home. Unless learning is viewed in this organized framework, being an intellectual animal may become drudgery.

Local naturalists observe nature without earning a livelihood from the observation. Their purpose is to fulfill their intellectual curiosity. They desire to gain a better knowledge of nature because they love the nature in their vicinity. Of course, most naturalists are ready to provide their materials, or the findings they have obtained through observation, when they realize that such materials are useful for compiling, for example, a red list. Even if it is learning pursued for personal pleasure, they will not retain the results for their own use when they could be useful for society.

However, in recent years, the number of successors to these non-professional naturalists has dramatically decreased, which may imply that the material- and energy-oriented mentality has become pervasive among youth. If the notion that learning must force students to only learn that which is useful for making a living becomes prevalent, the fundamental activity of humans to learn driven by curiosity will fade into obscurity. The result will be that the best and precious sense of happiness that humans have come to experience will remain unnoticed as it fades into oblivion.

6.2 Joy of Learning, Joy of Investigating

Is it scientifically correct to say with certainty that mankind is an intellectual animal? Although widely accepted as a rule of thumb, can it be proven scientifically? In recent years, the cognitive science of primates has advanced rapidly. The validity of viewing human intellectuality as a characteristic of living organisms has not yet been scientifically proven. Nevertheless, no one can deny the reality that human intellectual activities have resulted in building advanced cultures full of originality.

Putting aside whether the act of cognition has been scientifically ascertained, humans involved in intellectual activities do have intellectual curiosity. This is clearly illustrated in the history of Western culture by the fact that natural philosophy was already systematized, and the Academy began to provide education in the times of ancient Greece.

Human intellectual curiosity is directed towards a myriad of things and works to clarify unelucidated events. As a result, modern science has clarified various facts in a scientific way after more than 2,000 years of history. Many findings obtained through this process have applied to technology based on human intelligence and led to safety and affluence in human life, if at times erroneously harmful to humans. Through cultures, humans address the challenge of illuminating various unelucidated realities. This is the activity most unique to humans.

Intellectual curiosity develops in the world of science in a way specific to humans. In recent years, scientific achievements have been rated by their immediate application to technology. People do not put much weight on social science and tend to disregard the methods of natural history. This may reflect the fact that the essential nature of science driven by intellectual curiosity is fading into oblivion due to the emphasis on economic efficiency. Of course, it is not possible to ignore the real-world economic effects and efficiencies resulting from human activities. Nonetheless, theorizing about improved efficiency for future generations, without calculating only immediate benefits, may help direct human energy towards science in a different manner. This seems to be commonly understood among learned people. However, the real world has failed to catch up with this understanding, perhaps because of insufficient efforts made by those who strive devotedly to fulfill their intellectual curiosity. In recent years, the lesson learned through history that, from a long-term perspective, it has been none other than sincere approaches driven by intellectual curiosity that have continuously contributed to safety and affluence in human society, has been neglected.

My statement above may be subject to the criticism that I am excessively particular about real-world safety and affluence rather than living for the pure joy of learning. However, ignoring the real world and being one-sided in favor of mental richness will not gain the sympathies of people living in the real world.

I have described that activities pursued in the domain of natural history have various problems. Many times, I have been sarcastic with researchers who, without adequate achievements, uttered complaints and blamed outsiders for the problems. Nonetheless, I have avoided mentioning my own responsibility in those problems, which are more profound than theirs. While the perspective that natural history should contribute to scientific advances has profound implications, natural historians lack the necessary capacity to improve people's scientific literacy. Notwithstanding this awareness, I have failed to appeal to the public or to pursue activities intended to help the public understand it. I am honestly aware of this major responsibility. With this awareness, I would like to conclude this book by describing, once again, the significance of learning and the research work in the domain of natural history.

(1) Learning natural history

Upon encountering natural history, human, as an intellectual animal, experiences the greatest joy through direct stimulation through intellectual curiosity. Indeed, even infants who cannot adequately judge things are impressed by and feel joy from various phenomena that occur in the natural world. When fascinated by the motion of a wild animal or charmed by beautiful flowers, their eyes sparkle extraordinarily. At the fundamental point of learning, they show the most remarkable response to stimulation that inspires their intellectual curiosity.

However, as people grow and become guided by interest, they have fewer chances of being impressed simply by intellectual stimulation. Indeed, the number of people who fully enjoy this kind of pleasure seems to be decreasing. In addition, our social structure is not prepared to allow them this enjoyment.

When I see the behavior of people who participate in museum activities, I feel very sympathetic with the strength of their intellectual curiosity. At museums, the eyes of infants sparkle. Likewise, students and pupils taking part in museum activities voluntarily study natural phenomena, although their number is quite limited. Looking at them, I become confident that hopes for the future still rest with them, despite the wholesale criticism about children of today. Obviously, more educational institutions must be in place for young people interested in natural history to expand even slightly. For this, there are high expectations for the activities of existing facilities, despite being few.

Those who know, and fully enjoy, the pleasure of learning are not limited to a small number of young people. I am one of them. Although my interest in advanced research was sparked while seriously studying modern science when enrolled in a college, my eyes were opened to natural history through the opportunity to share the sincere impulse for exploration of nature with people driven by scientific curiosity. In retrospect, I was inspired by the scientific curiosity expressed by those who were attempting to confront the integrated whole of nature. Indeed, many around us inquire about the significance of living in harmony with nature sincerely driven by their scientific curiosity, as well as researchers who make the utmost effort for the growth of modern science. There exist people who devote their life to the joy of learning as expressed by the term *lifelong learning*. This is what really assures me.

If posed with the question of whether I fully enjoy the pleasure of learning experienced by naturalists that this book has described, I would have an odd feeling. In this book, I discuss naturalists as if they were unique to Japan.

However, when it comes to ferns, in addition to a large number of fern lovers in Japan, I have had direct contact with many enthusiastic fern lovers in Britain and the United States as well, despite the low economic value of the plant group. The British Pteridological Society is said to have been active since the Victorian era (I am not familiar with their activities in these preceding times). Likewise, in the United States, known as a country of practicality, fern lovers exist irrespective of their occupations. The American Fern Society is an organization with both professional researchers and non-professional naturalists as members. A group of people from the Society enlisting the cooperation of the Nippon Fernist Club once visited Japan to observe ferns. I also joined their observation tour organized to enjoy wild ferns.

I have stated on many occasions that my eyes were opened to ferns as a research material when I became acquainted with fern lovers who were not professional researchers. However, I am interested in ferns as a material for biodiversity research, rather than being fascinated by the plants as a devotee. I formed a science club at a newly established junior high school to explore ferns as an after-school activity. This was driven by a suggestion that ferns had many unelucidated points. It was not the same as being charmed by the beauty of ferns. When enrolled in graduate school, I chose to belong to a taxonomic course and embark on a fern research project because I considered that I would be able to explore what it is to be alive. (I still believe that it was the correct choice for me.) There may be people who love ferns very much and are happy when they are with them. I am somewhat different from them. Other naturalists whose curiosity is stimulated by the mystery of ferns may feel the same way; however, the tendency towards a quest for the truth driving the act of learning is likely uncommon. Of course, I am aware that it is common that even non-professional naturalists are in a quest for the truth driven by curiosity.

The joy of learning cannot be confined exclusively to the realm of mind. In a real-world context, materials gathered by non-professional naturalists through years of efforts have made major contributions to the compilation of a red list of vascular plants in Japan, as described earlier in this book.

Certainly, some people will probably offer a counterargument that in Japanese society people are busy making ends meet, and only those who have considerably higher incomes can speak about the joy of learning. Nevertheless, it is a precious characteristic that man as an intellectual animal feels the joy of learning without caring about gain or loss, once he or she begins to have interest in the structure of nature.

In the present-day context, I would like to additionally point out that the joy of learning serves to raise people's motivation for science and helps them improve scientific literacy. After all, people's spontaneous pursuit of the fundamental joy of learning will eventually lead to improved public scientific literacy and serve as a foundation for healthy growth of society. Through these extended arguments, it is correct to conclude that the joy of learning enables the most beneficial activities for society, rather than saying that there is a useful aspect to the joy of learning.

(2) Research pursued by natural historians

Using scientific methods, scientists analyze, and ascertain the reality of, various phenomena. They then develop new technologies based on the gained knowledge to contribute to people's safety and affluence. They are valued and feel happy when they are honored and find fortune because of their contributions.

In contrast, there are those who clarify problems guided by intellectual curiosity and build foundations for the further clarification of problems. Frequently, they are unable to apply their knowledge to technology and contribute to the affluence and safety of people living today. They may recognize that their findings may serve as a foundation for elucidating additional facts and feel joy from the results. Nonetheless, their results seldom attract public interest.

One typical example is the discovery of Mendel's laws. Mendel is said to have been confident in the results he obtained. However, while he was alive, his achievements did not have any impact on biology, nor was he honored by the public as a biologist. However, by being confident about his achievements, he hopefully felt pride as a scientist. He pursued research anticipating that he would make social contributions. He believed that to do this it was necessary to explore the universal principles underlying living organisms. It was a natural course of development that Mendel's research, intended to explore universal laws, was eventually acknowledged to state the basic principles of biology. Around a century after his death, genetics based on Mendel's research made numerous direct contributions to people's affluence and safety.

For Mendel, to learn was to correctly recognize the laws of nature. By applying analytical methods that went ahead of the times, he could correctly recognize laws underlying life that were known to nobody else. The joy of investigating the truth using the methods of natural history formed the base of his science.

Individuals driven by scientific curiosity who aim for social utility offer an illustrative example. In cultural development, it is normal that those with curiosity about science for science promote the growth of science and before long their achievements are applicable. The joy of pursuing natural history-related sciences, if developing from curiosity about science for science, will eventually serve as a guiding principle for promoting science, boosting cultural advances, and allowing human society to achieve healthy development. It stands to reason that experiencing the joy of pursuing science leads to these results.

Scientists were once described as people in honorable poverty conversant with scientific research. Nowadays, scientific achievements are commended and rewarded with large amounts of prize money. Results that demonstrate utility are promised to earn a large revenue from patent fees and provide opportunities for scientists to be economically successful. Successful scientists are eventually wealthy people. However, success is limited to a small group of blessed people, like with successful businessmen. Even if a scientist makes major scientific contributions and if the achievements are valuable in terms of the growth of science, there is no chance for the scientist's economic success unless the achievements are immediately applied to technology and demonstrate economic benefits.

Scientific achievements applied to technology are those that have been empirically verified. They are often facts elucidated for specific purposes. (The rapid evolution in certain fields of science during war serves as good evidence for this.) Indeed, financially supported research projects are often in fields easily applied to potentially useful technology. In connection with this, the reality is that research subsidies are often planned to fulfill political or economic needs and rarely based on scientific necessity. For this situation, the attitude of scientists driven purely by intellectual curiosity may bear some of the blame.

In practice, it may be difficult for basic sciences to be supported politically or economically. Nonetheless, as exemplified by the fact that institutions contributing to basic research, such as colleges and research institutes, are supported by society, social recognition of activities driven by intellectual curiosity remains in the real world. We should expect that the natural course of development of intellectual activities as a basic human function will be firmly recognized by society. As the self-proclaimed noblest of all animals, *Homo sapiens* should return to their original nature when pursuing activities.

Notwithstanding the above statement, this book is open to criticism as to whether natural history is conducive to activities for the proper development of human intelligence and whether this has been expressed in this book. At the same time, hopes in the field rest on natural historians making additional contributions to the production of concrete results.

Natural history can be described as a four-dimensional natural science.

By learning natural history, it is possible to recognize the significance of the historical background against which the present is taking place. In recent society, people seem to place importance solely on the current moment and have forgotten to live according to what the history teaches. Looking at them, I realize with some pleasure that man will be able to live with dignity as an intellectual animal through the joy of learning natural history. The joy of learning is always conducive to the pleasure of the quest for the truth. In addition to contributing to people's safety and sense of security in life, the quest for the truth brings the seeker the joy of being in quest for the truth. That said, after all, there are individuals who are in the quest for the truth from an independent standpoint on the one hand, and there are those who choose to live as a professional in quest for the truth on the other.

In this book I have attempted to organize the characteristics of natural history along with its historical development. Moreover, I have pointed out that science should be discussed from a global perspective, rather than from a local point of view. However, in the final part of this book, I would like to attract the interest of Japan's younger generations to natural history during what might be termed as today's scientific recession in Japan, although this may be a brief moment in history.

The scientific recession in Japan since the turn of the century is indicated by objective figures. Among learned people, it is widely known that this is because of the academic no-way-out situation resulting from science policies. Although one of the urgent issues for scientists is to strive for normalization, I would like to state my expectations for young Japanese people, rather than to complain about the situation, which is a reality.

Aside from Japan, efforts are also made on a global scale to advance science. Human society will not stand still in its eternal history, even with a policy recession in some country that may pose impediments. The absence of an outstanding contribution in science may slow down the pace of overall growth. However, it is probably not necessary to worry that a partial recession will result in an overall stoppage or recession. Although country in question may suffer a loss, it does so simply by its own responsibility.

In the academic no-way-out situation, there are those among young Japanese people who are unable to act spontaneously driven by scientific curiosity. The question arises as to what they are missing. I advise those who live in such a situation to return to a fundamental viewpoint of nature and to work on challenges driven honestly by curiosity, even if unable to compete in society. Fortunately, there is a mountain of natural history-related challenges on the Japanese archipelago. The archipelago's biodiversity is rich. Frequent natural disasters that occur on the Japanese archipelago present the opportunity for various experiments daily. There is a huge backlog of challenges to be addressed. Moreover, there are many naturalists who are interested in and in contact with these phenomena. The situation is now opportune to live with the joy of questing for the truth.

References

(Author names in *italics* show the literature in Japanese and with some comments for the non-Japanese readers.)

Aristotle. Complete works of Aristotle. [Japanese version in Iwanami Shoten Publ., 1968–1973.] For English version: the revised Oxford translation, 2 volumes, 1984.

Baldwin, E. 1952. Dynamic Aspect of Biochemistry. Cambridge Univ. Press, Cambridge.

Bell, P.R. 1960. The morphology and cytology of sporogenesis of *Trichomanes proliferum* Bl. New Phytol. 59: 53–59.

Bentham, G. & J.D.Hooker. 1862–83. Genera plantarum ad exemplaria imprimis in herbariis kewensibus servata definite, 3 vols. L Reeve & Co., London.

Botanical Society of Japan (ed.). 1982. Hundred Years History (in Japanese). Bot. Soc. Japan, Tokyo.

Botanical Society of Japan (ed.). 1887–1993. Botanical Magazine, Tokyo; 1993–. Journal of Plant Research. Bot. Soc. Japan, Tokyo.

Braithwaite, A.F. 1969. The cytology of some Hymenophyllaceae from the Solomon Islands. Brit. Fern Gaz. 10: 81–91.

Braithwaite, A.F. 1975. Cytotaxonomic observation on some Hymenophyllaceae from the New Hebrides, Fiji and New Caledonia. Bot. J. Linn. Soc. 71: 167–189.

Budge, E.A.W. 1978. Herb Doctors and Physicians in the Ancient World: The Divine Origin of the Craft of the Herbalist. Ares Publ., Chicago.

Buffon, G.-L.L. 1749–1804. L'Histoire Naturelle. L'Imprimerie royale, Paris.

Candolle, A.P.de. 1824–39. Prodromus Systematis Naturalis Regni Vegetabilis, 17 vols., continued by A.de Candolle, –1873. Paris.

Kurihara, N. (Choseisha-shujin) 1837. Iconography of *Psilotum*. Gyokusei-do.

Coen, E.S. & E.M.Meyerowitz. 1991. The war of the whorls: genetic interactions controlling flower development. Nature 353: 31–37.

Copeland, E.B. 1907. The comparative ecology of San Ramon Polypodiaceae. Phil. J. Sci. 2: 1–76.

Copeland, E.B. 1914 (3rd ed. 1931). Coconut. McMillan, London.

Copeland, E.B. 1924. Rice. McMillan, London.
Copeland, E.B. 1929. The oriental genera of Polypodiaceae. Univ. Calif. Publ. Bot. 16: 45–128.
Copeland, E.B. 1933. *Trichomanes*. Phil. J. Sci. 51: 119–280.
Copeland, E.B. 1937. *Hymenophyllum*. Phil. J. Sci. 64: 1–188.
Copeland, E.B. 1938. Genera Hymenophyllaceaea. Phil. J. Sci. 67: 1–110.
Copeland, E.B. 1940. Fern evolution in Antarctica. Phil. J. Sci. 70: 157–189.
Copeland, E.B. 1947. Genera Filicum: The Genera of Ferns. Chronica Bot., Waltham, MA.
Darnaedi, D., M.Kato & K.Iwatsuki. 1990. Electrophoretic evidence for the origin of *Dryopteris yakusilvicola* (Dryopteridaceae). Bot. Mag. Tokyo 103: 1–10.
Darwin, C. 1859. On the Origin of Species by Means of Natural Selection, or the Preservation of Favoured Races in the Struggle for Life. J.Murray, London.
Descartes, Rene. 1637. Discours de la method.
Diderot, D. & al. 1751–72. L'Encyclopédie.
Dioscorides, P. Later 1st century. De Materia Medica libri quinque.
Dodoens, R. 1554. Cruydeboeck.
Ebihara, A. (ed.). 2016–2017. The Standard of Ferns and Lyicophytes in Japan. I–II. Gakken Plus, Tokyo.
Endo, H. 2015. Idea of Hirokichi Saito. In *Ichinose, M. & H.Masaki* (ed.), A story of Todai-Hachiko, 149–158. Univ. Tokyo Press, Tokyo.
Engler, A. (ed.). 1900–. Das Pflanzenreich, regni vegetabilis conspectus, I-CVII. Verl. Wilhelm Engelmann, Leipzig.
Engler, A. & K.A.Prantl (eds.). 1887–1915. Die natürlichen Pflanzenfamilien, 23 vols.; 2nd ed. 1924–1980. 28 parts issued. Verl. Wilhelm Engelmann, Leipzig.
Farlow, W.G. 1874. An asexual growth from the prothallum of *Pteris cretica* var. *albo-lineata*. Quart. J. Micr. Sci. 14: 226–273.
Fukane, S. Around 918. Honzo-wamyo (Japanese names of plants in Herbalism); Revised and published by M.Taki in 1796.
Fuse, S. 2012. Herbarium specimens, repairing against salt damage. In *Iwatsuki, K. & A.Domoto* (ed.), A Catastrophic Disaster and Biodiversity, 82–85. Biodiversity Network Japan, Tokyo.
Gessner, C. 1551–1558. Historia Animalium.
Hara, H. (ed.). 1985. Origin and Evolution of Diversity in Plants and Plant Communities. Academia Scientific Book Inc., Tokyo.
Hara, T. 1995. Octopus, sea bream and human beings. Biohistory 9: 11.

Hasebe, M. 2015. Research of evolution by genome analysis. Gakken Plus, Tokyo.

Hasegawa, M., N.Minaka & T.Yahara. 1999. Reviving Darwin at present (in Japanese). Bun-ichi Publ. Co., Tokyo.

Hattori, T., N.Minamiyama & Y.Ogawa. 2010. Studies on vegetation in 8th Century based on the information read from Manyoshu. Veg. Sci. 27: 45–61.

Hayata, B. 1927. On the systematic importance of the stellar anatomy in the Filicales (in Japanese). Bot. Mag. Tokyo 41: 697–718.

Hearn, L. 1899. Insects Musicians. In Exotics and Retrospective. Little, Brown & Co., Boston.

Hennig, W. 1965. Phylogenetic Systematics. Ann. Rev. Entomol. 10: 97–116.

Heraclitus. 5–6th century BC. Fragments: The Collected Wisdom of Heraclitus. Viking, New York.

Hidaka, T. 2007–2008. Selected Works of HIDAKA Toshitaka, 8 vols. (in Japanese). Random House Kodansha Publ., Tokyo.

Hirase, S. 1895. Études sur la fécondation et l'embryogénie du *Ginkgo biloba* (1). J. Coll. Sci. Imp. Univ. Tokio 8: 307–322.

Hirase, S. 1896. Spermatozoid of *Ginkgo biloba* (in Japanese). Bot. Mag. Tokyo 10:171.

Hooke, R. 1665. Micrographia: or, Some physiological description of minute bodies made by magnifying glasses. J.Martyn & J.Allestry, London.

Hori, K., A.Ebihara & N.Murakami. 2018. Revised classification of the species within the *Dryopteris varia* complex (Dryopteridaceae) in Japan. Acta Phytotax. Geobot, 69: 77–108.

Hori, K., A.Tono, K.Fujimoto, J.Kato, A.Ebihara, Y.Watano & N. Murakami. 2014. Retoculate evolution in the apogamous *Dryopteris varia* complex (Dryopteridaceae, subgen. Erythrovariae, sect. Variae) and its related sexual species in Japan. J. Pl. Res. 127: 661–684.

Hu, Hsen-Hsu & W.-C. Cheng. 1948. On the new family Metasequoiaceae and on *Metasequoia glyptostroboides*, a living species of the genus Metasequoia found in Szechuan and Hupeh. Bull. Fan Mem. Inst. Biol. N.S. 1: 153–163.

Humbolt, F.H.A.von 1945–62. Kosmos: Entwurf einer physischen Weltbeschreibung, 5 vols.

Iinuma, Y. 1856–62. Soumoku-zusetsu (Iconography of Herbals and Trees of Japan): 20 volumes for Herbals. [10 volumes of Trees were edited by S.Kitamura and published in 1977 from Hoikusha Publ. Co., Ltd.,

Osaka.]

Ikeno, S. 1896. Das Spermatozoid von *Cycas revoluta* (in Japanese). Bot. Mag. Tokyo 10: 367–368.

International Organization of Plant Information (IOPI) (ed.). Species Plantarum: Flora of the World. 1999–2005. Introduction. 1999; 1. Irvingiaceae by D.J.Harris. 1999; 2. Stangeriaceae by E.M.A.Steyn, G.F.Smith & K.D.Hill. 1999; 3. Welwitchiaceae by E.M.A.Steyn & G.F.Smith. 1999; 4. Schisandraceae by R.M.K. Saunders. 2001; 5. Prioniaceae by S.L.Munro, J.Kirschner & H.P. Linder. 2001; 6–8. Juncaceae by J.Kirschner (ed.). 2002; 9–10 Chrysobalanaceae by G.T.Prance. 2003; 11. Saururaceae by A.R.Brach & Xia N.h. 2005.

Ito, M. 2013. Plant Systematics. Univ. Tokyo Press., Tokyo.

Ito, M., A.Soejima & M.Ono. 1998. Genetic diversity of the endemic plants of Bonin (Ogasawara Islands). In Stuessy, T.F. & M.Ono (eds.), Evolution and Speciation of Island Plants, 141–154. Cambridge Univ. Press. Cambridge.

Iwasaki, K. 1828. Honzou-zufu (Illustrated Flora of Japan).

Iwatsuki, K. 1958. Taxonomic studies of Pteridophyta II. Acta Phytotax. Geobot. 17: 161–166.

Iwatsuki, K. 1979. Distribution of filmy ferns in palaeotropics. In Larsen, K. & L.B.Holm-Nielsen (eds.), Tropical Botany, 309–314. Academic Press, London.

Iwatsuki, K. (ed.). 1992. Ferns and Fern Allies of Japan (in Japanese with color plates). Heibonsha Ltd., Tokyo.

Iwatsuki, K. 1993. Biology of Diversity (in Japanese). Iwanami Shoten Publ., Tokyo.

Iwatsuki, K., D.E.Boufford, T.Yamazaki & H.Ohba (ed.). 1993–2020. Flora of Japan. 8 items & 1 Gen. Index. Kodansha Sci. Co. Ltd., Tokyo.

Iwatsuki, K. & S.Mawatari (superv.). 1996–2008. Biodiversity Series, 7 volumes. Shokabo Co. Ltd., Tokyo. (1) *Iwatsuki, K. & S.Mawatari* (ed.). Species diversity. 1996. (2) *Kato, M.* (ed.). Plants: Their diversity and phylogeny. 1997. (3) *Chihara, M.* (ed.). Algae, their diversity and phylogeny. 1999. (4) *Sugiyama, J.* (ed.). Fungi, Bacteria, Virus: Their diversity and phylogeny. 2005. (5) *Shirayama, Y.* (ed.). Invertebrates, except for Insects: Their diversity and phylogeny. 2000. (6) *Ishikawa, R.* (ed.) Arthropoda, their diversity and phylogeny. 2008. (7) *Matsui, M.* (ed.) Vertebrates: their diversity and phylogeny. 2006.

Iwatsuki, K. 1997. Diversification of Plants under the Impact of Civilization (in Japanese). Univ. of Tokyo Press, Tokyo. [Chinese edition: 林蘇娟 (tr.). 2000. 文明与植物進化. 雲南科技出版社.]

Iwatsuki, K. 1999. Seimeikei, a new concept of biodiversity (in Japanese). Iwanami Shoten Publ., Tokyo. [English edition of the same subject: Iwatsuki, K. 2023. Spherophylon – The Integrated Lives of Earth's Diverse Organisms. Bookend Publ. Co., Ltd., Tokyo.]

Iwatsuki, K. 2002. Biology – on the basis of biodiversity (in Japanese). Shokabo Co. Ltd., Tokyo.

Iwatsuki, K. 2012. 30 Lectures on Evolution and Phylogeny (in Japanese). Asakura Publ. Co., Ltd., Tokyo.

Iwatsuki, K. 2012a. A journey to trace interrelationships among diverse organisms – An autobiography (in Japanese). Minervashobo Publ. Co., Kyoto.

Iwatsuki, K. 2017. Biodiversity (in Japanese). Environmental Research Quarterly 182: 68–73.

Iwatsuki, K., A.Ebihara & M.Kato. 2019. Taxonomic studies of pteridophytes of Ambon and Seram (Moluccas) collected on Indonesian-Japanese botanical expeditions 1983–1986. XIII. Hymenophyllaceae. PhytoKeys 119: 107–115.

Izawa, K. 2014. Primates in the New World, 2 vols. (in Japanese). Univ. Tokyo Press, Tokyo.

Jardine, N., J.A.Secord & E.C.Spary. 1996. Cultures of natural history. Cambridge Univ. Press, Cambridge.

Kagaku-Asahi (ed.). 1991. Genealogy of Lord-Biology (in Japanese). Asahi Shimbun Publ. Inc. Tokyo.

Kahneman, D. & A.Deaton. 2010. High income improves evaluation of life but not emotional well-being. Proc. Natn. Acad. Sci. 107: 16489–16493.

Kahneman, D. & N.Tomono (ed. by A.Yamauchi). 2011. Daniel Kahneman – His concept of Psychology and Economy (in Japanese). Rakkousha Inc., Tokyo.

Kaibara, E. 1709. Yamato-Honzo (Herbals in Japan).

Kämpfer, E. 1712. Amoenitatum exoticarum. Lemgo.

Kato, M. & K.Iwatsuki. 1985. An unusual submerged aquatic ecotype of *Asplenium unilaterale*. Amer. Fern J. 75: 73–76.

Kato, M., N.Nakato, S.Akiyama & K.Iwatsuki. 1990. The systematic position of *Asplenium cardiophyllum* (Aspleniaceae). Bot. Mag. Tokyo 105: 105–124.

Kawai, M. 1965. Newly-acquired pre-cultural behavior of the natural troop of Japanese monkeys on Koshima Islet. Primates 6: 1–30.

Kawai, M. 1969. Collection of Works of Masao Kawai. 13 volumes. Shogakukan Prod. Co. Ltd., Tokyo.

Kawamura, S. 1954–55. Colored Illustrations of Japanese Fungi, 8 vols. (in Japanese with color plates). Kazamashobo Publ., Tokyo.
Kihara, H. 1930. Genomanalyse bei *Triticum* und *Aegilops*. Cytologia 1: 263–270.
Kimura, M. 1968. Evolutionary rate at the molecular level. Nature 217: 624–626.
Kimura, M. 1983. The Neutral Theory of Molecular Evolution. Cambridge Univ. Press, Cambridge.
Kimura, Y. 1974. Natural History in Japan – Dutch studies and herbalism (in Japanese). Chuokoronsha, Inc., Tokyo.
Kusayama, M. (= Kawai, M.) 1997–2014. Book of Animals by Masao Kawai, 8 vols. Froebel-Kan Co. Ltd., Tokyo.
Lamarck, J.-B.P.A.M.C.de. 1809. Philosophie zoologique. Paris.
Lamarck, J.-B.P.A.M.C.de. 1815–22. Histoire naturelle des animaux sans vertèbres, présentant les caractères généraux et particuliers de ces animaux …, 7 vols.
Lawton, J. 1976. The structure of the arthropod community on bracken. Bot. J. Linn. Soc. 73: 187–216.
Li Shizhen. 1578–96. Compendium of Materia Medica (本草綱目), 52 vols. (in Chinese).
Lin, S.J., M.Kato & K.Iwatsuki. 1990. Sporogenesis, Reproductive Mode, and Cytotaxonomy of Some Species of *Sphenomeris*, *Lindsaea*, and *Tapeindium* (Lindsaeaceae). Amer. Fern J. 80: 97–109.
Lin, S.J., M.Kato & K.Iwatsuki. 1992. Diploid and triploid offspring of triploid agamosporous fern *Dryopteris pacifica*. Bot. Mag. Tokyo 105: 443–452.
Linnaeus, C. 1735. Systema naturae (1758, 10th ed.). Stockholm.
Linnaeus, C. 1751. Philosophia Botanica. Stockholm & Amsterdam.
Linnaeus, C. 1753. Species plantarum. Stockholm.
Linnaeus, C. 1754. Genera plantarum (5th ed.). Leiden.
Lovelock, J. 1972. Gaia: A New Look at Life on Earth. Oxford Univ. Press, Oxford. 3rd ed.: 2000.
Löve, A., D.Löve & R.E.G.Pichi Sermolli. 1977. Cytotaxonomical Atlas of the Pteridophyta. J.Cramer, Vadus.
Mayebara, K. 1931. Fl. Austrohigoensis. Hitoyoshi.
Makino, T. 1940. Makino's Illustrated Flora of Japan (in Japanese; many editions). Hokuryukan Co., Ltd., Tokyo.
Makino, T. 1956. Autobiography of MAKINO Tomitaro (in Japanese). Nagashima-shobo Co., Tokyo.
Makino, T. 1973–1974. Collections of Papers on Plants by MAKINO

Tomitaro (in Japanese; ed. by H.Nakamura). AKANE SHOBO Publ. Co., Ltd., Tokyo.

Mawatari, S. (ed.) 1995. Natural History of Animals (in Japanese). Hokkaido Univ, Press, Sapporo.

Mayr, E. 1942. Systematics and the origin of species from the viewpoint of a zoologist. Columbia Univ. Press, New York. [Reprint version: Harvard University Press. 1999]

Mendel, G.J. 1864. Versuche über Pflanzen-Hybriden. Verhand. nat. Ver. Bruenn, IV Abhandl. 3–47.

Miki, S. 1941. On the change of flora in Eastern Asia since Tertiary Period. The clay or lignate beds flora in Japan with special reference to the *Pinus trifolia* beds in Central Hondo. Jap. J. Bot. 11: 237–303.

Miki, S. 1953. Metasequoia – A living fossil (in Japanese). Geol. Soc. Jap., Tokyo.

Minakata, K. 1925. Curriculum vitae – A letter addressed to Y.Yabuki. (In *Masuda, K.* ed., 1994, MINAKATA Kumagusu's Esseys, 7–74. Chikumashobo Ltd., Tokyo.)

Minakata, K. 1971–75. Complete works of MINAKATA Kumagusu, 10 vols. with 2 supplements. Heibonsha Ltd., Tokyo.

Minamoto, S. 931. Wamyo-Ruijyushou (Encyclopedia indexed by Japanese names, including the part of Plant names).

Mitsuta, S., M.Kato & K.Iwatsuki. 1980. Steler structure of Aspleniaceae. Bot. Mag. Tokyo 93: 275–289.

Miyoshi, M. 1915. Naturdenkmal (in Japanese). Fuzambo Publ., Tokyo.

Museum of Nature and Human Activities, Hyogo (ed.). 2012. A trial of museum for everyone to enjoy learning (in Japanese). Kenseisha, Co., Tokyo.

Momose, S. 1967. Prothallia of Japanese Ferns (in Japanese with line drawings of prothallia). Univ. Tokyo Press, Tokyo.

Morse, E.S. 1879. Shell Mounds of Omori. Mem. Univ. Tokyo 1: part 1.

Morse, E.S. 1917. Japan Day by Day, 2 vols. Houghton Mifflin Co., Boston.

Mujica Cordano, J.A. 2012. Speech at the Earth Summit in Rio de Janeiro. (In *Kusaba, Y.* ed., *M.Nakagawa* illus., 2014, Speech of the world's poorest president (in Japanese). Choubunsha Publ. Co. Ltd., Tokyo.)

Murakami, N. 1995. Systematics and evolutionary biology of the fern genus *Hymenasplenium* (Aspleniaceae). J. Pl. Res. 108: 257–268.

Murakami, N. & S.-I.Hatanaka. 1988. A revised taxonomy of the *Asplenium unilaterale* complex in Japan and Taiwan. J. Fac. Sci. U. Tokyo III. 14: 183–199.

Murakami, N. & K.Iwatsuki. 1983. Observation on the variation of *Asplenium unilaterale* complex in Japan with special reference to apogamy. J. Jap. Bot. 58: 257–262.
Nakamura, K. 1991. From Life Science to Biohistory (in Japanese). Shogakukan Prod. Co. Ltd., Tokyo.
Nakamura, K. 2000. Biohistory (in Japanese). NHK Publ., Inc., Tokyo.
Needham, H.J.T.M. 1954–2004. Science and Civilization in China. 8 vols. Cambridge Univ. Press, Cambridge.
Nishimura, S. 1987. After unknown organisms – Three stars in Exploring Natural History (in Japanese). Heibonsha Ltd., Tokyo.
Nishimura, S. 1989. Linnaeus and his apostles – Dawn of Exploring Natural History (in Japanese). JinbunShoin Co., Kyoto.
Nishimura, S. 1999. Natural History under Civilization (in Japanese). 2 vols. Kinokuniya Co. Ltd., Tokyo.
Nitta, J.H., A.Ebihara & M.Ito. 2011. Reticulate evolution in the *Crepidomanes minutum* species complex (Hymenophyllaceae). Amer. J. Bot. 98: 1782–1800.
Ogura, Y. 1940. History of Department of Botany, Faculty of Science, Imperial University of Tokyo (in Japanese). Imp. Univ. Tokyo.
Okada, H., K.Ueda & Y.Kadono (ed.). 1994. Natural History of Plants – Evolution of Biodiversity (in Japanese). Hokkaido Univ. Press, Sapporo.
Okamura, K. 1936. Marine Algae of Japan (in Japanese). Uchida Rokakuho Publ. Co., Ltd., Tokyo.
Ono, R. 1803–06. Honzo-Komoku Keimou (Main Points and Details of Japanese Plants; in Japanese with many illustrations). 48 vols.
Orel, V. 1996. (Translated by Stephen Finn.) Gregor Mendel, the First Geneticist. Oxford Univ. Press, Oxford.
Pascal, B. 1669. Pensées.
Gaius Plinius. –77. Naturalis historia.
Ray, J. 1686–1704. Historia generalis plantarum, 3 vols.
Renard, J. 1896. Histoires naturelles.
Saitou, K. 1995. *Metaseqoia* – Emperor Showa interested in it (in Japanese). Chuokoronsha, Inc., Tokyo.
Siebold, P.F.von. 1832–82. Nippon. Leiden.
Siebold, P.F.von & J.G.von Zuccarini. 1835–70. Flora of Japan. Leiden.
Socal, R.R. & P.H.A.Sneath. 1963. Principles of Numerical Taxonomy. W.H.Freeman, San Francisco.
Special Research Committee to Evaluate Japanese Red Plants. 1989. Threatened Plants of Japan to be preserved urgently (in Japanese). WWF-Japan & NACS Japan.

Suzuki, T. & K.Iwatsuki. 1990. Genetic variation in agamosporous fern *Pteris cretica* L. in Japan. Heredity 65: 221–227.

Tagawa, M. 1959. Colored Illustration of the Japanese Pteridophyta (in Japanese with color plates). HOIKUSHA Publ., Co., Ltd., Osaka.

Takamiya, M. 1996. Index to Chromosomes of Japanese Pteridophyta (1910–1996). Jap. Pterid. Soc., Tokyo.

Tamba, Y. 984. Ishinpou (in Japanese; an encyclopedia of Medical science including topics of medicinal plants). 30 vols.

Tamura, N.M. & K.Suzuki. 2017. Newly founded Japanese journal issued by the Japanese Society for Plant Systematics, led by amalgamation of Japanese newsletters of JSPS and the Journal of Phytogeography and Taxonomy (in Japanese). Bunrui 17: 109–111.

Tanaka, J. 1986. Zhang Hua's 'Jiao-Liao Fu' as a self-recommendation (in Japanese). Chugoku Bungakuron-shu (Kyushu Univ.) 15:71–99.

Tao Hongjing. Around 500. 神農本草経集注 Shennong bencaojing jizhu (in Chinese).

Tardieu-Blot, M. 1932. Les Aspléniées du Tonkin. Thesis, prés. à la Fac. Sci. Paris ser. A no 1373 1re thèse.

Tardieu-Blot, M. & C.Christensen. 1939–51. Cryptogams vasculaires. In Lecomte, H. (ed.), Flore générale de l'Indo-Chine vol. VII-2. Masson et Cie, Paris.

Tashiro, K. (ed.) 1968–73. Diaries of TASHIRO Zentaro, 3 vols. (Meiji, Taisho & Showa; in Japaese). Sogensha, Inc. Osaka.

Theophrastos. Around 314 BC. Historia plantarum.

Thunberg, C.P. 1784. Flora Japonica. Lipsiae.

Thunberg, C.P. (ed. by L.-M.Langles). 1796. Voyage de C.P.Thunberg au Japan.

Tokuda, M. 1951. Evolution (in Japanese). Iwanami Shoten Publ., Tokyo.

Tokuda, M. 1952, 1956. Two Genetics; Two Genetics, sequel (in Japanese). Rironsha, Co., Tokyo.

Tournefort, J.P.de. 1694. Éléments de botanique.

Toyama, S., R.Yatabe & T.Inoue. 1882. Shintaishisho (New style poetries). Maruya-Zen'hichi Co., Tokyo.

Tsuji, Y. & N.Nakagawa (ed.). 2017. Primates in Japan – Studies on Japanese monkey as Primatology (in Japanese). Univ. Tokyo Press, Tokyo.

Tsumura Laboratory (ed.). 1916–. The Journal of Japanese Botany. Tsumura & Co., Tokyo.

Tsunewaki, K. 1993. Genome-plasmon interaction in wheat. Jap. J. Genetics 68: 1–34.

Tsunewaki, K. 1993. Phylogeny of Wheat determined by RFLP analysis of nuclear DNA 1. Einkorn wheat (in Japanese). Jap. J. Genetics 68: 73–79.

Tsurumi, K. 1979. MINAKATA Kumagusu – Comparative study in global conspectus (in Japanese). Nihon Minzoku Bunka Taikei vol. 4. Kodansha Ltd., Tokyo.

Udagawa, Y. 1822. Botanika-kyo (introduction to botany in the form of Bhuddism scripture [sutra]; in Japanese).

Udagawa, Y. 1835. Shokugaku-Keigen (Textbook of Botany; in Japanese), 3 vols.

Ueno, M. 1973. Nihon-Hakubutsugakushi (Records of Natural History in Japan). Heibonsha, Ltd., Tokyo.

Ueno, M. 1987. Nihon-Doubutsugakushi (History of Zoology in Japan). Yasaka Shobo Inc., Tokyo.

Ui, N. 1929. Flora of Kii Province (in Japanese). Kindai-Bungei Co., Ltd. Osaka.

Umesao, T. 1960. Nihon-tanken (Exploration in Japan). Chuokoron Shinsha, Inc., Tokyo.

Umesao, T. 2003. An Ecological View of History – Japanese Civilization in the World Context. Trans Pacific Press, Melbourne. (Translated from original Japanese book issued in 1957.)

University of Tokyo Press (ed.). 1993–2018. Natural History Series 50 vols. (in Japanese). Univ. Tokyo Press, Tokyo.

Wagner, W.H.Jr. 1954. Reticulate evolution in the Appalachian *Asplenium*s. Evolution 8: 103–118.

Wagner, W.H.Jr. 1964. Edwin Bingham Copeland (1873–1964) and his contributions to Pteridology. Amer. Fern J. 54: 177–188.

Watano, Y. & K.Iwatsuki. 1988. Genetic variation in the 'Japanese Apogamous Form' of the Fern *Asplenium unilaterale* Lam. Bot. Mag. Tokyo 101: 213–222.

Windelband, W. 1903. Lehrbuch der Geschichte der Philosophie, 3. Aufl. d. Geschichite der Philosophie. Tübingen.

Yamagiwa, J. 2012. Evolution of families (in Japanese). Univ. Tokyo Press, Tokyo.

Yamagiwa, J. 2015. Gorilla, 2nd ed. (in Japanese). Univ. Tokyo Press, Tokyo.

Yasugi, R. 1972. History of Modern Concept of Evolution (in Japanese). Chuokoron Shinsha, Inc., Tokyo.

Yatabe, R. 1890. A few words of explanation to European botanists. Bot. Mag. Tokyo 4: 305.

Yoroi, R. & K.Iwatsuki. 1977. An observation on the variation of *Trichomanes minutum* and allied species. Acta Phytotax. Geobot. 28: 152–159.

Yoshimura, F. 1987. Recovering the lost ground of Morphology – Beyond Molecular Biology (in Japanese). Academic Publishing Center, Inc., Tokyo.

Yoshino, Z. 1929. Flora of Bicchu Province (in Japanese; with suppl. 1–2, –1931). Yoshino Medical Shop, Okayama.

Zhang Hua. 3rd Century. (Revised by Chan Jing in 6th Century.) 博物志 Bowuzhi (in Chinese). [*K.Ozawa.* 2013. Zhang Hua's Bowuzhi (in Japanese). V2-Solution Books, Nagoya.]

Afterword

I first received a proposal about this book project in the spring of 2013. I was informed that by this book the Natural History series would be completed, that this last volume would be published with March 2018 as a target, and that the book would need to describe natural history pursued at Japanese universities, which had not been written in this series. After holding the first meeting on this book in June 2013, which set a basic direction for the work, I started writing.

This book was expected to describe natural history pursued at Japanese universities. Natural history at botanical gardens, zoos, and aquariums, as well as at natural history museums, was already described in the other books of the series. Consequently, I was told that this book would primarily portray the current state of natural history at Japanese universities, its fundamental role in the promotion of scientific research, and the challenges facing natural history.

However, to discuss the true nature of the pursuit of natural history, the discussion naturally led to generalities about what is expected of natural history at the moment.

When one attempts to outline the natural history topics above, numerous biodiversity-related examples come into view. I also view things from a perspective closely related to biodiversity. To know the current state of findings related to biodiversity, the seven-volume Biodiversity series (ed. IWATSUKI Kunio and MAWATARI Shunsuke, 1996–2008) produced through collaborative efforts of many researchers serves as a good guide.

I have been blessed with many opportunities to discuss science at lecture halls and in journal articles. However, it is difficult to summarize the integration within the field of science by using an example or two. For this purpose, I could not logically organize the story by focusing on a specific topic. Within a limited number of minutes or pages, it is inevitable to refer to bits of marginal issues. As a result, I sometimes naturally give the impression that my speech (or article) is disjointed, drawing criticism for being difficult to focus on and understand.

Obviously, a speech about an individual scientific achievement becomes more understandable when it focuses on specific examples. In these cases, it is likely that my delivery was not good or my wording less than excellent.

These factors may contribute to the criticism. Nevertheless, it is difficult to explain an integrative perspective in a short speech or article. Perhaps, a book is long enough to thoroughly narrate my point, although this depends on my ability to successfully exploit the valuable opportunity given to me.

In this book I have attempted to describe natural history primarily in terms of biodiversity. However, I am not completely sure if the use of these models for this description, as in science, is persuasive.

During the long process of writing this book, I was in close communication with the editor Kōmyō. In 2015, I presented the first draft to ŌTA Hidetoshi and AKIYAMA Hiroyuki of the Museum of Nature and Human Activities, Hyogo for a review. They furnished me with valuable comments.

I rewrote the first draft, incorporating their feedback. The draft was almost complete by the end of 2016. I had it reviewed and edited by KATŌ Noriko. In the final writing process, my manuscript, subject to errors due to my age, was polished for printing by two editors Kōmyō and Katō, renowned in the domain of natural history. Of course, I am responsible for the distinct ideas and descriptions in this book.

This book avoids the use of figures or tables, which would be helpful for reading the text, as suggested by Kōmyō. The sketches readers see at chapter breaks were drawn for use in treatises when I was young. The editor selected them from a limited collection, as I had lost my confidence in drawing at an early stage. The illustration on the cover is a painting drawn by AZUKI Mutsuko, a Sunday artist and a collaborative researcher working for the Museum of Nature and Human Activities.

The publication of this book owes much to many people, as usual. The text of this book is based on my research achievements, possible only through various forms of collaboration with many individuals. They include my long-acquainted research colleagues, teachers, senior researchers, and an extensive range of people from Japan and abroad who took part in research projects with me. Some of them appear in this book. In addition, many who have contributed to science policy have encouraged us and helped our research activities. Personally, in various ways, I received teachings from naturalists with whom I discussed science and ascertained facts, from those who did not have much interest in nature to the staff of organizations in which I was involved. I would like to take this opportunity to express my gratitude to all of these people. My only concern is how much I successfully expressed my experiences in this book. As I write this afterword, I am sincerely aware of the limits of my abilities.

The prize money that the International Cosmos Prize granted for my achievements was used to cover part of the expenses required for the publication of this book, as I believed that much of what I wanted to write in this book was in line with the purpose of the prize.

Nevertheless, the product is still unsatisfactory to me. Like a biological taxonomic chart appearing in a textbook in its hypothetic state is known by the author to be inconclusive, I publish this book knowing it is incomplete. This incompleteness is partly due to my incomplete knowledge in natural history. In addition, findings gained in natural history are still undeniably far from complete. Even if with godlike expressive power, I would be unable to describe the complete form of an incomplete fact. One needs not only to complain about one's insufficient competence, but also to see this truth.

I hope that this book, though published with my dissatisfaction, will draw the attention of those who have been indifferent to the real possibilities of natural history and that the proper development of natural history methods will facilitate the healthy growth of human society.

Lastly, I cannot finish without referring to the Natural History series, completed with this book and published by the University of Tokyo Press, and without mentioning the contributions from the editor KŌMYŌ Yoshifumi. While it is difficult to publish good books that have a small circulation, the fifty-volume series has taken quarter century to publish. The series provides precise descriptions of Japanese researchers' contributions to the field. Considering the whole picture of research, the series fails to cover all areas, and some excellent researchers were unfortunately unable to contribute to the series. However, it is not difficult to see the whole picture through the fifty volumes.

Contributions by natural historians are still far from sufficient, and consequently, the current state of science is biased, as I state in detail in this book. Researchers in this domain need to seriously consider this view and strive continuously to achieve further research results. Accordingly, I have a renewed recognition of the significance of what the Natural History series offers. I would like to praise the University of Tokyo Press afresh for their decision to work on the publication of the series.

I have heard that the editor KŌMYŌ Yoshifumi has already been honored by the Society of Evolutionary Studies, Japan; the Union of Japanese Societies for Natural History; and the Mammal Society of Japan. As a researcher in the domain of natural history, I would like to express my deep thanks for the profound and sincere contributions that he has made to the difficult editing work during the period of a quarter century.

Autumn of 2018
Iwatsuki Kunio

References

Iwatsuki, K. & al. (ed.). 1995. Flora of Japan, vol. 1.

Postscript to the English Version

It is, needless to say, quite difficult to replace the Japanese sentences with English, especially of the kind in this book. In addition to the general problems always encountered in translating into a different language, I have to comment here that there are at least two severe and complex problems in translating a Japanese book on natural history into English.

It is natural that we usually focus on topics related to Japanese affairs in books written in Japanese. The simple replacement of Japanese sentences with English, therefore, is not sufficiently helpful in letting English-speaking readers fully understand the contents of the original books. The contents themselves should be carefully explained for the English-speaking readers who are generally not familiar with the Japanese affairs, making it hardly possible to communicate the contents of such a book in a directly word-for-word translation of the original.

In apologizing for this fact with the expectation of filling the gap between the Japanese and English versions, I wish to make a brief complementary comment.

The natural history of Japan has been comprehensively surveyed; first in modern science by Mitsutaro Shirai (1891) and highlighted by Masuzo Ueno (1973). We are always aware of these and other excellent works when we take up historical topics in this field. In books written in Japanese, we usually refer to preceding research only by citing the literature in the bibliography. Without information given in the literature written in Japanese, therefore, it may barely be possible to fully follow discussions given in the books in Japanese, such as in my original 'Natural History.' In reading the English version of this 'Natural History,' no convenience to refer to the Japanese literature is offered. However, it is impossible to introduce most of the information given in previous research in this English version, and I have to request that readers of the English version follow the discussion to obtain the necessary information in other ways.

The original Japanese version of this book includes neither figures nor tables, as was suggested by the editor of the Series. In such a book mainly intended for the lifelong learning series, no original contributions were expected, but syntheses of published contributions to improve basic concepts on the topics in question were permitted. It was expected that the study and understanding of natural history would be introduced properly in considerable order from a historical perspective. Without tables and figures, it was possible to point out the main events in the developmental process of natural history, but in this English version it is difficult to introduce these facts only by referring to the literature in Japanese.

The original book in Japanese was the 50th and final volume of 50 issues of the Natural History Series offered by the UT Press, as noted in the Foreword to this volume. It was suggested by the editor of the Series that contributions to natural history by universities were to be included in this volume. In writing the 50th and final volume, it was natural that I always had in my mind the contributions included in the preceding 49 volumes. They are all fully in Japanese, and it is hardly possible for English-speaking readers to learn from the comprehensive information included in those volumes. Knowing all of these difficulties, I dare to offer this English version to the public, expecting the readers to generously understand the circumstances.

The second problem to be faced in the original version was the situation where I was at that time a pure researcher focused on a particular field and expected to contribute to a general audience as an academic expert.

I had been in the services of four universities in my career, two of which were national universities, one a private university and one an open university. In a university setting, each constituent member usually contributes to basic research work in one's particular field of expertise as well as to administering to one's educational duties in that field, and I followed this general lifestyle as a university staff member. In the two national universities, Kyoto and Tokyo, my position was professor of botany. In the University of Tokyo, an additional duty required me to serve as director of the Koishikawa Botanical Gardens in addition to administrative requirements.

All researchers are usually involved in the activities of the academic societies pertaining to their interests. Board members of such societies are generally chosen from among the members through elections. I was elected and served on the boards of various academic societies. Experts with particular areas of interest are often invited to serve in the governmental and non-governmental organizations and to provide information in their specific fields. I have been involved in various activities in such organizations, both national and international, especially on biodiversity issues, which have been hotly discussed these days in our society.

While I was at the University of Tokyo, I was appointed to director of the botanical gardens, and was involved in the activities of both Japanese and international associations of botanical gardens. I also worked on lifelong learning initiatives in addition to conducting pure scientific research within the university. For the Open University of Japan, for lifelong learning through remote education, my duties expanded during the five years I spent at this University. In addition to my experiences in research and education within the universities, I also became involved in a natural history museum in my later years, as director of the Museum of Nature and Human Activities in Hyogo. Based on the experiences I gained through my associations with these institutions, I was involved in various services in the field of lifelong learning in addition to conducting my own pure scientific investigation.

As a university staff member specializing in botany, I worked on systematic botany and species diversity in plants, especially of the pteridophytes. When I started my research career on species diversity in the 1950s, it was a field of pure science. Soon, biodiversity loss grew to become one of the major environmental issues, and I was involved in various activities as a specialist in this particular field. (Strictly speaking, I am not a specialist in biodiversity loss, but of species diversity of vascular plants. In that position I came to realize the recent phenomenon called biodiversity loss. I may better note here that my recent principal scientific contribution is a monograph of the Hymenophyllaceae for 'Flora Malesiana' published in 2023. I also co-edited the 'Japanese Species on the Edge of Survival' issued one year earlier.)

Based on such experiences, my research in the field of natural history expanded and became closely connected to the environmental issues faced by modern society. In this 'Natural History,' I expected to integrate the basic research on plant species diversity and environmental issues under the keyword, 'biodiversity,' although it was terribly difficult to do and I do not know if I succeeded in this book, at least partly or not at all.

Despite the difficult problems encountered in my situation, which made me think that the content might be rather immature in the original Japanese version, I arranged to translate the original Japanese version directly and dared to publish it in English. My main objective was to explain the traditional Japanese concept of the natural environment. Such an idea is less popular even within Japanese society, especially among the younger, modern generation. (It is not necessary to note here that the elders usually complain that the younger generation does not follow the traditional concepts. Such attitudes were recorded even 10 centuries ago, as noted in the classical diary of a Heian aristocrat.)

Recently, it is rather easy to understand the diversity of concepts developed in various areas. Even traditional Japanese concepts are introduced globally through various channels. However, the particular Japanese concept and nature is sometimes introduced erroneously, or is inaccurately accepted even by modern Japanese. I had to comment on these traditional concepts in the historical summary of the natural history research in Japan.

Traditional concepts of the Japanese life within the natural environment are in general wrongly understood today, for the sense of life is different between our ancestors and modern people. The most distinct difference is traced in the sense of the monetary system. In recent times, the value of every object is based on its monetary price. The continuous growth of the economic system is expected to maintain the development of this monetary evaluation of everything far into the future.

The endless growth of the economy in its present form cannot be expected. Such a system is applied only for the time being and knowing that it will continue to develop over time. (I do not know how many years it will be, several decades, a few centuries, or still longer; anyway, few or no people living today will see the outcome.) In contrast, in discussing environmental issues, our target is on the scale of centuries, for the life of forests is usually counted in this time scale.

In translating each word or phrase into its equivalent in another language, we always meet a difficulty in finding what portion of the meaning of the word in question is to be correctly translated. I mean that each word or phrase expresses a certain range of meaning, and the expected equivalent in the other language in general encompasses more or less a different range of meanings. A word expresses a concept and contains a broad range of meaning, especially in the case of abstract thoughts. Every word, therefore, presents its own complex problem when it is translated into any other language.

In taking up the case of the Japanese equivalent of 'nature' as an example, I wish to show its difficulty in translation. The European expression, 'nature,' was first introduced to Japan in a Dutch-Japanese dictionary (Inamura S. & al. 1896), where Dutch 'natuur' was given as the equivalent of the Japanese word 'shizen.' At that time, the two Chinese characters that formed shizen were pronounced 'jinen' and understood to mean something like 'automatic' without artificial influences. (Now, it is explained to mean 'naturally.')

Nature in Japan at that time was recognized as lush greenery, and that understanding is the same presently. Thus, a green site, such as a terrace field, is often called 'secondary nature.'

I then realized that the usage by Japanese conservationists of 'secondary nature' is self-contradictory, or a combination of two words that contradict each other. 'Nature' is, strictly speaking, a word that describes a native site without artificial impacts; 'secondary' means a site with artificial influences. In Japan, greenery sites are 'shizen.' There are various greenery sites, even those with considerable artificial influences. Japanese often use the phrase 'secondary nature' to describe such places. (This is further commented on relation to the note on *satoyama* below.)

We use the botanical term secondary vegetation with the understanding that secondary vegetation is considered to be a part of nature. Such use is generally accepted, and even in English it is hardly possible to clearly differentiate between the natural and man-made world.

Another definition in Japanese is 'nature in urban area.' For instance, parks with greenery are described as natural sites, although a park is a kind of man-made feature. Various wild animals even live in urban areas and wild plants are growing in big cities. In Japan they are often referred to as 'nature in the cities,' but they may be better referred to as 'pockets of nature' (Boufford, pers. comm.). Wild organisms are naturally the elements of nature, even when they live in urban (man-made) areas.

In understanding the word 'natural,' we even apply this to less greenery sites. Deserts and summit areas of higher mountains (above the tree line) are less green, but we usually describe them as harsh natural environments. Less greenery sites may be natural, and in this 'nature' means fully primitive sites.

When we replace the Japanese word 'shizen' with the English 'nature,' we usually do not explain to which meaning the word in question is applied. In that sense, the reader of the translated word can read the meaning of the word in one's own way. Very often, the true meaning will be missed and the real concept of the author cannot be transferred to the reader.

To explain the above discussion further, I will take up some particular Japanese words that are now more or less popular in global conspectus. They are particular Japanese words that are difficult to translate into other languages, as the words express particular concepts of traditional Japanese. They are explained variously, but are in most cases not completely introduced even within Japan, for they are often carelessly applied to the modern lifeforms while forgetting their original use. (I know that language is alive and changing day by day, but still I expect to express the concept correctly by respectfully applying the traditional usage to the word.)

In translating the original Japanese version into English, I am afraid that some words are hardly possible to translate literally word for word (from Japanese to English). I comment here on three particular words, apologizing for the limited possibility of direct translation. The examples of Japanese words taken up here are '*kyousei* – 共生,' '*satoyama*' and '*mottainai.*' For those words, I have also given short explanations in another of my books: 'Spherophylon' (Iwatsuki, 2023).

The word ***kyousei*** is composed of two Chinese characters, 共生, but is pure Japanese, and I suspect that Chinese people cannot understand the real Japanese concept from the combination of these two characters. This Japanese word has been used for a long time, and was once applied in the teaching of Buddhism doctrines by Honen (1133–1212; a Buddhist saint). Buddhist thought greatly influenced the principal ideas of the general Japanese populace, as was the influence of Christian philosophy on western people.

Kyousei is composed of two Chinese characters. The first character, 'kyou' means together, both (for two), all (for more than two), side by side, syn-, and so on. The second, 'sei,' living, life, crude, unprocessed, and many other meanings (and as a character for raw meat, fresh vegetables, draft beer, dry-cured ham, the condition of buckwheat noodles, etc.). By combining these two characters, *kyousei* expresses a distinct meaning rather than simply summing up the meaning of the two characters.

It may be noted here that the same word *kyousei* with these two characters was introduced, some 100 years ago as an equivalent in translation of the biological term *symbiosis.* In spite of the same characters and pronunciation, the biological term is defined carefully in science (as a homonym) and differs in meaning from the popular (and Buddhist) word *kyousei* used in Japan over a long period of time. Symbiosis covers biological phenomena, i.e., mutualism, commensalism and parasitism.

I was awarded the International Cosmos Prize in 2016. This Prize was founded after the success of the International Garden and Greenery Exposition in Osaka, Japan, in 1990. In awarding the Prize, it was intended to promote the principal concepts of the exposition, and that concept, symbolized in Japanese by the word *kyousei*, was expressed in English as 'Harmonious co-existence of nature and humankind'(cf. Tea Time 7). As an awardee of the Prize, I am responsible for promoting the principal concept of the exposition. The original Japanese slogan includes the word *kyousei*, which is translated into English as *harmonious co-existence.* From this English phrase, however, I do not recall accurately the philosophy of *kyousei*, as this Japanese concept is really Japanese.

The traditional Japanese concept of nature is distinct in considering humankind as an element of nature. All the other components of nature live just in parallel with humankind, having an equivalent worth of existence with each other. This concept is different from that of the general concept of western people, who consider that the humankind is the lord of creation and is allowed to utilize all other natural elements as resources for exploitation. (I note that it is not a strictly binary opposite, but a general tendency.) The basic Japanese concept is still alive in the Japanese tradition, even at present, as is shown by the reality that many Japanese pray at shrines, especially on New Year's Day (not as a religious, but as a traditional event). It is natural that younger generations do not practice such traditions, and it is regretful that various good customs have been abandoned in recent times in parallel with the disappearance of tradition.

In the traditional Japanese concept, all natural objects are considered to harbor deities and to possess a spirit. Japanese people have a respect for all natural objects as precious colleague. When we take organisms as food, we are always grateful to them. When we take foods three times a day, we give thanks each time to the deities, 'I give thanks (to the deities) for this meal (いただきます – *itadakimasu*)' at the beginning and 'Thank you (to the deities) for giving us a good meal (ごちそうさまでした – *gochisousamadeshita*)' at the end. Every object in nature is our colleague, not just a resource for our life, but as our equivalent with all others in nature. We therefore expect to practice *kyousei* (harmonious co-existence) with all the objects in nature within which the deities dwell.

Under the influence of his Buddhism teacher, celebrated architect Kisho Kurokawa (1934–2007), developed his philosophy under his personal concept of *kyousei*. He published several papers and books in English and French on this philosophy (Kurokawa 1994). His ideas were really unique in introducing the traditional Japanese concept along with his own interpretations. He used the word symbiosis as the English equivalent of *kyousei*, and extended his ideas to the co-existence of organisms. His discussions developed beyond Buddhism, but were hardly concerned with the traditional Japanese concept in relation to the existence of eight million deities. (It may be noted here that the Japanese Buddhism is really the Japanese concept under a syncretism of introduced Buddhism and traditional Shinto.) Still, I should note that Kurokawa's philosophy provides a variety of information about the Japanese concept of *kyousei*.

The Japanese archipelago is composed of steep terrain and is mostly covered by mountains with little spaces for plains. It is rather difficult to extend the agricultural fields more widely, and even moderate slopes have been developed into terrace fields blessed by rich precipitation. In utilizing natural resources effectively, the Japanese people have additionally continued a system of hunting and collecting in the backyard of their villages and have enlarged ***satoyama*** areas throughout the archipelago.

Following traditional religious views, Japanese people considered remote, secluded mountains as sacred places, where numerous (the number is often given as eight million; eight million usually means a great many or even infinite number) deities dwelled. The soul of humans was believed to go there after they passed away. *Okuyama* (奥山 – mountain recesses) was usually considered to be such a sacred place.

Satoyama (里山 – village mountains) was developed as the area between *okuyama* and the village area (*hitozato* – 人里 – human village or *satochi* – 里地 – countryside) even at the time of the New Stone Age when cultivation was initiated, villages were being developed and people could settle down. Regional divisions into *okuyama*, *satoyama*, and *hitozato* were developed throughout the Japanese archipelago since the New Stone Age (although there were no words to express these areas and concepts at the time). This regional division grew subconsciously as part of the lifestyle of the general people without governmental direction by the authorities. It was never planned as a unifying idea, but formed as part of the practices of the inhabitants of the villages without governmental planning. *Satoyama* thus developed independently in various areas based on independent ideas of the local people concerned, with the result of more or less different but similar landscapes.

Okuyama, with reduced influence from human activities, has thus been a happy place in nature for wild organisms. Although *Satoyama* was utilized by the people, biodiversity developed there in particular ways under artificial influences. Wild animals enjoyed the natural resources, but did not like to encounter humans or to have unexpected conflicts. Usually, the human population worked in *satoyama* in the daytime. Wild animals, to avoid meeting people, shifted their activities in the *satoyama* areas to dawn, to the late evening and at night. Although it is natural for wild animals to be active even in daytime, they are careful to avoid people working there, and good at hiding from humans, when they unwillingly meet them. When I was young, I often made field excursions alone, and on such occasions I sometimes met populations of wild animals there. When they noticed me, they stopped what they were doing and remained quiet. I realized that the wild animals were probably holding their breath and waiting for me to go away.

Satoyama has been an area shared between humans and wild animals in such a way, and has been kept beautiful for many years. Wild animals live in *okuyama* and go about their business in both in *okuyama* and *satoyama*, while humans go about their business in *hitozato* and often also in *satoyama*. Thus, in *satoyama* wild animals and humans form a good *kyousei* (harmonious co-existence) relationship and avoid useless conflicts. Wild animals learn not to come to the *hitozato* area beyond the *satoyama*. Habitat segregation has developed in an ideal way and co-existence has remained harmoniously.

After the 1960s, however, the lifestyle of the Japanese people changed as petroleum became the main energy source even in rural areas. Petroleum became available at reasonable prices throughout the country, and various petroleum-dependent machines were produced and widely distributed, even in private homes. People then became less dependent on collecting firewood and/or producing charcoals in *satoyama*, although it became necessary for incomes to increase to be able to purchase petroleum. A self-sufficient life-style could not be maintained, even in local areas, as monetary economy developed and people needed a greater income. Thus, *satoyama* was abandoned almost simultaneously throughout the Japanese archipelago since the 1960s. Now, half a century later, the former *satoyama* areas are mostly abandoned and have become degraded.

Satoyama is now a popular term globally, but the definition of *satoyama* differs according to the user. The Japanese term was used sporadically even several centuries ago, but it became a popular word as was proposed by the late Professor Tsunahide Shidei (1911–2009; 2006) in the 1960s to express the concept of *satoyama* forests that covered the backyard of villages where the local people collected natural resources, mostly the sources of energy, such as firewood and charcoal.

Satoyama functioned in the Japanese archipelago as a buffer zone between *okuyama* (a more or less core natural area) and *hitozato* (artificially developed region) for many years. But *satoyama* was abandoned and has become rough through recent changes in lifestyle in general. *Satoyama* itself was blessed with rich biodiversity in addition to its role as a buffer zone between *okuyama* and *hitozato*, especially in relation to environmental issues. Elimination of *satoyama* has greatly influenced the human environment in the Japanese archipelago, especially in its biodiversity.

The Environmental Agency of Japan (at that time, now Ministry of the Environment) took up the topics of *satoyama*, and intended to spread information about *satoyama*. They asked a native English-speaking adviser to translate the word *satoyama*, and he proposed 'countryside' as an English equivalent of *satoyama*. Actually, in the 1970s, pamphlets issued by the Environmental Agency explained *satoyama* using 'countryside' as the English equivalent of *satoyama*. However, countryside refers to the area around villages, or *hitozato*. In fact, in the 1970s when *satoyama* was taken up as popular word, terrace fields were exemplified as a typical component of *satoyama*.

I received the suggestion from Dr. David Boufford that woodlot in the western world is more similar to the Japanese concept of *satoyama* than countryside. I do not have enough understanding of the word woodlot, and am not able to compare these two words. I would note here that *satoyama* is a natural transitional area between *hitozato* and *okuyama*, and has functioned ecologically as a buffer between these two areas. I do not know if woodlot has such a function in keeping the remote and primitive mountains as paradise of wild organisms.

In these days, artificially developed and fully green areas, such as paddy fields, were brought to the attention of conservationists. In Japan, peculiar Japanese English was used in translation to describe them as 'secondary nature.' As the equivalent term for nature in Japanese is 'shizen,' this word was used (at least by the conservationists) to mean fully green places. So, artificially developed but fully green areas such as *satoyama* (in the strict sense) as well as paddy fields are collectively understood as being secondary 'shizen,' and are called *satoyama* (in a broader sense). (In this broader sense, *satoyama* may be translated as 'countryside.') The English word 'nature' is used for the places without artificial influence, while *satoyama* and terrace fields are not quite natural sites, even in a 'secondary' state. [I have learned such an English expression as 'Habit is second nature,' but this meaning of 'nature' does not apply to the natural environment, but to a characteristic of human beings.]

I view the importance for biodiversity issues in the agricultural sites in Japan, including terrace fields, as fully green zones, even though they are artificially developed. These artificial green zones are rich in biodiversity and better for the human environment. However, *satoyama* (forest) in the strict sense has another particular meaning and functions as the buffer zone between *okuyama* and *hitozato*, based on the traditional Japanese concept. In this sense, *satoyama*, in its strict sense as defined by Professor Shidei, should be a focus when environmental issues are under consideration.

In 2010 when the biodiversity COP 10 was held at Nagoya, Japan, the Japanese government proposed an international project named SATOYAMA Initiative for maintaining green even in artificially developed areas, and at that time the term *satoyama* for this particular project was written in full capital letters. It is difficult to explain it, but SATOYAMA, as used in this Initiative, was applied in its broadest sense (which is now a popular application of this word in general). It is true that the term *satoyama* is used to describe various landscapes nowadays, even within Japan. The promotion of growing green in artificially developed areas is an important perspective as an environmental issue, and the SATOYAMA Initiative is to be promoted globally.

Satoyama in the strict sense is to be recognized as a buffer zone between *okuyama* and *satochi*, and a zone based on the daily lives of the people in general, or as the fruit borne from the natural evolution of Japanese people. This evolution succeeded as the result of the general concept of Japanese in considering *okuyama* to be a sacred site with eight million deities and all natural objects having the same value with the human beings themselves and living in harmonious co-existence with each other. In this sense, the concept of *satoyama* (in the strict sense) may be recognized only when the *kyousei* lifestyle is correctly understood. The strict definition of a buffer zone in a scientific sense should separate the core (natural) zone and the developed (artificial) zone and should be perfectly continuous; *satoyama* is not such a zone in its constitution. I now focus on a discussion of the buffer between natural and artificial areas, as it has functioned through human history in the Japanese archipelago.

This is a brief explanation why I would use here the word *satoyama* in its original and strict sense against the general and broader use of the word as interpreted in recent days. And I do not wish to cause confusion in discussions due to different understandings of the word *satoyama*.

Mottainai (もったいない) is a traditional Japanese word recently introduced globally by the Nobel Laureate Wangari Maathai (1940–2011; 2009) and others. In its recent use in the area of conservation activities, this word is related to conservation activities aimed at reducing, reusing and recycling.

The meaning of this Japanese word is, as noted in dictionaries under their second or third explanation: 'It is regretful that the materials which may still be useful are abandoned in vain.' For this usage, the word *mottainai* is translated into English as: 'too good to waste,' 'What a waste!' 'That is a waste,' 'It is wasteful with ...,' 'You do not know what you are missing,' and so on. This meaning usually explained in recent days, is understood economically as 'repairing is less expensive than buying new.'

The original use of this meaning is, as will be noted below, in accordance with the traditional concept of the Japanese: everything on the earth co-exists with the deities, and we should make use of anything with sincere respect, never waste in vain. In applying the word in the recent way of thinking, principally based on an economic (or monetary) concept, cheaper materials may easily be abandoned after calculation of the monetary advantage of such a decision.

We are currently facing great food loss and waste, which are serious problems at the moment. Discarding food is deeply regretted. These wastefully abandoned foods are usually fully regretted and said 'it is *mottainai* (wasteful) to have such a great food loss and waste, even though such food could be used to save many hungry people.'

In Japanese word, *mottai* (勿体 – the thing itself) was originally applied to any materials existing in nature. As noted before, all natural objects co-exist with deities according to the traditional concepts of ordinary Japanese people. This meant that all natural objects should be thoughtfully respected as being with deities and must not be treated lightly or in vain. Respect for all materials does not depend on whether it is useful or not, even if it is expensive or has zero market value.

Recent usage of this word depends on the modern preference to judge all materials based on their commercial value. Expensive materials are highly respected. However, even during the Yedo Era, which ended some 150 years ago, the Japanese people said that they did not wish to become rich overnight. They earned money through their daily labor and usually spent it within that same day, enjoying everyday life as much as possible. It was not necessary to save money for the future, since the general population had close mutual help within their society. Worried people who could not work and earn money for that day by any particular means were helped by others in their society. No one was really wealthy in these days. Their daily supplies were often barely sufficient, but still all of them were able to survive through a mutual collaboration system established through their own determination, not by the system controlled by the government. This mutual help system came from their concept that all material things (*mottai*) belonged to the deities and were not their private property, even when they purchased something through their earnings. Everything is for everyone, and everyone should live on materials in harmony with the deities.

People at that time used materials very carefully. Their belongings were repaired repeatedly until they finally became useless. Actually, resources are not abundant in the Japanese archipelago, and people carefully used their belongings as long as they were useful.

The story is that when the clothes (Japanese kimono) at that time became dirty after being worn for many days, the stitches were removed and the article of clothing was reduced to its individual parts so that they could be washed carefully and individually. The cleaned fabrics were repaired, if necessary, and sewn back together again so that the clean, repaired kimono could be worn as if it was new. After several cleanings, however, stitch zone became damaged and was unable to be sewn again. The damaged zone was cut off and the narrow, remaining pieces of fabric were used to make clothes for children. After several washings, again, the children's kimono could not be sewn up. The fabric fragments could then be used as patches, such as for the covers of futon, Japanese bedding. They were used again and again until they were no longer able to be used for patches and repairs. The fragments of fabric were then gathered together and made into mops to clean the rooms and furniture. When the scraps were thoroughly worn out after various uses, they were dried and used as fuel. The ashes produced were taken to the farms to be used as fertilizer. Thus, there was no rubbish to be thrown out and the original clothes were fully utilized as valuable materials with the deities. Such activities were performed not strictly for economic purpose, but just for fully appreciating every object on earth. In the Yedo Era, numerous types of laborers worked to repair various household items.

The concept of *mottainai* went beyond the monetary value of an object and resulted in a lifestyle that promoted harmonious co-existence between nature and mankind. Such a concept cannot be established without sincere respect for every object in nature. The concept of *kyousei* is, thus, related to the idea of *mottainai*, as both of them are based on the traditional Japanese concept of understanding that all objects in nature are associated with a deity and are to be respected.

Since the above three words are hardly possible to translate word for word into English, I hope that the explanations above will make it possible to understand the real meaning of each. Without referring to either of the other two words, any one of the three cannot be understood correctly and sufficiently. The above three words are taken up as examples that are closely related to environmental issues. In the case of this book, there are many other words that are hardly translatable, literally word for word, into English. Any particular English word in dictionary has a range of meanings more or less out of alignment with the responding Japanese word; typical examples of such words are *nature*, *education*, *science and technology*, *evolution*, and so on.

I usually read books translated from foreign languages, especially in the fields of philosophy, bibliography, novels and poems, essays and so on, and owe much to the comprehensive contributions of translators. In comparing the translated version of my book with the original, I have some complicated feelings on each word translated. Still, I hope to communicate my thoughts and intentions to those who cannot read Japanese and must rely on the translated version of my sentences, I hope that the main content of the descriptions will be conveyed without serious misunderstandings.

References to the Postscript:

Inamura, S., G.Udagawa & H.Okada. 1796. Halma Wage. [F.Halma: Nederduits Woordenboek; Dutch-Japanese Dictionary, following to the style of Halma's Dutch-Francois Dictionary ed. 2, 1729.]

Iwatsuki, K. 2023. Spherophylon – The Integrated Lives of Earth's Diverse Organisms. Bookend Publ. Co., Tokyo.

Iwatsuki, K. & A.Ebihara. 2023. Hymenophyllaceae, in Flora Malesiana series II, volume 5. Naturalis, Biodiversity Centre, Leiden.

Kurokawa, K. 1994. The Philosophy of Symbiosis. Academy Editions, London.

Maathai, W.M. 2009. Statement by Prof. W. Maathai, Nobel Peace Laureate, on behalf of Civil Society. United Nations. Cited in E.M.Siniawer, 2014, "Affluence of the Heart": Wastefulness and the Search for Meaning in Millennial Japan, J. Asian Stud. 73(1): 177.

Shidei, T. 2006. Shinrin-wa-moriyahayashi-dewanai – Watashino Shinrin-ron (Forests are not mori nor hayashi – My concept on forest) (in Japanese). Nakanishiya Publ. Co., Kyoto.

Shirai, M. 1891. Nihon Hakubutsugaku Nempyo (Chronological Table for Natural History in Japan) (in Japanese). Maruzen, Tokyo.

Ueno, M. 1973. Nihon Hakubutsugakushi (History of Natural History in Japan) (in Japanese). Heibonsha, Tokyo.

INDEX

A

B

C

D

E

F

G

H

I

J

K

L

M

N

O

P

R

V

W

X

Y

Z